Fundamentals of Mathematics

Denny Burzynski
Wade Ellis, Jr.

West Valley College
Saratoga, California

SAUNDERS COLLEGE PUBLISHING

Philadelphia   Fort Worth   Chicago
San Francisco   Montreal   Toronto
London   Sydney   Tokyo

Text Typeface: Century Schoolbook
Compositor: Progressive Typographers, Inc.
Acquisitions Editor: Robert B. Stern
Developmental Editor: Ellen Newman
Project Editor: Janet B. Nuciforo
Copy Editor: Charlotte Nelson
Managing Editor: Carol Field
Art Director: Carol Bleistine
Art Assistant: Doris Bruey
Text Designer: Emily Harste
Cover Designer: Lawrence R. Didona
Text Artwork: ANCO/Boston, Inc. and Grafacon, Inc.
Production Manager: Merry Post

Cover Credit: Coonley Playhouse Clerestory No. 2 stained glass window © 1988 The Frank Lloyd Wright Foundation. Oakbrook-Esser Studios, exclusive Frank Lloyd Wright art glass licensee.

Printed in the United States of America

FUNDAMENTALS OF MATHEMATICS

0-03-063901-8

Library of Congress Catalog Card Number: 88-043412

901   071   987654321

To the next generation of explorers: Kristi, BreAnne, Lindsey, Randi, Piper, Meghan, Wyatt, Lara, Mason, and Sheanna.

Preface

Fundamentals of Mathematics is a work text that covers the traditional topics studied in a modern prealgebra course, as well as the topics of estimation, elementary analytic geometry, and introductory algebra. It is intended for students who (1) have had a previous course in prealgebra, (2) wish to meet the prerequisite of a higher level course such as elementary algebra, and (3) need to review fundamental mathematical concepts and techniques.

This text will help the student develop the insight and intuition necessary to master arithmetic techniques and manipulative skills. It was written with the following main objectives: (1) to provide the student with an understandable and usable source of information, (2) to provide the student with the maximum opportunity to see that arithmetic concepts and techniques are logically based, (3) to instill in the student the understanding and intuitive skills necessary to know how and when to use particular arithmetic concepts in subsequent material, courses, and nonclassroom situations, and (4) to give the student the ability to correctly interpret arithmetically obtained results. We have tried to meet these objectives by presenting material dynamically, much the way an instructor might present the material visually in a classroom. (See the development of the concept of addition and subtraction of fractions in Section 5.2, for example.) Intuition and understanding are some of the keys to creative thinking; we believe that the material presented in this text will help the student realize that mathematics is a creative subject.

This text can be used in standard lecture or self-paced classes. To help meet our objectives and to make the study of prealgebra a pleasant and rewarding experience, *Fundamentals of Mathematics* is organized as follows.

▢ PEDAGOGICAL FEATURES

The work text format gives the student space to practice mathematical skills with ready reference to sample problems. The chapters are divided into sections, and each section is a complete treatment of a particular topic, which includes the following features: Section Overview, Sample Sets, Practice Sets, Section Exercises, Exercises for Review, and Answers to Practice Sets. The chapters begin with Objectives and end with a Summary of Key Concepts, an Exercise Supplement, and a Proficiency Exam.

Objectives

Each chapter begins with a set of objectives identifying the material to be covered. Each section begins with an overview that repeats the objectives for that particular section. Sections are divided into subsections that correspond to the section objectives, which makes for easier reading.

Sample Sets

Fundamentals of Mathematics contains examples that are set off in boxes for easy reference. The examples are referred to as Sample Sets for two reasons: (1) They serve as a representation to be imitated, which we believe will foster understanding of mathematical concepts and provide experience with mathematical techniques. (2) Sample Sets also serve as a preliminary representation of problem-solving techniques that may be used to solve more general and more complicated problems. The examples have been carefully chosen to illustrate and develop concepts and techniques in the most instructive, easily remembered way. Concepts and techniques preceding the examples are introduced at a level below that normally used in similar texts and are thoroughly explained, assuming little previous knowledge.

Practice Sets

A parallel Practice Set follows each Sample Set, which reinforces the concepts just learned. There is adequate space for the student to work each problem directly on the page.

Answers to Practice Sets

The Answers to Practice Sets are given at the end of each section and can be easily located by referring to the page number, which appears after the last Practice Set in each section.

Section Exercises

The exercises at the end of each section are graded in terms of difficulty, although they are not grouped into categories. There is an ample number of problems, and after working through the exercises, the student will be capable of solving a variety of challenging problems.

The problems are paired so that the odd-numbered problems are equivalent in kind and difficulty to the even-numbered problems. Answers to the odd-numbered problems are provided at the back of the book.

Exercises for Review

This section consists of five problems that form a cumulative review of the material covered in the preceding sections of the text and is not limited to material in that chapter. The exercises are keyed by section for easy reference. Since these exercises are intended for review only, no work space is provided.

Summary of Key Concepts

A summary of the important ideas and formulas used throughout the chapter is included at the end of each chapter. More than just a list of terms, the summary is a valuable tool that reinforces concepts in preparation for the Proficiency Exam at the end of the chapter, as well as future exams. The summary keys each item to the section of the text where it is discussed.

Exercise Supplement

In addition to numerous section exercises, each chapter includes approximately 100 supplemental problems, which are referenced by section. Answers to the odd-numbered problems are included in the back of the book.

Proficiency Exam

Each chapter ends with a Proficiency Exam that can serve as a chapter review or evaluation. The Proficiency Exam is keyed to sections, which enables the student to refer back to the text for assistance. Answers to all the problems are included in the Answer Section at the end of the book.

❏ CONTENT

The writing style used in *Fundamentals of Mathematics* is informal and friendly, offering a straightforward approach to prealgebra mathematics. We have made a deliberate effort not to write another text that minimizes the use of words because we believe that students can best study arithmetic concepts and understand arithmetic techniques by using words *and* symbols rather than symbols alone. It has been our experience that students at the prealgebra level are not nearly experienced enough with mathematics to understand symbolic explanations alone; they need literal explanations to guide them through the symbols.

We have taken great care to present concepts and techniques so they are understandable and easily remembered. After concepts have been developed, students are warned about common pitfalls. We have tried to make the text an information source accessible to prealgebra students.

Chapter 1 — Addition and Subtraction of Whole Numbers This chapter includes the study of whole numbers, including a discussion of the Hindu-Arabic numeration and the base ten number systems. Rounding whole numbers is also presented, as are the commutative and associative properties of addition.

Chapter 2 — Multiplication and Division of Whole Numbers The operations of multiplication and division of whole numbers are explained in this chapter. Multiplication is described as repeated addition. Viewing multiplication in this way may provide students with a visualization of the meaning of algebraic terms such as $8x$ when they start learning algebra. The chapter also includes the commutative and associative properties of multiplication.

Chapter 3 — Exponents, Roots, and Factorizations of Whole Numbers The concept and meaning of the word *root* is introduced in this chapter. A method of reading root notation and a method of determining some common roots, both mentally and by calculator, is then presented. We also present grouping symbols and the order of operations, prime factorization of whole numbers, and the greatest common factor and least common multiple of a collection of whole numbers.

Chapter 4 — Introduction to Fractions and Multiplication and Division of Fractions We recognize that fractions constitute one of the foundations of problem solving. We have, therefore, given a detailed treatment of the operations of multiplication and division of fractions and the logic behind these operations. We believe that the logical treatment and many practice exercises will help students retain the information presented in this chapter and enable them to use it as a foundation for the study of rational expressions in an algebra course.

Chapter 5 — Addition and Subtraction of Fractions, Comparing Fractions, and Complex Fractions A detailed treatment of the operations of addition and subtraction of fractions and the logic behind these operations is given in this chapter. Again, we believe that the logical treatment and many practice exercises will help students retain the information, thus enabling them to use it in the study of rational expressions in an algebra course. We have tried to make explanations dynamic. A method for comparing fractions is introduced, which gives the student another way of understanding the relationship between the words *denominator* and *denomination*. This method serves to show the student that it is sometimes possible to compare two different types of quantities. We also study a method of simplifying complex fractions and of combining operations with fractions.

Chapter 6 — Decimals The student is introduced to decimals in terms of the base ten number system, fractions, and digits occurring to the right of the units position. A method of converting a fraction to a decimal is discussed. The logic behind the standard methods of operating on decimals is presented and many examples of how to apply the methods are given. The word *of* as related to the operation of multiplication is discussed. Nonterminating divisions are examined, as are combinations of operations with decimals and fractions.

Chapter 7 — Ratios and Rates We begin by defining and distinguishing the terms *ratio* and *rate*. The meaning of proportion and some applications of proportion problems are described. Proportion problems are solved using the "Five-Step Method." We hope that by using this method the student will discover the value of introducing a variable as a first step in problem solving and the power of organization. The chapter concludes with discussions of percent, fractions of one percent, and some applications of percent.

Chapter 8 — Techniques of Estimation One of the most powerful problem-solving tools is a knowledge of estimation techniques. We feel that estimation is so important that we devote an entire chapter to its study. We examine three estimation techniques: estimation by rounding, estimation by clustering, and estimation by rounding fractions. We also include a section on the distributive property, an important algebraic property.

Chapter 9 — Measurement and Geometry This chapter presents some of the techniques of measurement in both the United States system and the metric system. Conversion from one unit to another (in a system) is examined in terms of unit fractions. A discussion of the simplification of denominate numbers is also included. This discussion helps the student understand more clearly the association between pure numbers and dimensions. The chapter concludes with a study of perimeter and circumference of geometric figures and area and volume of geometric figures and objects.

Chapter 10 — Signed Numbers   A look at algebraic concepts and techniques is begun in this chapter. Basic to the study of algebra is a working knowledge of signed numbers. Definitions of variables, constants, and real numbers are introduced. We then distinguish between positive and negative numbers, learn how to read signed numbers, and examine the origin and use of the double-negative property of real numbers. The concept of absolute value is presented both geometrically (using the number line) and algebraically. The algebraic definition is followed by an interpretation of its meaning and several detailed examples of its use. Addition, subtraction, multiplication, and division of signed numbers are presented first using the number line, then with absolute value.

Chapter 11 — Algebraic Expressions and Equations   The student is introduced to some elementary algebraic concepts and techniques in this final chapter. Algebraic expressions and the process of combining like terms are discussed in Sections 11.1 and 11.2. The method of combining like terms in an algebraic expression is explained by using the interpretation of multiplication as a description of repeated addition (as in Chapter 2).

❑ ANCILLARY PACKAGE

Users of *Fundamentals of Mathematics* will receive an extensive set of ancillary items that substantially assist the instructor in presenting the course as well as motivating the student. This package includes the following:

Math Review Pak   With every copy of the text, students will receive a package of approximately 150 index cards containing important ideas and formulas from the Summary of Key Concepts at the end of each chapter. Each index card lists a key word on one side with the definition or explanation on the reverse side; the key word is referenced to the section of the text in which it is discussed. This is a useful device to help students remember the concepts learned in the text and to study for exams.

Instructor's Manual   The printed manual contains chapter-by-chapter objectives, lecture suggestions, and recommendations for using the entire ancillary package, as well as:

• Answers to both even- and odd-numbered exercises
• Pretests and answers for each chapter of the text

Prepared Tests   This set of tests contains six written tests for each of the chapters in the book. Half of the tests have open-ended questions and the other half have multiple-choice questions. Thus, instructors have greater flexibility in testing. The answers to all of the questions are included. There is also a final examination and a Diagnostic Test (with answers included) that can be used for placing students in the proper course level.

Computerized Test Bank   Available for the Apple and IBM PC, this bank contains 2,500 questions, both open-ended and multiple choice, enabling the instructor to create many unique tests. The instructor is able to edit the questions as well as add new questions. The software solves each problem and prints the answer in a grading key on a separate sheet.

Test Bank   This printed bank contains five tests for each chapter of the book that were generated from the Computerized Test Bank and bound as another source of tests. Answers to the tests are included on separate grading keys.

Maxis Interactive Software   This program disk contains practice problems from each chapter of the text. Use of the software will provide the student with an alternative way to learn the material and, at the same time, with individualized attention. The program automatically advances to the next level of difficulty once the student has successfully solved a few problems; the student may also ask to see the solution to check his or her understanding of the process used. Once the student has completed a section, a printout is available showing how many problems were answered correctly. The software is keyed to the textbook and refers the student to the appropriate section of the text if an incorrect answer is input. A useful tool to check skills and to identify and correct any difficulties in finding solutions, this software is available for the Apple II and IBM microcomputers.

Videotapes   A complete set of videotapes (15 hours) is free to adopters and will give added assistance or serve as a quick review of the book. Keyed to the text, the videotapes "walk" the student through each section of the book and provide another approach toward mastery of the given topic. Prepared by the authors of the text, the videotapes often include a short summary of the theory used in solving problems before working some of the odd-numbered problems in the text. The tapes include a list of topics covered and the amount of time spent on each section, as well as suggestions for using the videotapes in conjunction with the text.

Student Solutions Manual and Study Guide   This guide contains step-by-step solutions to one fourth of the problems in the Exercise Sets (every other odd-numbered problem) in addition to providing the student with a short summary of the important concepts in each chapter. This will help the student practice the techniques used in solving problems.

☐ ACKNOWLEDGMENTS

Many extraordinarily talented people are responsible for helping to create this text. We wish to acknowledge the efforts and skill of the following mathematicians. Their contributions have been invaluable.

Barbara Conway, Berkshire Community College
Bill Hajdukiewicz, Miami-Dade Community College
Virginia Hamilton, Shawnee State University
David Hares, El Centro College
Norman Lee, Ball State University
Ginger Y. Manchester, Hinds Junior College
John R. Martin, Tarrant County Junior College
Shelba Mormon, Northlake College
Lou Ann Pate, Pima Community College
Gus Pekara, Oklahoma City Community College
David Price, Tarrant County Junior College
David Schultz, Virginia Western Community College
Sue S. Watkins, Lorain County Community College
Elizabeth M. Wayt, Tennessee State University
Prentice E. Whitlock, Jersey City State College
Thomas E. Williamson, Montclair State College

Special thanks to the following individuals for their careful accuracy reviews of manuscript, galleys, and page proofs: Steve Blasberg, West Valley College; Wade Ellis, Sr., University of Michigan; John R. Martin, Tarrant County Junior College; and Jane Ellis. We would also like to thank Amy Miller and Guy Sanders, Branham High School.

Our sincere thanks to Debbie Wiedemann for her encouragement, suggestions concerning psychobiological examples, proofreading much of the manuscript, and typing many of the section exercises; Sandi Wiedemann for collating the annotated reviews, counting the examples and exercises, and untiring use of "white-out"; and Jane Ellis for solving and typing all of the exercise solutions.

We thank the following people for their excellent work on the various ancillary items that accompany *Fundamentals of Mathematics:* Steve Blasberg, West Valley

College; Wade Ellis, Sr., University of Michigan; and Jane Ellis (Instructor's Manual); John R. Martin, Tarrant County Junior College (Student Solutions Manual and Study Guide); Virginia Hamilton, Shawnee State University (Computerized Test Bank); Patricia Morgan, San Diego State University (Prepared Tests); and George W. Bergeman, Northern Virginia Community College (Maxis Interactive Software).

We also thank the talented people at Saunders College Publishing whose efforts made this text run smoothly and less painfully than we had imagined. Our particular thanks to Bob Stern, Mathematics Editor, Ellen Newman, Developmental Editor, and Janet Nuciforo, Project Editor. Their guidance, suggestions, open minds to our suggestions and concerns, and encouragement have been extraordinarily helpful. Although there were times we thought we might be permanently damaged from rereading and rewriting, their efforts have improved this text immensely. It is a pleasure to work with such high-quality professionals.

San Jose, California   **Denny Burzynski**
December 1988   **Wade Ellis, Jr.**

I would like to thank Doug Campbell, Ed Lodi, and Guy Sanders for listening to my frustrations and encouraging me on. Thanks also go to my cousin, David Raffety, who long ago in Sequoia National Forest told me what a differential equation is.

Particular thanks go to each of my colleagues at West Valley College. Our everyday conversations regarding mathematics instruction have been of the utmost importance to the development of this text and to my teaching career.

D.B.

Contents

1 ☐ Addition and Subtraction of Whole Numbers   1

    1.1 Whole Numbers   2
    1.2 Reading and Writing Whole Numbers   7
    1.3 Rounding Whole Numbers   12
    1.4 Addition of Whole Numbers   18
    1.5 Subtraction of Whole Numbers   29
    1.6 Properties of Addition   44
    SUMMARY OF KEY CONCEPTS   48
    EXERCISE SUPPLEMENT   50
    PROFICIENCY EXAM   53

2 ☐ Multiplication and Division of Whole Numbers   55

    2.1 Multiplication of Whole Numbers   56
    2.2 Concepts of Division of Whole Numbers   65
    2.3 Division of Whole Numbers   71
    2.4 Some Interesting Facts about Division   81
    2.5 Properties of Multiplication   85
    SUMMARY OF KEY CONCEPTS   89
    EXERCISE SUPPLEMENT   90
    PROFICIENCY EXAM   93

3 ☐ Exponents, Roots, and Factorizations of Whole Numbers   95

    3.1 Exponents and Roots   96
    3.2 Grouping Symbols and the Order of Operations   101
    3.3 Prime Factorization of Natural Numbers   110
    3.4 The Greatest Common Factor   117
    3.5 The Least Common Multiple   121
    SUMMARY OF KEY CONCEPTS   127
    EXERCISE SUPPLEMENT   129
    PROFICIENCY EXAM   131

4 ☐ Introduction to Fractions and Multiplication and Division of Fractions   135

    4.1 Fractions of Whole Numbers   136
    4.2 Proper Fractions, Improper Fractions, and Mixed Numbers   145
    4.3 Equivalent Fractions, Reducing Fractions to Lowest Terms, and Raising Fractions to Higher Terms   154
    4.4 Multiplication of Fractions   165
    4.5 Division of Fractions   175
    4.6 Applications Involving Fractions   181
    SUMMARY OF KEY CONCEPTS   189
    EXERCISE SUPPLEMENT   192
    PROFICIENCY EXAM   197

5 ☐ Addition and Subtraction of Fractions, Comparing Fractions, and Complex Fractions   201

5.1 Addition and Subtraction of Fractions with Like Denominators   202
5.2 Addition and Subtraction of Fractions with Unlike Denominators   207
5.3 Addition and Subtraction of Mixed Numbers   215
5.4 Comparing Fractions   221
5.5 Complex Fractions   225
5.6 Combinations of Operations with Fractions   230
SUMMARY OF KEY CONCEPTS   238
EXERCISE SUPPLEMENT   239
PROFICIENCY EXAM   243

6 ☐ Decimals   247

6.1 Reading and Writing Decimals   248
6.2 Converting a Decimal to a Fraction   254
6.3 Rounding Decimals   257
6.4 Addition and Subtraction of Decimals   260
6.5 Multiplication of Decimals   265
6.6 Division of Decimals   274
6.7 Nonterminating Divisions   284
6.8 Converting a Fraction to a Decimal   288
6.9 Combinations of Operations with Decimals and Fractions   294
SUMMARY OF KEY CONCEPTS   298
EXERCISE SUPPLEMENT   299
PROFICIENCY EXAM   301

7 ☐ Ratios and Rates   304

7.1 Ratios and Rates   304
7.2 Proportions   309
7.3 Applications of Proportions   316
7.4 Percent   325
7.5 Fractions of One Percent   330
7.6 Applications of Percents   334
SUMMARY OF KEY CONCEPTS   347
EXERCISE SUPPLEMENT   348
PROFICIENCY EXAM   351

8 ☐ Techniques of Estimation   355

8.1 Estimation of Rounding   356
8.2 Estimation of Clustering   363
8.3 Mental Arithmetic — Using the Distributive Property   366
8.4 Estimation by Rounding Fractions   370
SUMMARY OF KEY CONCEPTS   374
EXERCISE SUPPLEMENT   375
PROFICIENCY EXAM   379

9 ☐ Measurement and Geometry   383

9.1 Measurement and the United States System   384
9.2 The Metric System of Measurement   389
9.3 Simplification of Denominate Numbers   394
9.4 Perimeter and Circumference of Geometric Figures   402
9.5 Area and Volume of Geometric Figures and Objects   410

SUMMARY OF KEY CONCEPTS   421
EXERCISE SUPPLEMENT   423
PROFICIENCY EXAM   427

10 ▣ Signed Numbers   433

10.1 Variables, Constants, and Real Numbers   434
10.2 Signed Numbers   439
10.3 Absolute Value   442
10.4 Addition of Signed Numbers   445
10.5 Subtraction of Signed Numbers   451
10.6 Multiplication and Division of Signed Numbers   454
SUMMARY OF KEY CONCEPTS   464
EXERCISE SUPPLEMENT   465
PROFICIENCY EXAM   467

11 ▣ Algebraic Expressions and Equations   469

11.1 Algebraic Expressions   470
11.2 Combining Like Terms Using Addition and Subtraction   477
11.3 Solving Equations of the Form $x + a = b$ and $x - a = b$ 480
11.4 Solving Equations of the Form $ax = b$ and $x/a = b$ 487
11.5 Applications I: Translating Words to Mathematical Symbols   495
11.6 Applications II: Solving Problems   500
SUMMARY OF KEY CONCEPTS   516
EXERCISE SUPPLEMENT   518
PROFICIENCY EXAM   523

Answers to Selected Exercises   A-1
Index   I-1

1

Addition and Subtraction of Whole Numbers

After completing this chapter, you should

Section 1.1 Whole Numbers
- know the difference between numbers and numerals
- know why our number system is called the Hindu-Arabic numeration system
- understand the base ten positional number system
- be able to identify and graph whole numbers

Section 1.2 Reading and Writing Whole Numbers
- be able to read and write a whole number

Section 1.3 Rounding Whole Numbers
- understand that rounding is a method of approximation
- be able to round a whole number to a specified position

Section 1.4 Addition of Whole Numbers
- understand the addition process
- be able to add whole numbers
- be able to use the calculator to add one whole number to another

Section 1.5 Subtraction of Whole Numbers
- understand the subtraction process
- be able to subtract whole numbers
- be able to use a calculator to subtract one whole number from another whole number

Section 1.6 Properties of Addition
- understand the commutative and associative properties of addition
- understand why 0 is the additive identity

1.1 Whole Numbers

Section
Overview

❑ **NUMBERS AND NUMERALS**
❑ **THE HINDU-ARABIC NUMERATION SYSTEM**
❑ **THE BASE TEN POSITIONAL NUMBER SYSTEM**
❑ **WHOLE NUMBERS**
❑ **GRAPHING WHOLE NUMBERS**

❑ NUMBERS AND NUMERALS

We begin our study of introductory mathematics by examining its most basic building block, the **number.**

Number

> A **number** is a concept. It exists only in the mind.

The earliest concept of a number was a thought that allowed people to mentally picture the size of some collection of objects. To write down the number being conceptualized, a **numeral** is used.

Numeral

> A **numeral** is a symbol that represents a number.

In common usage today we do not distinguish between a number and a numeral. In our study of introductory mathematics, we will follow this common usage.

☆ SAMPLE SET A

The following are numerals. In each case, the first represents the number four, the second represents the number one hundred twenty-three, and the third, the number one thousand five. These numbers are represented in different ways.

(a) Hindu-Arabic numerals

    4,   123,   1005

(b) Roman numerals

    IV,   CXXIII,   MV

(c) Egyptian numerals

    | | | |,   ꝯ ∩∩ | | |,   ⚇ | | | | |
    Strokes   Coiled rope,   Lotus flower
               heel bones,   and strokes
               and strokes

★ PRACTICE SET A

Do the phrases "four," "one hundred twenty-three," and "one thousand five" qualify as numerals? _____

                                                    yes/no

❑ THE HINDU-ARABIC NUMERATION SYSTEM

Hindu-Arabic Numeration
System

Our society uses the **Hindu-Arabic numeration system.** This system of numeration began shortly before the third century when the Hindus invented the numerals

0 1 2 3 4 5 6 7 8 9

Leonardo Fibonacci

About a thousand years later, in the thirteenth century, a mathematician named Leonardo Fibonacci of Pisa introduced the system into Europe. It was then popularized by the Arabs. Thus, the name, Hindu-Arabic numeration system.

☐ THE BASE TEN POSITIONAL NUMBER SYSTEM

Digits

The Hindu-Arabic numerals 0 1 2 3 4 5 6 7 8 9 are called **digits.** We can form any number in the number system by selecting one or more digits and placing them in certain positions. Each position has a particular value. The Hindu mathematician who devised the system about A.D. 500 stated that "from place to place each is ten times the preceding." It is for this reason that our number system is called a **positional** number system with **base ten.**

Base Ten Positional System
Commas

When numbers are composed of more than three digits, **commas** are sometimes used to separate the digits into groups of three. These groups of three are called **periods** and they greatly simplify reading numbers.

Periods

In the Hindu-Arabic numeration system, a period has a value assigned to each of its three positions, and the values are the same for each period. The position values are

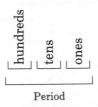

Thus, each period contains a position for the values of one, ten, and hundred. Notice that, in looking from right to left, the value of each position is ten times the preceding. Each period has a particular name.

As we continue from right to left, there are more periods. The five periods listed above are the most common, and in our study of introductory mathematics, they are sufficient.

The following diagram illustrates our positional number system to trillions. (There are, to be sure, other periods.)

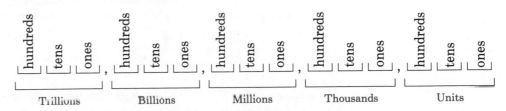

In our positional number system, the **value of a digit** is determined by its *position* in the number.

☆ **SAMPLE SET B**

1. Find the value of 6 in the number 7,261.

Since 6 is in the tens position of the units period, its value is 6 tens.

6 tens = 60

Continued

2. Find the value of 9 in the number 86,932,106,005.

   Since 9 is in the hundreds position of the millions period, its value is 9 hundred millions.

   9 hundred millions = 9 hundred million

3. Find the value of 2 in the number 102,001.

   Since 2 is in the ones position of the thousands period, its value is 2 one thousands.

   2 one thousands = 2 thousand

★ PRACTICE SET B

1. Find the value of 5 in the number 65,000.

2. Find the value of 4 in the number 439,997,007,010.

3. Find the value of 0 in the number 108.

❑ WHOLE NUMBERS

Numbers that are formed using only the digits

0 1 2 3 4 5 6 7 8 9

Whole Numbers

are called **whole numbers.** They are

0, 1, 2, 3, 4, 5, 6, 7, 8, 9, 10, 11, 12, 13, 14, 15, . . .

The three dots at the end mean "and so on in this same pattern."

❑ GRAPHING WHOLE NUMBERS

Number Line

Origin

Whole numbers may be visualized by constructing a **number line.** To construct a number line, we simply draw a straight line and choose any point on the line and label it 0. This point is called the **origin.** We then choose some convenient length, and moving to the right, mark off consecutive intervals (parts) along the line starting at 0. We label each new interval endpoint with the next whole number.

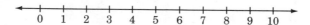

Graphing

We can visually display a whole number by drawing a closed circle at the point labeled with that whole number. Another phrase for visually displaying a whole number is **graphing** the whole number. The word **graph** means to "visually display."

☆ SAMPLE SET C

1. Graph the following whole numbers: 3, 5, 9.

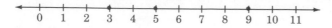

2. Specify the whole numbers that are graphed on the following number line. The break in the number line indicates that we are aware of the whole numbers between 0 and 106, and 107 and 872, but we are not listing them due to space limitations.

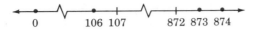

The numbers that have been graphed are

0,   106,   873,   874

★ PRACTICE SET C

1. Graph the following whole numbers: 46, 47, 48, 325, 327.

2. Specify the whole numbers that are graphed on the following number line.

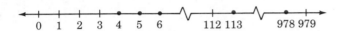

Answers to Practice Sets are on p. 6.

A **line** is composed of an endless number of points. Notice that we have labeled only some of them. As we proceed, we will discover new types of numbers and determine their location on the number line.

Section 1.1 EXERCISES

1. What is a number?

2. What is a numeral?

3. Does the word "eleven" qualify as a numeral?

4. How many different digits are there?

5. Our number system, the Hindu-Arabic number system, is a _____ number system with base _____.

6. Numbers composed of more than three digits are sometimes separated into groups of three by commas. These groups of three are called _____.

7. In our number system, each period has three values assigned to it. These values are the same for each period. From right to left, what are they?

8. Each period has its own particular name. From right to left, what are the names of the first four?

9. In the number 841, how many tens are there?

10. In the number 3,392, how many ones are there?

11. In the number 10,046, how many thousands are there?

12. In the number 779,844,205, how many ten millions are there?

13. In the number 65,021, how many hundred thousands are there?

For problems 14–17, give the value of the indicated digit in the given number.

14. 5 in 599

15. 1 in 310,406

16. 9 in 29,827

17. 6 in 52,561,001,100

18. Write a two-digit number that has an eight in the tens position.

19. Write a four-digit number that has a one in the thousands position and a zero in the ones position.

20. How many two-digit whole numbers are there?

21. How many three-digit whole numbers are there?

22. How many four-digit whole numbers are there?

23. Is there a smallest whole number? If so, what is it?

24. Is there a largest whole number? If so, what is it?

25. Another term for "visually displaying" is _____.

26. The whole numbers can be visually displayed on a _____.

27. Graph (visually display) the following whole numbers on the number line below: 0, 1, 31, 34.

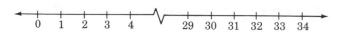

28. Construct a number line in the space provided below and graph (visually display) the following whole numbers: 84, 85, 901, 1006, 1007.

29. Specify, if any, the whole numbers that are graphed on the following number line.

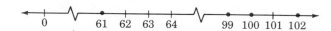

30. Specify, if any, the whole numbers that are graphed on the following number line.

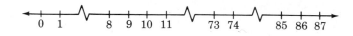

★ **Answers to Practice Sets (1.1)**

A. Yes. Letters are symbols. Taken as a collection (a written word), they represent a number.

B. **1.** five thousand **2.** four hundred billion **3.** zero tens, or zero

C. **1.**

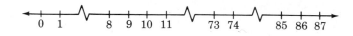

2. 4, 5, 6, 113, 978

1.2 Reading and Writing Whole Numbers

**Section
Overview**
☐ **READING WHOLE NUMBERS**
☐ **WRITING WHOLE NUMBERS**

Because our number system is a positional number system, reading and writing whole numbers is quite simple.

☐ READING WHOLE NUMBERS

To convert a number that is formed by digits into a verbal phrase, use the following method:

1. Beginning at the right and working right to left, separate the number into distinct periods by inserting commas every three digits.
2. Beginning at the left, read each period individually, saying the period name.

☆ SAMPLE SET A

Write the following numbers as words.

1. Read 42958.

1. Beginning at the right, we can separate this number into distinct periods by inserting a comma between the 2 and 9.

42,958

2. Beginning at the left, we read each period individually:

```
 ⌊_⌋⌊4⌋⌊2⌋ , ⟶ Forty-two thousand
 ⌊_____⌋
 Thousands period
```

```
 ⌊9⌋⌊5⌋⌊8⌋ ⟶ nine hundred fifty-eight
 ⌊_____⌋
 Units period
```

Forty-two thousand, nine hundred fifty-eight.

2. Read 307991343.

1. Beginning at the right, we can separate this number into distinct periods by placing commas between the 1 and 3 and the 7 and 9.

307,991,343

2. Beginning at the left, we read each period individually.

```
 ⌊3⌋⌊0⌋⌊7⌋ , ⟶ Three hundred seven million,
 ⌊_____⌋
 Millions period
```

```
 ⌊9⌋⌊9⌋⌊1⌋ , ⟶ nine hundred ninety-one thousand,
 ⌊_____⌋
 Thousands period
```

Continued

$\underbrace{\boxed{3} \, \boxed{4} \, \boxed{3}}_{\text{Units period}} \longrightarrow$ three hundred forty-three

Three hundred seven million, nine hundred ninety-one thousand, three hundred forty-three.

3. Read 36000000000001.

   1. Beginning at the right, we can separate this number into distinct periods by placing commas.

      36,000,000,001

   2. Beginning at the left, we read each period individually.

   $\underbrace{\boxed{} \, \boxed{3} \, \boxed{6}}_{\text{Trillions period}} , \longrightarrow$ Thirty-six trillion,

   $\underbrace{\boxed{0} \, \boxed{0} \, \boxed{0}}_{\text{Billions period}} , \longrightarrow$ zero billion,

   $\underbrace{\boxed{0} \, \boxed{0} \, \boxed{0}}_{\text{Millions period}} , \longrightarrow$ zero million,

   $\underbrace{\boxed{0} \, \boxed{0} \, \boxed{0}}_{\text{Thousands period}} , \longrightarrow$ zero thousand,

   $\underbrace{\boxed{0} \, \boxed{0} \, \boxed{1}}_{\text{Units period}} \longrightarrow$ one

   Thirty-six trillion, one.

★ PRACTICE SET A

Write each number in words.

1. 12,542   2. 101,074,003   3. 1,000,008

☐ WRITING WHOLE NUMBERS

To express a number in digits that is expressed in words, use the following method:

1. Notice first that a number expressed as a verbal phrase will have its periods set off by commas.
2. Starting at the beginning of the phrase, write each period of numbers individually.
3. Using commas to separate periods, combine the periods to form one number.

☆ **SAMPLE SET B**

Write each number using digits.

1. Seven thousand, ninety-two.

Using the comma as a period separator, we have

Seven thousand , ⟶ 7,

ninety-two ⟶ 092

7,092

2. Fifty billion, one million, two hundred thousand, fourteen.

Using the commas as period separators, we have

Fifty billion , ⟶ 50,

one million , ⟶ 001,

two hundred thousand , ⟶ 200,

fourteen ⟶ 014

50,001,200,014

3. Ten million, five hundred twelve.

The comma sets off the periods. We notice that there is no thousands period. We'll have to insert this ourselves.

Ten million , ⟶ 10,

zero thousand , ⟶ 000,

five hundred twelve ⟶ 512

10,000,512

★ **PRACTICE SET B**

Express each number using digits.

1. One hundred three thousand, twenty-five.

2. Six million, forty thousand, seven.

3. Twenty trillion, three billion, eighty million, one hundred nine thousand, four hundred two.

4. Eighty billion, thirty-five.

Answers to Practice Sets are on p. 12.

Section 1.2 EXERCISES

For problems 1–32, write all numbers in words.

1. 912

2. 84

3. 1491

4. 8601

5. 35,223

6. 71,006

7. 437,105

8. 201,040

9. 8,001,001

10. 16,000,053

11. 770,311,101

12. 83,000,000,007

13. 106,100,001,010

14. 3,333,444,777

15. 800,000,800,000

16. A particular community college has 12,471 students enrolled.

17. A person who watches 4 hours of television a day spends 1460 hours a year watching T.V.

18. Astronomers believe that the age of the earth is about 4,500,000,000 years.

19. Astronomers believe that the age of the universe is about 20,000,000,000 years.

20. There are 9690 ways to choose four objects from a collection of 20.

21. If a 412 page book has about 52 sentences per page, it will contain about 21,424 sentences.

22. In 1980, in the United States, there was $1,761,000,000,000 invested in life insurance.

23. In 1979, there were 85,000 telephones in Alaska and 2,905,000 telephones in Indiana.

24. In 1975, in the United States, it is estimated that 52,294,000 people drove to work alone.

25. In 1980, there were 217 prisoners under death sentence that were divorced.

26. In 1979, the amount of money spent in the United States for regular-session college education was $50,721,000,000,000.

27. In 1981, there were 1,956,000 students majoring in business in U.S. colleges.

28. In 1980, the average fee for initial and follow up visits to a medical doctors office was about $34.

29. In 1980, there were approximately 13,100 smugglers of aliens apprehended by the Immigration border patrol.

30. In 1980, the state of West Virginia pumped 2,000,000 barrels of crude oil, whereas Texas pumped 975,000,000 barrels.

31. The 1981 population of Uganda was 12,630,000 people.

32. In 1981, the average monthly salary offered to a person with a Master's degree in mathematics was $1,685.

For problems 33–45, write each number using digits.

33. Six hundred eighty-one

34. Four hundred ninety

35. Seven thousand, two hundred one

36. Nineteen thousand, sixty-five

37. Five hundred twelve thousand, three

38. Two million, one hundred thirty-three thousand, eight hundred fifty-nine

39. Thirty-five million, seven thousand, one hundred one

40. One hundred million, one thousand

41. Sixteen billion, fifty-nine thousand, four

42. Nine hundred twenty billion, four hundred seventeen million, twenty-one thousand

43. Twenty-three billion

44. Fifteen trillion, four billion, nineteen thousand, three hundred five

45. One hundred trillion, one

EXERCISES FOR REVIEW

(1.1) **46.** How many digits are there?

(1.1) **47.** In the number 6,641, how many tens are there?

(1.1) **48.** What is the value of 7 in 44,763?

(1.1) **49.** Is there a smallest whole number? If so, what is it?

(1.1) **50.** Write a four-digit number with a 9 in the tens position.

★ **Answers to Practice Sets (1.2)**

A. **1.** Twelve thousand, five hundred forty-two
2. One hundred one million, seventy-four thousand, three **3.** One million, eight

B. **1.** 103,025 **2.** 6,040,007 **3.** 20,003,080,109,402 **4.** 80,000,000,035

1.3 Rounding Whole Numbers

Section Overview	☐ ROUNDING AS AN APPROXIMATION ☐ THE METHOD OF ROUNDING WHOLE NUMBERS

☐ ROUNDING AS AN APPROXIMATION

A primary use of whole numbers is to keep count of how many objects there are in a collection. Sometimes we're only interested in the approximate number of objects in the collection rather than the precise number. For example, there are *approximately* 20 symbols in the collection below.

The *precise* number of symbols in the above collection is 18.

Rounding

We often approximate the number of objects in a collection by mentally seeing the collection as occurring in groups of tens, hundreds, thousands, etc. This process of approximation is called **rounding**. Rounding is very useful in estimation. We will study estimation in Chapter 8.

When we think of a collection as occurring in groups of tens, we say we're *rounding to the nearest ten*. When we think of a collection as occurring in groups of hundreds, we say we're *rounding to the nearest hundred*. This idea of rounding continues through thousands, ten thousands, hundred thousands, millions, etc.

The process of rounding whole numbers is illustrated in the following examples.

1. Round 67 to the nearest ten.

On the number line, 67 is more than halfway from 60 to 70. The digit immediately to the right of the tens digit, the round-off digit, is the indicator for this.

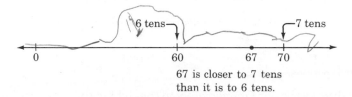

67 is closer to 7 tens
than it is to 6 tens.

Thus, 67, rounded to the nearest ten, is 70.

2. Round 4,329 to the nearest hundred.

On the number line, 4,329 is less than halfway from 4,300 to 4,400. The digit to the immediate right of the hundreds digit, the round-off digit, is the indicator.

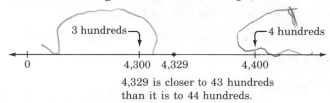

4,329 is closer to 43 hundreds
than it is to 44 hundreds.

Thus, 4,329, rounded to the nearest hundred is 4,300.

3. Round 16,500 to the nearest thousand.

On the number line, 16,500 is exactly halfway from 16,000 to 17,000.

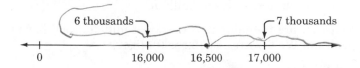

By convention, when the number to be rounded is *exactly halfway* between two numbers, it is rounded to the *higher* number.

Thus, 16,500, rounded to the nearest thousand, is 17,000.

4. A person whose salary is $41,450 per year might tell a friend that she makes $41,000 per year. She has rounded 41,450 to the *nearest* thousand. The number 41,450 is closer to 41,000 than it is to 42,000.

☑ THE METHOD OF ROUNDING WHOLE NUMBERS

Rounding Whole Numbers

From the observations made in the preceding examples, we can use the following method to **round a whole number** to a particular position.

1. Mark the position of the round-off digit.
2. Note the digit to the immediate right of the round-off digit.
   (a) If it is less than 5, replace it and all the digits to its right with zeros. Leave the round-off digit unchanged.
   (b) If it is 5 or larger, replace it and all the digits to its right with zeros. Increase the round-off digit by 1.

☆ SAMPLE SET A

Use the method of rounding whole numbers to solve the following problems.

1. Round 3,426 to the nearest ten.

1. We are rounding to the tens position. Mark the digit in the tens position

   3,426
   ↑
   tens position

2. Observe the digit immediately to the right of the tens position. It is 6. Since 6 is greater than 5, we *round up* by replacing 6 with 0 and adding 1 to the digit in the tens position (the round-off position): $2 + 1 = 3$.

   3,430

Thus, 3,426 rounded to the nearest ten is 3,430.

2. Round 9,614,018,007 to the nearest ten million.

1. We are rounding to the nearest ten million.

   9,614,018,007
   ↑
   ten millions position

2. Observe the digit immediately to the right of the ten millions position. It is 4. Since 4 is less than 5, we *round down* by replacing 4 and all the digits to its right with zeros.

   9,610,000,000

Thus, 9,614,018,007 rounded to the nearest ten million is 9,610,000,000.

3. Round 148,422 to the nearest million.

1. Since we are rounding to the nearest million, we'll have to *imagine* a digit in the millions position. We'll write 148,422 as 0,148,422.

   0,148,422
   ↑
   millions position

2. The digit immediately to the right is 1. Since 1 is less than 5, we'll *round down* by replacing it and all the digits to its right with zeros.

   0,000,000

   This number is 0.

Thus, 148,422 rounded to the nearest million is 0.

4. Round 397,000 to the nearest ten thousand.

1. We are rounding to the nearest ten thousand.

   397,000
   ↑
   ten thousand position

2. The digit immediately to the right of the ten thousand position is 7. Since 7 is greater than 5, we round up by replacing 7 and all the digits to its right with zeros and adding 1 to the digit in the ten thousands position. But $9 + 1 = 10$ and we must carry the 1 to the next (the hundred thousands) position.

400,000

Thus, 397,000 rounded to the nearest ten thousand is 400,000.

★ **PRACTICE SET A**

Use the method of rounding whole numbers to solve each problem.

1. Round 3387 to the nearest hundred.

2. Round 26,515 to the nearest thousand.

3. Round 30,852,900 to the nearest million.

4. Round 39 to the nearest hundred.

5. Round 59,600 to the nearest thousand.

Answers to the Practice Set are on p. 17.

Section 1.3 EXERCISES

For problems 1–23, complete the table by rounding each number to the indicated positions.

	hundred	thousand	ten thousand	million
1. 1,642				
2. 5,221				
3. 91,803				
4. 106,007				
5. 208				

	hundred	thousand	ten thousand	million
6. 199	_____	_____	_____	_____
7. 863	_____	_____	_____	_____
8. 794	_____	_____	_____	_____
9. 925	_____	_____	_____	_____
10. 909	_____	_____	_____	_____
11. 981	_____	_____	_____	_____
12. 965	_____	_____	_____	_____
13. 551,061,285	_____	_____	_____	_____
14. 23,047,991,521	_____	_____	_____	_____
15. 106,999,413,206	_____	_____	_____	_____
16. 5,000,000	_____	_____	_____	_____
17. 8,006,001	_____	_____	_____	_____
18. 94,312	_____	_____	_____	_____
19. 33,486	_____	_____	_____	_____
20. 560,669	_____	_____	_____	_____
21. 388,551	_____	_____	_____	_____
22. 4,752	_____	_____	_____	_____
23. 8,209	_____	_____	_____	_____

24. In 1950, there were 5,796 cases of diphtheria reported in the United States. Round to the nearest hundred.

25. In 1979, 19,309,000 people in the United States received federal food stamps. Round to the nearest ten thousand.

26. In 1980, there were 1,105,000 people between 30 and 34 years old enrolled in school. Round to the nearest million.

27. In 1980, there were 29,100,000 reports of aggravated assaults in the United States. Round to the nearest million.

For problems 28–36, round the numbers to the position you think is most reasonable for the situation.

28. In 1980, for a city of one million or more, the average annual salary of police and firefighters was $16,096.

29. The average percentage of possible sunshine in San Francisco, California, in June is 73%.

30. In 1980, in the state of Connecticut, $3,777,000,000 in defense contract payroll was awarded.

31. In 1980, the federal government paid $5,463,000,000 to Viet Nam veterans and dependants.

32. In 1980, there were 3,377,000 salespeople employed in the United States.

33. In 1948, in New Hampshire, 231,000 popular votes were cast for the president.

34. In 1970, the world production of cigarettes was 2,688,000,000,000.

35. In 1979, the total number of motor vehicle registrations in Florida was 5,395,000.

36. In 1980, there were 1,302,000 registered nurses in the United States.

EXERCISES FOR REVIEW

(1.1) **37.** There is a term that describes the visual displaying of a number. What is the term?

(1.1) **38.** What is the value of 5 in 26,518,206?

(1.2) **39.** Write 42,109 as you would read it.

(1.2) **40.** Write "six hundred twelve" using digits.

(1.2) **41.** Write "four billion eight" using digits.

★ **Answers to Practice Set (1.3)**

A. **1.** 3,400   **2.** 27,000   **3.** 31,000,000   **4.** 0   **5.** 60,000

1.4 Addition of Whole Numbers

**Section
Overview**

- ❑ **ADDITION**
- ❑ **ADDITION VISUALIZED ON THE NUMBER LINE**
- ❑ **THE ADDITION PROCESS**
- ❑ **ADDITION INVOLVING CARRYING**
- ❑ **CALCULATORS**

❑ ADDITION

Suppose we have two collections of objects that we combine together to form a third collection. For example,

⠿ is combined with ⠇ to yield ⠿⠿

We are combining a collection of four objects with a collection of three objects to obtain a collection of seven objects.

> The process of combining two or more objects (real or intuitive) to form a third, the total, is called **addition.**

Addition

Addends or Terms
Sum/+
=

In addition, the numbers being added are called **addends** or **terms,** and the total is called the **sum.** The **plus symbol** (+) is used to indicate addition, and the **equal symbol** (=) is used to represent the word "equal." For example, $4 + 3 = 7$ means "four added to three equals seven."

❑ ADDITION VISUALIZED ON THE NUMBER LINE

Addition is easily visualized on the number line. Let's visualize the addition of 4 and 3 using the number line.

To find $4 + 3$,

1. Start at 0.
2. Move to the right 4 units. We are now located at 4.
3. From 4, move to the right 3 units. We are now located at 7.

Thus, $4 + 3 = 7$.

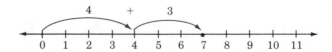

❑ THE ADDITION PROCESS

We'll study the process of addition by considering the sum of 25 and 43.

$$
\begin{array}{ll}
25 & \text{means} \\
+43 &
\end{array}
\qquad
\begin{array}{l}
2 \text{ tens} + 5 \text{ ones} \\
+4 \text{ tens} + 3 \text{ ones} \\
\hline
6 \text{ tens} + 8 \text{ ones}
\end{array}
$$

We write this as 68.

We can suggest the following procedure for adding whole numbers using this example.

The Process of Adding
Whole Numbers

> ### To **add whole numbers,**
>
> | | *Example:* |
> |---|---|
> | *The process:* | 25 |
> | 1. Write the numbers vertically, placing corresponding positions in the same column. | $+43$ |
> | 2. Add the digits in each column. Start at the right (in the ones position) and move to the left, placing the sum at the bottom. | 25 |
> | | $+43$ |
> | | 68 |

CAUTION: Confusion and incorrect sums can occur when the numbers are *not aligned* in columns properly. Avoid writing such additions as

$$\begin{array}{r} 25 \\ +\ \ 43 \\ \hline \end{array} \qquad \begin{array}{r} 25 \\ +43 \\ \hline \end{array}$$

☆ SAMPLE SET A

1. Add 276 and 103.

$$\begin{array}{r} 276 \\ +103 \\ \hline 379 \end{array}$$
    $6 + 3 = 9.$
    $7 + 0 = 7.$
    $2 + 1 = 3.$

2. Add 1459 and 130.

$$\begin{array}{r} 1459 \\ +\ \ 130 \\ \hline 1589 \end{array}$$
    $9 + 0 = 9.$
    $5 + 3 = 8.$
    $4 + 1 = 5.$
    $1 + 0 = 1.$

In each of these examples, each individual sum does not exceed 9. We will examine individual sums that exceed 9 in the next section.

★ PRACTICE SET A

Perform each addition. Show the expanded form in problems 1 and 2.

1. Add 63 and 25.   **2.** Add 4,026 and 1,501.   **3.** Add 231,045 and 36,121.

⬜ ADDITION INVOLVING CARRYING

It often happens in addition that the sum of the digits in a column will exceed 9. This happens when we add 18 and 34. We show this in expanded form as follows.

This sum exceeds 9.

$$\begin{array}{l} 18 = 1 \text{ ten } + \ 8 \text{ ones} \qquad\qquad\qquad 12 \text{ ones} \\ +34 = 3 \text{ tens} + \ 4 \text{ ones} \\ \hline \qquad\quad 4 \text{ tens} + 12 \text{ ones} = 4 \text{ tens} + 1 \text{ ten} + 2 \text{ ones} \\ \qquad\qquad\qquad\qquad\quad = 5 \text{ tens} + 2 \text{ ones} \\ \qquad\qquad\qquad\qquad\quad = 52 \end{array}$$

Notice that when we add the 8 ones to the 4 ones we get 12 ones. We then convert the 12 ones to 1 ten and 2 ones. In vertical addition, we show this conversion by **carrying** the ten to the tens column. We write a 1 at the top of the tens column to indicate the carry. This same example is shown in a shorter form as follows:

$$
\begin{array}{r}
1 \\
18 \\
+34 \\
\hline
52
\end{array}
\quad
\begin{array}{l}
8 + 4 = 12 \\
\text{Write 2, carry 1 ten to the top of the next column to the left.}
\end{array}
$$

☆ SAMPLE SET B

Perform the following additions. Use the process of carrying when needed.

1. Add 1875 and 358.

$$
\begin{array}{r}
111 \\
1875 \\
+\ 358 \\
\hline
2233
\end{array}
$$

$5 + 8 = 13.$	Write 3, carry 1 ten.
$1 + 7 + 5 = 13.$	Write 3, carry 1 hundred.
$1 + 8 + 3 = 12.$	Write 2, carry 1 thousand.
$1 + 1 = 2.$	

The sum is 2233.

2. Add 89,208 and 4,946.

$$
\begin{array}{r}
11\ \ 1 \\
89,208 \\
+\ 4,946 \\
\hline
94,154
\end{array}
$$

$8 + 6 = 14.$	Write 4, carry 1 ten.
$1 + 0 + 4 = 5.$	Write the 5 (nothing to carry).
$2 + 9 = 11.$	Write 1, carry one thousand.
$1 + 9 + 4 = 14.$	Write 4, carry one ten thousand.
$1 + 8 = 9.$	

The sum is 94,154.

3. Add 38 and 95.

$$
\begin{array}{r}
11 \\
38 \\
+\ 95 \\
\hline
133
\end{array}
$$

$8 + 5 = 13.$	Write 3, carry 1 ten.
$1 + 3 + 9 = 13.$	Write 3, carry 1 hundred.
$1 + 0 = 1.$	

As you proceed with the addition, it is a good idea to keep in mind what is actually happening.

$$
\begin{array}{r}
38 \\
+95
\end{array}
\quad\text{means}
$$

		3 tens	+ 8 ones
		+ 9 tens	+ 5 ones
		12 tens	+13 ones
=		12 tens + 1 ten +	3 ones
=		13 tens	+ 3 ones
= 1 hundred +		3 tens	+ 3 ones
= 133			

The sum is 133.

4. Find the sum 2648, 1359, and 861.

$$
\begin{array}{r}
111 \\
2648 \\
1359 \\
+\ 861 \\
\hline
4868
\end{array}
$$

$8 + 9 + 1 = 18.$	Write 8, carry 1 ten.
$1 + 4 + 5 + 6 = 16.$	Write 6, carry 1 hundred.
$1 + 6 + 3 + 8 = 18.$	Write 8, carry 1 thousand.
$1 + 2 + 1 = 4.$	

The sum is 4,868.

Numbers other than 1 can be carried as illustrated in problem 5.

5 Find the sum of the following numbers.

```
132 1
878016     6 + 5 + 1 + 7 = 19.      Write 9, carry the 1.
  9905     1 + 1 + 0 + 5 + 1 = 8.   Write 8.
 38951     0 + 9 + 9 + 8 = 26.      Write 6, carry the 2.
+ 56817    2 + 8 + 9 + 8 + 6 = 33.  Write 3, carry the 3.
983689     3 + 7 + 3 + 5 = 18.      Write 8, carry the 1.
           1 + 8 = 9.               Write 9.
```

The sum is 983,689.

6. The number of students enrolled at Riemann College in the years 1984, 1985, 1986, and 1987 was 10,406, 9,289, 10,108, and 11,412, respectively. What was the total number of students enrolled at Riemann College in the years 1985, 1986, and 1987?

We can determine the total number of students enrolled by adding 9,289, 10,108, and 11,412, the number of students enrolled in the years 1985, 1986, and 1987.

```
 1  11
  9,289
 10,108
+11,412
 30,809
```

The total number of students enrolled at Riemann College in the years 1985, 1986, and 1987 was 30,809.

★ PRACTICE SET B

Perform each addition. For problems 1–3, show the expanded form.

1. Add 58 and 29.   **2.** Add 476 and 85.   **3.** Add 27 and 88.

4. Add 67,898 and 85,627.

For problems 5–7, find the sums.

5.	**6.**	**7.**
57	847	16,945
26	825	8,472
84	796	387,721
		21,059
		629

❏ CALCULATORS

Calculators provide a very simple and quick way to find sums of whole numbers. For the two problems in Sample Set C, assume the use of a calculator that does not require the use of an ENTER key (such as many Hewlett-Packard calculators).

☆ **SAMPLE SET C**

Use a calculator to find each sum.

1. 34 + 21   **Display Reads**

Type	34	34
Press	+	34
Type	21	21
Press	=	55

The sum is 55.

2. 106 + 85 + 322 + 406   **Display Reads**

Type	106	106	The calculator keeps a running subtotal.
Press	+	106	
Type	85	85	
Press	+	191	⟵ **106 + 85**
Type	322	322	
Press	+	513	⟵ **191 + 322**
Type	406	406	
Press	=	919	⟵ **513 + 406**

The sum is 919.

★ **PRACTICE SET C**

Use a calculator to find the following sums.

1. 62 + 81 + 12   **2.** 9,261 + 8,543 + 884 + 1,062   **3.** 10,221 + 9,016 + 11,445

Answers to Practice Sets are on p. 28.

Section 1.4 EXERCISES

For problems 1–40, perform the additions. If you can, check each sum with a calculator.

1. 14 + 5

2. 12 + 7

3. 46 + 2

4. 83 + 16

5. 77 + 21

6.   321
      + 42

7.   916
       + 62

8.   104
       +561

9.   265
       +103

10. 552 + 237

11. 8,521 + 4,256

12.   16,408
        + 3,101

13.   16,515
        +42,223

14. 616,702 + 101,161

15. 43,156,219 + 2,013,520

16. 17 + 6

17. 25 + 8

18.   84
        + 7

19.   75
        + 6

20. 36 + 48

21. 74 + 17

22. 486 + 58

23. 743 + 66

24. 381 + 88

25.   687
        +175

26.   931
        +853

27. 1,428 + 893

28. 12,898 + 11,925

29.   631,464
        +509,740

30.   805,996
        + 98,516

31.   39,428,106
        +522,936,005

32. 5,288,423,100 + 16,934,785,995

33. 98,876,678,521,402 + 843,425,685,685,658

34. 41 + 61 + 85 + 62

35. $21 + 85 + 104 + 9 + 15$

36.
```
   116
    27
   110
   110
+   8
```

37.
```
 75,206
  4,152
+16,007
```

38.
```
  8,226
    143
 92,015
      8
487,553
  5,218
```

39.
```
   50,006
    1,005
  100,300
   20,008
1,000,009
  800,800
```

40.
```
       616
    42,018
     1,687
       225
 8,623,418
12,506,508
        19
     2,121
   195,643
```

For problems 41–50, perform the additions and round to the nearest hundred.

41.
```
1,468
2,183
```

42.
```
928,725
 15,685
```

43.
```
   82,006
3,019,528
```

44.
```
18,621
 5,059
```

45.
```
92
48
```

46.
```
16
37
```

47.
```
21
16
```

48.
```
11,172
22,749
12,248
```

49.
```
240
280
210
310
```

50.
```
  9,573
101,279
122,581
```

For problems 51–55, replace the letter m with the whole number that will make the addition true.

51.
```
  62
+ m
  67
```

52.
```
  106
+ m
  113
```

53.
```
  432
+ m
  451
```

54.
```
  803
+ m
  830
```

55.
```
1,893
+   m
1,981
```

56. The number of nursing and related care facilities in the United States in 1971 was 22,004. In 1978, the number was 18,722. What was the total number of facilities for both 1971 and 1978?

57. The number of persons on food stamps in 1975, 1979, and 1980 was 19,179,000, 19,309,000, and 22,023,000, respectively. What was the total number of people on food stamps for the years 1975, 1979, and 1980?

58. The enrollment in public and nonpublic schools in the years 1965, 1970, 1975, and 1984 was 54,394,000, 59,899,000, 61,063,000, and 55,122,000, respectively. What was the total enrollment for those years?

59. The area of New England is 3,618,770 square miles. The area of the Mountain states is 863,563 square miles. The area of the South Atlantic is 278,926 square miles. The area of the Pacific states is 921,392 square miles. What is the total area of these regions?

60. In 1960, the IRS received 1,188,000 corporate income tax returns. In 1965, 1,490,000 returns were received. In 1970, 1,747,000 returns were received. In 1972–1977, 1,890,000; 1,981,000; 2,043,000; 2,100,000; 2,159,000; and 2,329,000 returns were received, respectively. What was the total number of corporate tax returns received by the IRS during the years 1960, 1965, 1970, 1972–1977?

61. Find the total number of scientists employed in 1974.

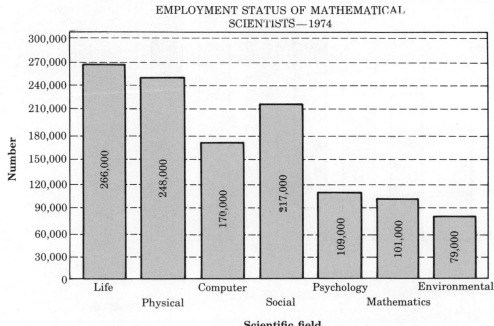

EMPLOYMENT STATUS OF MATHEMATICAL SCIENTISTS—1974

62. Find the total number of sales for space vehicle systems for the years 1965–1980.

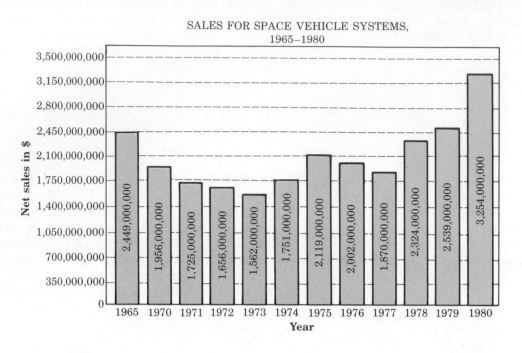

SALES FOR SPACE VEHICLE SYSTEMS, 1965–1980

63. Find the total baseball attendance for the years 1960–1980.

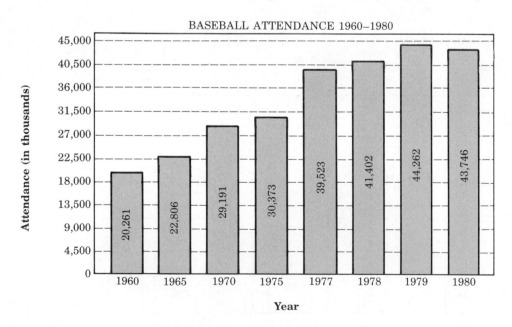

BASEBALL ATTENDANCE 1960–1980

64. Find the number of prosecutions of federal officials for 1970–1980.

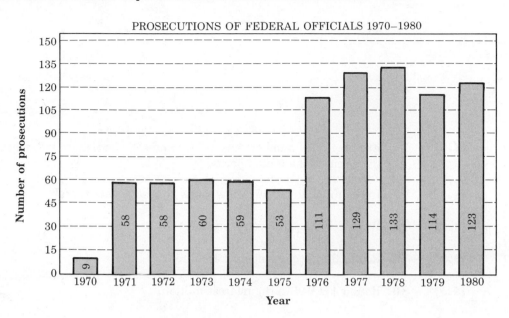

PROSECUTIONS OF FEDERAL OFFICIALS 1970–1980

For problems 65–75, try to add the numbers mentally.

65. 5
5
3
7

66. 8
2
6
4

69. 6
4
3
1
6
7
9
4

70. 20
30

67. 9
1
8
5
2

68. 5
2
5
8
3
7

71. 15
35

72. 16
14

73. 23
    27

74. 82
    18

75. 36
    14

EXERCISES FOR REVIEW

(1.1) **76.** Each period of numbers has its own name. From right to left, what is the name of the fourth period?

(1.1) **77.** In the number 610,467, how many thousands are there?

(1.2) **78.** Write 8,840 as you would read it.

(1.3) **79.** Round 6,842 to the nearest hundred.

(1.3) **80.** Round 431,046 to the nearest million.

★ Answers to Practice Sets (1.4)

A. **1.** 88,
  6 tens + 3 ones
  +2 tens + 5 ones
  8 tens + 8 ones

2. 5,527,
  4 thousands + 0 hundreds + 2 tens + 6 ones
  +1 thousand + 5 hundreds + 0 tens + 1 one
  5 thousands + 5 hundreds + 2 tens + 7 ones

3. 267,166

B. **1.** 87,
  5 tens + 8 ones
  +2 tens + 9 ones
  7 tens + 17 ones
  = 7 tens + 1 ten + 7 ones
  = 8 tens + 7 ones
  = 87

2. 561,
  4 hundreds + 7 tens + 6 ones
  +   8 tens + 5 ones
  4 hundreds + 15 tens + 11 ones
  = 4 hundreds + 15 tens + 1 ten + 1 one
  = 4 hundreds + 16 tens + 1 one
  = 4 hundreds + 1 hundred + 6 tens + 1 one
  = 5 hundreds + 6 tens + 1 one
  = 561

3. 115,
  2 tens + 7 ones
  + 8 tens + 8 ones
  10 tens + 15 ones
  = 10 tens + 1 ten + 5 ones
  = 11 tens + 5 ones
  = 1 hundred + 1 ten + 5 ones
  = 115

4. 153,525 **5.** 167 **6.** 2,468 **7.** 434,826

C. **1.** 155 **2.** 19,750 **3.** 30,682

DANIEL

1.5 Subtraction of Whole Numbers

**Section
Overview**

- ☑ SUBTRACTION
- ☑ SUBTRACTION AS THE OPPOSITE OF ADDITION
- ☑ THE SUBTRACTION PROCESS
- ☑ SUBTRACTION INVOLVING BORROWING
- ☑ BORROWING FROM ZERO
- ☑ CALCULATORS

☑ SUBTRACTION

Subtraction

> **Subtraction** is the process of determining the remainder when part of the total is removed.

Suppose the sum of two whole numbers is 11, and from 11 we remove 4. Using the number line to help our visualization, we see that if we are located at 11 and move 4 units to the left, and thus remove 4 units, we will be located at 7. Thus, 7 units remain when we remove 4 units from 11 units.

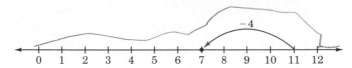

The Minus Symbol

Minuend

Subtrahend
Difference

The **minus symbol** ($-$) is used to indicate subtraction. For example, $11 - 4$ indicates that 4 is to be subtracted from 11. The number immediately in front of or above the minus symbol is called the **minuend,** and it represents the *original* number of units. The number immediately following or below the minus symbol is called the **subtrahend,** and it represents the number of units *to be removed.* The *result* of the subtraction is called the **difference** of the two numbers. For example, in $11 - 4 = 7$, 11 is the minuend, 4 is the subtrahend, and 7 is the difference.

☑ SUBTRACTION AS THE OPPOSITE OF ADDITION

Subtraction can be thought of as the opposite of addition. We show this in the problems in Sample Set A.

DANIEL

☆ SAMPLE SET A

$$8 - 5 = 3 \quad \text{since} \quad 3 + 5 = 8.$$
$$9 - 3 = 6 \quad \text{since} \quad 6 + 3 = 9.$$

★ PRACTICE SET A

Complete the following statements.

1. $7 - 5 =$ _____   since   _____ $+ 5 = 7$.

2. $9 - 1 =$ _____   since   _____ $+ 1 = 9$.

3. $17 - 8 =$ _____   since   _____ $+ 8 = 17$.

❑ THE SUBTRACTION PROCESS

We'll study the process of the subtraction of two whole numbers by considering the difference between 48 and 35.

$$
\begin{array}{r}
48 \\
-35 \\
\end{array}
\qquad \text{means} \qquad
\begin{array}{r}
4 \text{ tens} + 8 \text{ ones} \\
-3 \text{ tens} - 5 \text{ ones} \\
\hline
1 \text{ ten} + 3 \text{ ones} \\
\end{array}
$$

which we write as 13.

The Process of Subtracting Whole Numbers

To subtract two whole numbers,	
The process	*Example*
1. Write the numbers vertically, placing corresponding positions in the same column.	$\begin{array}{r} 48 \\ -35 \end{array}$
2. Subtract the digits in each column. Start at the right, in the ones position, and move to the left, placing the difference at the bottom.	$\begin{array}{r} 48 \\ -35 \\ \hline 13 \end{array}$

☆ SAMPLE SET B

Perform the following subtractions.

1.
$$\begin{array}{r} 275 \\ -142 \\ \hline 133 \end{array}$$
$5 - 2 = 3.$
$7 - 4 = 3.$
$2 - 1 = 1.$

2.
$$\begin{array}{r} 46{,}042 \\ -\ 1{,}031 \\ \hline 45{,}011 \end{array}$$
$2 - 1 = 1.$
$4 - 3 = 1.$
$0 - 0 = 0.$
$6 - 1 = 5.$
$4 - 0 = 4.$

3. Find the difference between 977 and 235.

Write the numbers vertically, placing the larger number on top. Line up the columns properly.

$$\begin{array}{r} 977 \\ -235 \\ \hline 742 \end{array}$$

The difference between 977 and 235 is 742.

4. In Keys County in 1987, there were 809 cable television installations. In Flags County in 1987, there were 1,159 cable television installations. How many more cable television installations were there in Flags County than in Keys County in 1987?

We need to determine the difference between 1,159 and 809.

$$\begin{array}{r} \overset{1\ 1}{} \\ 1{,}159 \\ -\ \ 809 \\ \hline 350 \end{array}$$

There were 350 more cable television installations in Flag County than in Keys County in 1987.

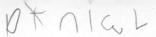

★ **PRACTICE SET B**

Perform the following subtractions.

1. 534
    −203

2. 857
    − 43

3. 95,628
    −34,510

4. 11,005
    − 1,005

5. Find the difference between 88,526 and 26,412.

In each of these problems, each bottom digit is less than the corresponding top digit. This may not always be the case. We will examine the case where the bottom digit is greater than the corresponding top digit in the next section.

☑ SUBTRACTION INVOLVING BORROWING

Minuend
Subtrahend

It often happens in the subtraction of two whole numbers that a digit in the **minuend** (top number) will be less than the digit in the same position in the **subtrahend** (bottom number). This happens when we subtract 27 from 84.

   84
 −27

We do not have a name for 4 − 7. We need to rename 84 in order to continue. We'll do so as follows:

   84 = 8 tens + 4 ones
 −27 = 2 tens + 7 ones

    7 tens + 1 ten + 4 ones
    2 tens   + 7 ones

    7 tens + 10 ones + 4 ones
    2 tens   + 7 ones

Our new name for 84 is 7 tens + 14 ones.

    7 tens + 14 ones
    2 tens + 7 ones
    5 tens + 7 ones
    = 57

Notice that we converted 8 tens to 7 tens + 1 ten, and then we converted the 1 ten to 10 ones. We then had 14 ones and were able to perform the subtraction.

Borrowing

The process of **borrowing** (converting) is illustrated in the problems of Sample Set C.

☆ SAMPLE SET C

```
        714
1.    8̸4        1. Borrow 1 ten from the 8 tens. This leaves 7 tens.
    − 27         2. Convert the 1 ten to 10 ones.
      57         3. Add 10 ones to 4 ones to get 14 ones.
```

```
        517
2.    6̸72       1. Borrow 1 hundred from the 6 hundreds. This leaves 5 hundreds.
    −  91        2. Convert the 1 hundred to 10 tens.
      581        3. Add 10 tens to 7 tens to get 17 tens.
```

★ PRACTICE SET C

Perform the following subtractions. Show the expanded form for problems 1, 2, and 3.

1. 53	2. 76	3. 872	4. 441	5. 775	6. 5,663
−35	−28	−565	−356	− 66	−2,559

Borrowing More Than Once

Sometimes it is necessary to borrow more than once. This is shown in the problems in Sample Set D.

☆ SAMPLE SET D

Perform the subtractions. Borrowing more than once if necessary.

```
        513
        3̸11
1.    6̸4̸1       1. Borrow 1 ten from the 4 tens. This leaves 3 tens.
    −  358       2. Convert the 1 ten to 10 ones.
      283        3. Add 10 ones to 1 one to get 11 ones. We can now perform 11 − 8.
                 4. Borrow 1 hundred from the 6 hundreds. This leaves 5 hundreds.
                 5. Convert the 1 hundred to 10 tens.
                 6. Add 10 tens to 3 tens to get 13 tens.
                 7. Now 13 − 5 = 8.
                 8. 5 − 3 = 2.
```

```
        12
        4̸2̸14
2.    5̸3̸4       1. Borrow 1 ten from the 3 tens. This leaves 2 tens.
    −   85       2. Convert the 1 ten to 10 ones.
      449        3. Add 10 ones to 4 ones to get 14 ones. We can now perform 14 − 5.
                 4. Borrow 1 hundred from the 5 hundreds. This leaves 4 hundreds.
                 5. Convert the 1 hundred to 10 tens.
                 6. Add 10 tens to 2 tens to get 12 tens. We can now perform 12 − 8 = 4.
                 7. Finally, 4 − 0 = 4.
```

3.   71529
  − 6952

After borrowing, we have

$$\begin{array}{r} 10 \\ 14 \\ 6\cancel{0}\!\!\not{4}12 \\ 71529 \\ -\ \ 6952 \\ \hline 64577 \end{array}$$

★ **PRACTICE SET D**

Perform the following subtractions.

1.   526
  −358

2.   63,419
  − 7,779

3.   4,312
  −3,123

☐ **BORROWING FROM ZERO**

It often happens in a subtraction problem that we have to borrow from one or more zeros. This occurs in problems such as

(1)   503   and   (2)   5000
   − 37   −   37

We'll examine each case.

1. Borrowing from a single zero.

   Consider the problem   503
                         − 37

   Since we do not have a name for $3 - 7$, we must borrow from 0.

   $$\begin{array}{l} 503 = 5\ \text{hundreds} + 0\ \text{tens} + 3\ \text{ones} \\ -\ 37 \hspace{2.5cm} 3\ \text{tens} + 7\ \text{ones} \\ \hline \end{array}$$

   Since there are no tens to borrow, we must borrow 1 hundred. One hundred = 10 tens.

   $$\begin{array}{l} 4\ \text{hundreds} + 10\ \text{tens} + 3\ \text{ones} \\ \hspace{2.3cm} 3\ \text{tens} + 7\ \text{ones} \\ \hline \end{array}$$

   We can now borrow 1 ten from 10 tens (leaving 9 tens). One ten = 10 ones and 10 ones + 3 ones = 13 ones.

   $$\begin{array}{l} 4\ \text{hundreds} + 9\ \text{tens} + 13\ \text{ones} \\ \hspace{2.3cm} 3\ \text{tens} + \ 7\ \text{ones} \\ \hline 4\ \text{hundreds} + 6\ \text{tens} + \ \ 6\ \text{ones} = 466 \end{array}$$

Now we can suggest the following method for borrowing from a single zero.

Borrowing from a Single Zero

To borrow from a single zero,

1. Decrease the digit to the immediate left of zero by one.
2. Draw a line through the zero and make it a 10.
3. Proceed to subtract as usual.

☆ **SAMPLE SET E**

Perform this subtraction.

$$\begin{array}{r} 503 \\ -\ 37 \\ \hline \end{array}$$

The number 503 contains a single zero

1. The number to the immediate left of 0 is 5. Decrease 5 by 1.

$$5 - 1 = 4$$

$$\begin{array}{r} ^{4\,10} \\ 5\cancel{0}3 \\ -\ 37 \\ \hline \end{array}$$

2. Draw a line through the zero and make it a 10.
3. Borrow from the 10 and proceed.

1 ten = 10 ones
10 ones + 3 ones = 13 ones

$$\begin{array}{r} ^{9} \\ ^{4\,\cancel{10}\,13} \\ 5\cancel{0}\cancel{3} \\ -\ 37 \\ \hline 466 \end{array}$$

★ **PRACTICE SET E**

Perform each subtraction.

1. $\begin{array}{r} 906 \\ -\ 18 \\ \hline \end{array}$ **2.** $\begin{array}{r} 5102 \\ -\ 559 \\ \hline \end{array}$ **3.** $\begin{array}{r} 9055 \\ -\ 386 \\ \hline \end{array}$

2. Borrowing from a group of zeros.

Consider the problem $\begin{array}{r} 5000 \\ -\ 37 \\ \hline \end{array}$

In this case, we have a group of zeros.

$\begin{array}{rl} 5000 = & 5 \text{ thousands} + 0 \text{ hundred} + 0 \text{ tens} + 0 \text{ ones} \\ -\ 37 = & \underline{\hspace{5cm} 3 \text{ tens} + 7 \text{ ones}} \end{array}$

Since we cannot borrow any tens or hundreds, we must borrow 1 thousand. One thousand = 10 hundreds.

$\begin{array}{l} 4 \text{ thousands} + 10 \text{ hundreds} + 0 \text{ tens} + 0 \text{ ones} \\ \underline{\hspace{6cm} 3 \text{ tens} + 7 \text{ ones}} \end{array}$

We can now borrow 1 hundred from 10 hundreds. One hundred = 10 tens.

4 thousands + 9 hundreds + 10 tens + 0 ones
                             3 tens + 7 ones
_____

We can now borrow 1 ten from 10 tens. One ten = 10 ones.

4 thousands + 9 hundreds + 9 tens + 10 ones
                             3 tens + 7 ones
_____
4 thousands + 9 hundreds + 6 tens + 3 ones = 4,963

From observations made in this procedure we can suggest the following method for borrowing from a group of zeros.

Borrowing from a Group of Zeros

> To borrow from a group of zeros,
>
> 1. Decrease the digit to the immediate left of the group of zeros by one.
> 2. Draw a line through each zero in the group and make it a 9, except the rightmost zero, make it 10.
> 3. Proceed to subtract as usual.

☆ SAMPLE SET F

Perform each subtraction.

1.   40,000
       −   125

The number 40,000 contains a group of zeros.

1. The number to the immediate left of the group is 4. Decrease 4 by 1.

   $4 - 1 = 3$

$$\begin{array}{r} 3\,9\ 9\,9\,10 \\ 4\cancel{0},\cancel{0}\cancel{0}\cancel{0} \\ -\quad\ \ 125 \\ \end{array}$$

2. Make each 0, except the rightmost one, 9. Make the rightmost 0 a 10.
3. Subtract as usual.

$$\begin{array}{r} 3\,9\ 9\,9\,10 \\ 4\cancel{0},\cancel{0}\cancel{0}\cancel{0} \\ -\quad\ \ 125 \\ \hline 39{,}875 \end{array}$$

2.   8,000,006
       −   41,107

The number 8,000,006 contains a group of zeros.

1. The number to the immediate left of the group is 8. Decrease 8 by 1.

   $8 - 1 = 7$

$$\begin{array}{r} 7\ 9\,9\,9\ 9\,1\,0 \\ \cancel{8},\cancel{0}\cancel{0}\cancel{0},\cancel{0}\cancel{0}6 \\ -\quad\ \ 41{,}107 \\ \end{array}$$

2. Make each zero, except the rightmost one, 9. Make the rightmost 0 a 10.
3. To perform the subtraction, we'll need to borrow from the ten.

   1 ten = 10 ones
   10 ones + 6 ones = 16 ones

$$\begin{array}{r} 9 \\ 7\ 9\,9\,9\ \cancel{9}\cancel{1}\cancel{0}16 \\ \cancel{8},\cancel{0}\cancel{0}\cancel{0},\cancel{0}\cancel{0}6 \\ -\quad\ \ 41{,}107 \\ \hline 7{,}958{,}899 \end{array}$$

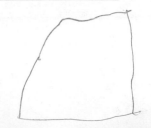

★ PRACTICE SET F

Perform each subtraction.

1.   21,007
      − 4,873

2.   10,004
      − 5,165

3.   16,000,000
      −   201,060

☐ CALCULATORS

In practice, calculators are used to find the difference between two whole numbers.

☆ SAMPLE SET G

Find the difference between 1006 and 284.

		Display Reads
Type	1006	1006
Press	−	1006
Type	284	284
Press	=	722

The difference between 1006 and 284 is 722.

(What happens if you type 284 first and then 1006? We'll study such numbers in Chapter 10.)

★ PRACTICE SET G

1. Use a calculator to find the difference between 7338 and 2809.

2. Use a calculator to find the difference between 31,060,001 and 8,591,774.

Answers to Practice Sets are on p. 43.

Section 1.5 EXERCISES

For problems 1–35, perform the subtractions. You may check each difference with a calculator.

1.   15
     − 8

2.   19
     − 8

3.   11
     − 5

4.   14
     − 6

5. $\begin{array}{r} 12 \\ -\ 9 \\ \hline \end{array}$ 6. $\begin{array}{r} 56 \\ -12 \\ \hline \end{array}$

7. $\begin{array}{r} 74 \\ -33 \\ \hline \end{array}$ 8. $\begin{array}{r} 80 \\ -61 \\ \hline \end{array}$

9. $\begin{array}{r} 350 \\ -141 \\ \hline \end{array}$ 10. $\begin{array}{r} 800 \\ -650 \\ \hline \end{array}$

11. $\begin{array}{r} 35,002 \\ -14,001 \\ \hline \end{array}$ 12. $\begin{array}{r} 5,000,566 \\ -2,441,326 \\ \hline \end{array}$

13. $\begin{array}{r} 400,605 \\ -121,352 \\ \hline \end{array}$ 14. $\begin{array}{r} 46,400 \\ -\ 2,012 \\ \hline \end{array}$

15. $\begin{array}{r} 77,893 \\ -\ \ \ 421 \\ \hline \end{array}$ 16. $\begin{array}{r} 42 \\ -18 \\ \hline \end{array}$

17. $\begin{array}{r} 51 \\ -27 \\ \hline \end{array}$ 18. $\begin{array}{r} 622 \\ -\ 88 \\ \hline \end{array}$

19. $\begin{array}{r} 261 \\ -\ 73 \\ \hline \end{array}$ 20. $\begin{array}{r} 242 \\ -158 \\ \hline \end{array}$

21. $\begin{array}{r} 3,422 \\ -1,045 \\ \hline \end{array}$ 22. $\begin{array}{r} 5,565 \\ -3,985 \\ \hline \end{array}$

23. $\begin{array}{r} 42,041 \\ -15,355 \\ \hline \end{array}$ 24. $\begin{array}{r} 304,056 \\ -\ 20,008 \\ \hline \end{array}$

25. $\begin{array}{r} 64,000,002 \\ -\ \ \ 856,743 \\ \hline \end{array}$ 26. $\begin{array}{r} 4,109 \\ -\ 856 \\ \hline \end{array}$

27. $\begin{array}{r} 10,113 \\ -\ 2,079 \\ \hline \end{array}$ 28. $\begin{array}{r} 605 \\ -\ 77 \\ \hline \end{array}$

29. $\begin{array}{r} 59 \\ -26 \\ \hline \end{array}$ 30. $\begin{array}{r} 36,107 \\ -\ 8,314 \\ \hline \end{array}$

31. $\begin{array}{r} 92,526,441,820 \\ -59,914,805,253 \\ \hline \end{array}$ 32. $\begin{array}{r} 1,605 \\ -\ 881 \\ \hline \end{array}$

33. $\begin{array}{r} 30,000 \\ -26,062 \\ \hline \end{array}$ 34. $\begin{array}{r} 600 \\ -216 \\ \hline \end{array}$

35. $\begin{array}{r} 9,000,003 \\ -\ \ 726,048 \\ \hline \end{array}$

For problems 36–58, perform each subtraction.

36. Subtract 63 from 92.
(*Hint:* The word "from" means "beginning at." Thus, 63 from 92 means beginning at 92, or 92 − 63.)

37. Subtract 35 from 86.

38. Subtract 382 from 541.

39. Subtract 1,841 from 5,246.

40. Subtract 26,082 from 35,040.

41. Find the difference between 47 and 21.

42. Find the difference between 1,005 and 314.

43. Find the difference between 72,085 and 16.

44. Find the difference between 7,214 and 2,049.

45. Find the difference between 56,108 and 52,911.

46. How much bigger is 92 than 47?

47. How much bigger is 114 than 85?

48. How much bigger is 3,006 than 1,918?

49. How much bigger is 11,201 than 816?

50. How much bigger is 3,080,020 than 1,814,161?

51. In Wichita, Kansas, the sun shines about 74% of the time in July and about 59% of the time in November. How much more of the time (in percent) does the sun shine in July than in November?

52. The lowest temperature on record in Concord, New Hampshire in May is 21°F, and in July it is 35°F. What is the difference in these lowest temperatures?

53. In 1980, there were 83,000 people arrested for prostitution and commercialized vice and 11,330,000 people arrested for driving while intoxicated. How many more people were arrested for drunk driving than for prostitution?

54. In 1980, a person with a bachelor's degree in accounting received a monthly salary offer of $1,293, and a person with a marketing degree a monthly salary offer of $1,145. How much more was offered to the person with an accounting degree than the person with a marketing degree?

55. In 1970, there were about 793 people per square mile living in Puerto Rico, and 357 people per square mile living in Guam. How many more people per square mile were there in Puerto Rico than Guam?

56. The 1980 population of Singapore was 2,414,000 and the 1980 population of Sri Lanka was 14,850,000. How many more people lived in Sri Lanka than in Singapore in 1980?

57. In 1977, there were 7,234,000 hospitals in the United States and 64,421,000 in Mainland China. How many more hospitals were there in Mainland China than in the United States in 1977?

58. In 1978, there were 3,095,000 telephones in use in Poland and 4,292,000 in Switzerland. How many more telephones were in use in Switzerland than in Poland in 1978?

For problems 59–64, use the corresponding graphs to solve the problems.

59. How many more life scientists were there in 1974 than mathematicians?

60. How many more social, psychological, mathematical, and environmental scientists were there than life, physical, and computer scientists?

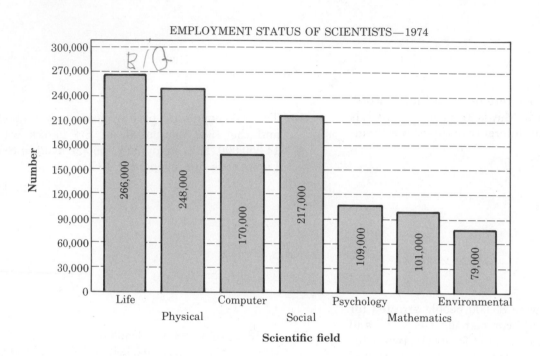

EMPLOYMENT STATUS OF SCIENTISTS—1974

61. How many more prosecutions were there in 1978 than in 1974?

62. How many more prosecutions were there in 1976–1980 than in 1970–1975?

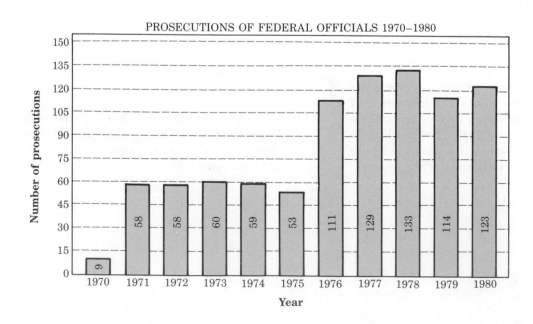

PROSECUTIONS OF FEDERAL OFFICIALS 1970–1980

63. How many more dry holes were drilled in 1960 than in 1975?

64. How many more dry holes were drilled in 1960, 1965, and 1970 than in 1975, 1978 and 1979?

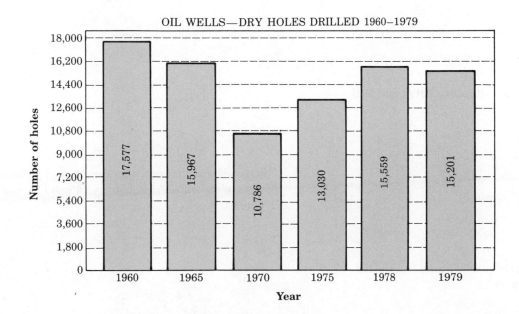

OIL WELLS—DRY HOLES DRILLED 1960–1979

For problems 65–69, replace the ▨ with the whole number that will make the subtraction true.

65.
$$\begin{array}{r} 14 \\ - \; ▨ \\ \hline 3 \end{array}$$

66.
$$\begin{array}{r} 21 \\ - \; ▨ \\ \hline 14 \end{array}$$

67.
$$\begin{array}{r} 35 \\ - \; □ \\ \hline 25 \end{array}$$

68.
$$\begin{array}{r} 16 \\ - \; ▨ \\ \hline 9 \end{array}$$

69.
$$\begin{array}{r} 28 \\ - \; □ \\ \hline 16 \end{array}$$

For problems 70–79, find the solutions.

70. Subtract 42 from the sum of 16 and 56.

71. Subtract 105 from the sum of 92 and 89.

72. Subtract 1,127 from the sum of 2,161 and 387.

73. Subtract 37 from the difference between 263 and 175.

74. Subtract 1,109 from the difference between 3,046 and 920.

75. Add the difference between 63 and 47 to the difference between 55 and 11.

76. Add the difference between 815 and 298 to the difference between 2,204 and 1,016.

77. Subtract the difference between 78 and 43 from the sum of 111 and 89.

78. Subtract the difference between 18 and 7 from the sum of the differences between 42 and 13, and 81 and 16.

79. Find the difference between the differences of 343 and 96, and 521 and 488.

EXERCISES FOR REVIEW

(1.1)   **80.** In the number 21,206, how many hundreds are there?

(1.1)   **81.** Write a three-digit number that has a zero in the ones position.

(1.1)   **82.** How many three-digit whole numbers are there?

(1.3)   **83.** Round 26,524,016 to the nearest million.

(1.4)   **84.** Find the sum of $846 + 221 + 116$.

★ **ANSWERS TO PRACTICE SETS (1.5)**

A. **1.** $7 - 5 = 2$ since $2 + 5 = 7$ **2.** $9 - 1 = 8$ since $8 + 1 = 9$ **3.** $17 - 8 = 9$ since $9 + 8 = 17$

B. **1.** 331   **2.** 814   **3.** 61,118   **4.** 10,000   **5.** 62,114

C. **1.** 18, $\begin{array}{l} 5 \text{ tens} + 3 \text{ ones} \\ - \quad 3 \text{ tens} + 5 \text{ ones} \\ \hline 4 \text{ tens} + 1 \text{ ten} + 3 \text{ ones} \\ - \quad 3 \text{ tens} \qquad\quad + 5 \text{ ones} \\ \hline 4 \text{ tens} + 13 \text{ ones} \\ - \quad 3 \text{ tens} + \ 5 \text{ ones} \\ \hline 1 \text{ ten} + \ 8 \text{ ones} \\ = 18 \end{array}$

   2. 48, $\begin{array}{l} 7 \text{ tens} + 6 \text{ ones} \\ - \quad 2 \text{ tens} + 8 \text{ ones} \\ \hline 6 \text{ tens} + 1 \text{ ten} + 6 \text{ ones} \\ - \quad 2 \text{ tens} \qquad\quad + 8 \text{ ones} \\ \hline 6 \text{ tens} + 16 \text{ ones} \\ - \quad 2 \text{ tens} + \ 8 \text{ ones} \\ \hline 4 \text{ tens} + \ 8 \text{ ones} \\ = 48 \end{array}$

   3. 307, $\begin{array}{l} 8 \text{ hundreds} + 7 \text{ tens} + 2 \text{ ones} \\ - \quad 5 \text{ hundreds} + 6 \text{ tens} + 5 \text{ ones} \\ \hline 8 \text{ hundreds} + 6 \text{ tens} + 1 \text{ ten} + 2 \text{ ones} \\ - \quad 5 \text{ hundreds} + 6 \text{ tens} \qquad\quad + 5 \text{ ones} \\ \hline 8 \text{ hundreds} + 6 \text{ tens} + 12 \text{ ones} \\ - \quad 5 \text{ hundreds} + 6 \text{ tens} + \ 5 \text{ ones} \\ \hline 3 \text{ hundreds} + 0 \text{ tens} + \ 7 \text{ ones} \\ = 307 \end{array}$

   4. 85   **5.** 709   **6.** 3,104

D. **1.** 168   **2.** 55,640   **3.** 1,189

E. **1.** 888   **2.** 4,543   **3.** 8,669

F. **1.** 16,134   **2.** 4,839   **3.** 15,789,940

G. **1.** 4,520   **2.** 22,468,227

1.6 Properties of Addition

Section
Overview
☐ **THE COMMUTATIVE PROPERTY OF ADDITION**
☐ **THE ASSOCIATIVE PROPERTY OF ADDITION**
☐ **THE ADDITIVE IDENTITY**

We now consider three simple but very important properties of addition.

Commutative Property of
Addition

☐ THE COMMUTATIVE PROPERTY OF ADDITION

If two whole numbers are added in any order, the sum will not change.

☆ SAMPLE SET A

Add the whole numbers

$$
\begin{array}{l}
8 \\
5
\end{array}
\qquad
\begin{array}{l}
8 + 5 = 13 \\
5 + 8 = 13
\end{array}
$$

The numbers 8 and 5 can be added in any order. Regardless of the order they are added, the sum is 13.

★ PRACTICE SET A

1. Use the commutative property of addition to find the sum of 12 and 41 in two different ways.

$$
\begin{array}{l}
12 \\
41
\end{array}
$$

2. Add the whole numbers

$$
\begin{array}{l}
837 \\
1{,}958
\end{array}
$$

☐ THE ASSOCIATIVE PROPERTY OF ADDITION

Associative Property of
Addition

If three whole numbers are to be added, the sum will be the same if the first two are added first, then that sum is added to the third, or, the second two are added first, and that sum is added to the first.

Using Parentheses

It is a common mathematical practice to **use parentheses** to show which pair of numbers we wish to combine first.

☆ **SAMPLE SET B**

Add the whole numbers.

43
16
27

43 and 16 are associated.
$(43 + 16) + 27 = 59 + 27 = 86.$
$43 + (16 + 27) = 43 + 43 = 86.$
16 and 27 are associated.

★ **PRACTICE SET B**

Use the associative property of addition to add the following whole numbers two different ways.

1.
17
32
25

2.
1,629
806
429

☐ THE ADDITIVE IDENTITY

0 Is the Additive Identity

The whole number 0 is called the **additive identity,** since when it is added to any whole number, the sum is identical to that whole number.

☆ **SAMPLE SET C**

Add the whole numbers.

29
0

$29 + 0 = 29.$
$0 + 29 = 29.$

Zero added to 29 does not change the identity of 29.

★ **PRACTICE SET C**

Add the following whole numbers.

1.
8
0

2.
0
5

3. Suppose we let the letter x represent a choice for some whole number. For problems (a) and (b), find the sums. For problem (c), find the sum provided we now know that x represents the whole number 17.

(a)
```
       0
  x
```

(b)
```
  0
      x
```

(c)
```
  x
      0
```

Answers to Practice Sets are on p. 47.

Section 1.6 EXERCISES

For problems 1–15, add the numbers in two ways.

1.
```
  8
     29
```

2.
```
  36
     12
```

9.
```
   32
       8
  5
```

10.
```
   16
  18
       14
```

3.
```
      36
  48
```

4.
```
        26
  117
```

11.
```
    52
  10
       38
```

12.
```
   84
        7
  36
```

5.
```
  456
     112
```

6.
```
  1,096
  4,251
```

13.
```
    114
  17
        425
```

14.
```
      1019
  11
          586
```

7.
```
  73,205
     49,118
```

8.
```
  265,094
     32,508
```

15.
```
   37,728
       1,261
  4,472
```

For problems 16–19, show that the pairs of quantities yield the same sum.

16. $(11 + 27) + 9$ and $11 + (27 + 9)$

17. $(80 + 52) + 6$ and $80 + (52 + 6)$

18. $(114 + 226) + 108$ and $114 + (226 + 108)$

19. $(731 + 256) + 171$ and $731 + (256 + 171)$

20. The fact that

(a first number + a second number) + third number = a first number + (a second number + a third number)

is an example of the _____ property of addition.

21. The fact that

0 + any number = that particular number

is an example of the _____ property of addition.

22. The fact that

a first number + a second number = a second number + a first number

is an example of the _____ property of addition.

23. Use the numbers 15 and 8 to illustrate the commutative property of addition.

24. Use the numbers 6, 5, and 11 to illustrate the associative property of addition.

25. The number zero is called the additive identity. Why is the term identity so appropriate?

EXERCISES FOR REVIEW

(1.1) **26.** How many hundreds in 46,581?

(1.2) **27.** Write 2,218 as you would read it.

(1.3) **28.** Round 506,207 to the nearest thousand.

(1.4) **29.** Find the sum of $\begin{array}{r} 482 \\ + \ 68 \end{array}$

(1.5) **30.** Find the difference: $\begin{array}{r} 3,318 \\ - \ 429 \end{array}$

★ **ANSWERS TO PRACTICE SETS (1.6)**

A. **1.** $12 + 41 = 53$ and $41 + 12 = 53$
2. $837 + 1,958 = 2,795$ and $1,958 + 837 = 2,795$

B. **1.** $(17 + 32) + 25 = 49 + 25 = 74$ and $17 + (32 + 25) = 17 + 57 = 74$
2. $(1,629 + 806) + 429 = 2,435 + 429 = 2,864$
$1,629 + (806 + 429) = 1,629 + 1,235 = 2,864$

C. **1.** 8   **2.** 5   **3.** (a) x (b) x (c) 17

Number / Numeral (1.1)

A *number* is a concept. It exists only in the mind. A *numeral* is a symbol that represents a number. It is customary not to distinguish between the two (but we should remain aware of the difference).

Hindu-Arabic Numeration System (1.1)

In our society, we use the *Hindu-Arabic* numeration system. It was invented by the Hindus shortly before the third century and popularized by the Arabs about a thousand years later.

Digits (1.1)

The numbers 0, 1, 2, 3, 4, 5, 6, 7, 8, 9 are called *digits*.

Base Ten Positional System (1.1)

The Hindu-Arabic numeration system is a positional number system with *base ten*. Each position has value that is ten times the value of the position to its right.

Commas / Periods (1.1)

Commas are used to separate digits into groups of three. Each group of three is called a *period*. Each period has a name. From right to left, they are ones, thousands, millions, billions, etc.

Whole Numbers (1.1)

A *whole number* is any number that is formed using only the digits (0, 1, 2, 3, 4, 5, 6, 7, 8, 9).

Number Line (1.1)

The *number line* allows us to visually display the whole numbers.

Graphing (1.1)

Graphing a whole number is a term used for visually displaying the whole number. The graph of 4 appears below.

Reading Whole Numbers (1.2)

To express a whole number as a verbal phrase:

1. Begin at the right and, working right to left, separate the number into distinct periods by inserting commas every three digits.
2. Begin at the left, and read each period individually.

Writing Whole Numbers (1.2)

To rename a number that is expressed in words to a number expressed in digits:

1. Notice that a number expressed as a verbal phrase will have its periods set off by commas.
2. Start at the beginning of the sentence, and write each period of numbers individually.
3. Use commas to separate periods, and combine the periods to form one number.

Rounding (1.3)

Rounding is the process of approximating the number of a group of objects by mentally "seeing" the collection as occurring in groups of tens, hundreds, thousands, etc.

Addition (1.4)

Addition is the process of combining two or more objects (real or intuitive) to form a new, third object, the total, or sum.

Addends / Sum (1.4)

In addition, the numbers being added are called *addends* and the result, or total, the *sum*.

Subtraction (1.5)

Subtraction is the process of determining the remainder when part of the total is removed.

Minuend / Subtrahend Difference **(1.5)**	$18 - 11 = 7$

minuend   subtrahend   difference

Commutative Property of Addition (1.6)

If two whole numbers are added in either of two orders, the sum will not change.

$3 + 5 = 5 + 3$

Associative Property of Addition (1.6)

If three whole numbers are to be added, the sum will be the same if the first two are added and that sum is then added to the third, or if the second two are added and the first is added to that sum.

$(3 + 5) + 2 = 3 + (5 + 2)$

Parentheses in Addition (1.6)

Parentheses in addition indicate which numbers are to be added first.

Additive Identity (1.6)

The whole number 0 is called the *additive identity* since, when it is added to any particular whole number, the sum is identical to that whole number.

$0 + 7 = 7$
$7 + 0 = 7$

EXERCISE SUPPLEMENT

For problems 1–35, find the sums and differences.

1. 908
    + 29

2. 529
    +161

3. 549
    + 16

4. 726
    +892

5. 390
    +169

6. 166
    +660

7. 391
    +951

8. 48
    +36

9. 1,103
    + 898

10. 1,642
    + 899

11. 807
    +1,156

12. 80,349
    + 2,679

13. 70,070
    + 9,386

14. 90,874
    + 2,945

15. 45,292
    +51,661

16. 1,617
    +54,923

17. 702,607
    + 89,217

18. 6,670,006
    + 2,495

19. 267
    +8,034

20. 7,007
    +11,938

21. 131,294
    + 9,087

22. 5,292
    + 161

23. 17,260
    +58,964

24. 7,006
    −5,382

25. 7,973
    −3,018

26. 16,608
    − 1,660

27. 209,527
    − 23,916

28. 584
    −226

29. 3,313
    −1,075

30. 458
    −122

31. 1,007
    + 331

32. 16,082
    + 2,013

33. 926
    − 48

34. 736
    +5,869

35. 676,504
    − 58,277

For problems 36–39, add the numbers.

36. 769
    795
    298
    746

37. 554
    184
    883

38. 30,188
    79,731
    16,600
    66,085
    39,169
    95,170

39. 2,129
    6,190
    17,044
    30,447
    292
    41
    428,458

For problems 40–50, combine the numbers as indicated.

40. 2,957 + 9,006

41. 19,040 + 813

42. 350,212 + 14,533

43. 970 + 702 + 22 + 8

44. 3,704 + 2,344 + 429 + 10,374 + 74

45. 874 + 845 + 295 − 900

46. 904 + 910 − 881

47. $521 + 453 - 334 + 660$

48. $892 - 820 - 9$

49. $159 + 4,085 - 918 - 608$

50. $2,562 + 8,754 - 393 - 385 - 910$

For problems 51–63, add and subtract as indicated.

51. Subtract 671 from 8,027.

52. Subtract 387 from 6,342.

53. Subtract 2,926 from 6,341.

54. Subtract 4,355 from the sum of 74 and 7,319.

55. Subtract 325 from the sum of 7,188 and 4,964.

56. Subtract 496 from the difference of 60,321 and 99.

57. Subtract 20,663 from the difference of 523,150 and 95,225.

58. Add the difference of 843 and 139 to the difference of 4,450 and 839.

59. Add the difference of 997,468 and 292,513 to the difference of 22,140 and 8,617.

60. Subtract the difference of 8,412 and 576 from the sum of 22,140 and 8,617.

61. Add the sum of 2,273, 3,304, 847, and 16 to the difference of 4,365 and 864.

62. Add the sum of 19,161, 201, 166,127, and 44 to the difference of the sums of 161, 2,455, and 85, and 21, 26, 48, and 187.

63. Is the sum of 626 and 1,242 the same as the sum of 1,242 and 626? Justify your claim.

1. _WHATYOU_

1. **(1.1)** What is the largest digit?

2. _NUMBER_

2. **(1.1)** In the Hindu-Arabic number system, each period has three values assigned to it. These values are the same for each period. From right to left, what are they?

3. _X_

3. **(1.1)** In the number 42,826, how many hundreds are there?

4. _X_

4. **(1.1)** Is there a largest whole number? If so, what is it?

5. _X_

5. **(1.1)** Graph the following whole numbers on the number line: 2, 3, 5.

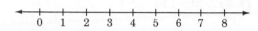

6. _X_

6. **(1.2)** Write the number 63,425 as you would read it aloud.

7. _✓_

7. **(1.2)** Write the number eighteen million, three hundred fifty-nine thousand, seventy-two.

8. _✓_

8. **(1.3)** Round 427 to the nearest hundred.

9. _✓_

9. **(1.3)** Round 18,995 to the nearest ten.

10. _X_

10. **(1.3)** Round to the most reasonable digit: During a semester, a mathematics instructor uses 487 pieces of chalk.

11. _X_

For problems 11–17, find the sums and differences.

11. **(1.4)**   627
              + 48

12. _X_

12. **(1.4)** 3106 + 921

53

13. _____

13. (1.4) $\begin{array}{r} 152 \\ + \ 36 \end{array}$

14. _____

14. (1.4) $\begin{array}{r} 5{,}189 \\ 6{,}189 \\ 4{,}122 \\ +8{,}001 \end{array}$

15. _____

15. (1.4) $21 + 16 + 42 + 11$

16. _____

16. (1.5) $520 - 216$

17. _____

17. (1.5) $\begin{array}{r} 80{,}001 \\ - \ \ 9{,}878 \end{array}$

18. _____

18. (1.5) Subtract 425 from 816.

19. _____

19. (1.5) Subtract 712 from the sum of 507 and 387.

20. _____

20. (1.6) Is the sum of 219 and 412 the same as the sum of 412 and 219? If so, what makes it so?

2

Multiplica-tion and Divi-sion of Whole Numbers

After completing this chapter, you should

Section 2.1 Multiplication of Whole Numbers
- understand the process of multiplication
- be able to multiply whole numbers
- be able to simplify multiplications with numbers ending in zero
- be able to use a calculator to multiply one whole number by another

Section 2.2 Concepts of Division of Whole Numbers
- understand the process of division
- understand division of a nonzero number into zero
- understand why division by zero is undefined
- be able to use a calculator to divide one whole number by another

Section 2.3 Division of Whole Numbers
- be able to divide a whole number by a single or multiple digit divisor
- be able to interpret a calculator statement that a division results in a remainder

Section 2.4 Some Interesting Facts about Division
- be able to recognize a whole number that is divisible by 2, 3, 4, 5, 6, 8, 9, or 10

Section 2.5 Properties of Multiplication
- understand and appreciate the commutative and associative properties of multiplication
- understand why 1 is the multiplicative identity

2.1 Multiplication of Whole Numbers

Section Overview

- ☐ **MULTIPLICATION**
- ☐ **THE MULTIPLICATION PROCESS WITH A SINGLE DIGIT MULTIPLIER**
- ☐ **THE MULTIPLICATION PROCESS WITH A MULTIPLE DIGIT MULTIPLIER**
- ☐ **MULTIPLICATIONS WITH NUMBERS ENDING IN ZERO**
- ☐ **CALCULATORS**

☐ MULTIPLICATION

Multiplication is a description of repeated addition.

In the addition of

$5 + 5 + 5$

the number 5 is repeated **3 times.** Therefore, we say we have **three times five** and describe it by writing

3×5

Thus,

$3 \times 5 = 5 + 5 + 5$

Multiplicand

Multiplier

In a multiplication, the repeated addend (number being added) is called the **multiplicand.** In 3×5, the 5 is the multiplicand. Also, in a multiplication, the number that records the number of times the multiplicand is used is called the **multiplier.** In 3×5, the 3 is the multiplier.

☆ SAMPLE SET A

Express each repeated addition as a multiplication. In each case, specify the multiplier and the multiplicand.

1. $7 + 7 + 7 + 7 + 7 + 7$

   __6 × 7__.   Multiplier is ___6___. Multiplicand is ___7___.

2. $18 + 18 + 18$

   __3 × 18__.   Multiplier is ___3___. Multiplicand is ___18___.

★ PRACTICE SET A

Express each repeated addition as a multiplication. In each case, specify the multiplier and the multiplicand.

1. $12 + 12 + 12 + 12$

   _____.   Multiplier is _____. Multiplicand is _____.

2. $36 + 36 + 36 + 36 + 36 + 36 + 36 + 36$

   _____.   Multiplier is _____. Multiplicand is _____.

3. $0 + 0 + 0 + 0 + 0$

_____. Multiplier is _____. Multiplicand is _____.

4. $\underbrace{1847 + 1847 + \cdots + 1847}$

       12,000 times

_____. Multiplier is _____. Multiplicand is _____.

Factors
Product

In a multiplication, the numbers being multiplied are also called **factors.** The result of a multiplication is called the **product.** In $3 \times 5 = 15$, the 3 and 5 are not only called the multiplier and multiplicand, but they are also called factors. The product is 15.

Indicators of Multiplication
$\times, \cdot, (\ \ )$

The multiplication symbol ($\times$) is not the only symbol used to indicate multiplication. Other symbols include the dot ($\cdot$) and pairs of parentheses ($\ \ $). The expressions

$$3 \times 5, \quad 3 \cdot 5, \quad 3(5), \quad (3)5, \quad (3)(5)$$

all represent the same product.

❑ THE MULTIPLICATION PROCESS WITH A SINGLE DIGIT MULTIPLIER

Since multiplication is repeated addition, we should not be surprised to notice that **carrying** can occur. Carrying occurs when we find the product of 38 and 7.

$$\begin{array}{r} 5 \\ 38 \\ \times\ \ 7 \\ \hline 266 \end{array}$$

First, we compute $7 \times 8 = 56$. Write the 6 in the ones column. Carry the 5. Then take $7 \times 3 = 21$. Add to 21 the 5 that was carried: $21 + 5 = 26$. The product is 266.

☆ SAMPLE SET B

Find the following products.

1.
$$\begin{array}{r} 1 \\ 64 \\ \times\ \ 3 \\ \hline 192 \end{array}$$

$3 \times 4 = 12.$ Write the 2, carry the 1.

$3 \times 6 = 18.$ Add to 18 the 1 that was carried: $18 + 1 = 19.$

The product is 192.

2.
$$\begin{array}{r} 13 \\ 526 \\ \times\ \ \ 5 \\ \hline 2,630 \end{array}$$

$5 \times 6 = 30.$ Write the 0, carry the 3.

$5 \times 2 = 10.$ Add to 10 the 3 that was carried: $10 + 3 = 13.$ Write the 3, carry the 1.

$5 \times 5 = 25.$ Add to 25 the 1 that was carried: $25 + 1 = 6.$

The product is 2,630.

Continued

3. $\quad\begin{array}{r}{}^{7\ 3}\\ 1{,}804\\ \times\quad 9\\ \hline 16{,}236\end{array}$

$9 \times 4 = 36.$ Write the 6, carry the 3.
$9 \times 0 = 0.$ Add to the 0 the 3 that was carried. $0 + 3 = 3$. Write the 3.
$9 \times 8 = 72.$ Write the 2, carry the 7.
$9 \times 1 = 9.$ Add to the 9 the 7 that was carried. $9 + 7 = 16$. Since there are no more multiplications to perform, write both the 1 and 6.

The product is 16,236.

★ PRACTICE SET B

Find the following products.

1. $\begin{array}{r}37\\ \times\ 5\end{array}$ 2. $\begin{array}{r}78\\ \times\ 8\end{array}$ 3. $\begin{array}{r}536\\ \times\ 7\end{array}$ 4. $\begin{array}{r}40{,}019\\ \times\quad 8\end{array}$ 5. $\begin{array}{r}301{,}599\\ \times\quad 3\end{array}$

☐ THE MULTIPLICATION PROCESS WITH A MULTIPLE DIGIT MULTIPLIER

In a multiplication in which the multiplier is composed of two or more digits, the *multiplication must take place in parts*. The process is as follows:

First Partial Product

Part 1:   Multiply the multiplicand by the ones digit of the multiplier. This product is called the **first partial product.**

Second Partial Product

Part 2:   Multiply the multiplicand by the tens digit of the multiplier. This product is called the **second partial product.** Since the tens digit is used as a factor, the second partial product is written below the first partial product so that its rightmost digit appears in the tens column.

Part 3:   If necessary, continue this way finding partial products. Write each one below the previous one so that the rightmost digit appears in the column directly below the digit that was used as a factor.

Total Product

Part 4:   Add the partial products to obtain the **total product.**

Note: It may be necessary to carry when finding each partial product.

☆ SAMPLE SET C

1. Multiply 326 by 48.

Part 1: $\begin{array}{r}{}^{2\ 4}\\ 326\\ \times\ 48\\ \hline 2608\end{array}$ ⟵ **First partial product.**

Part 2: $\begin{array}{r}{}^{1\ 2}\\ {}^{2\ 4}\\ 326\\ \times\ 48\\ \hline 2608\\ 1304\end{array}$ ⟵ **Second partial product.**

Part 3:   This step is unnecessary since all of the digits in the multiplier have been used.

Part 4:   Add the partial products to obtain the total product.

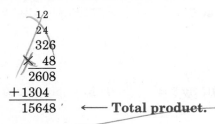

$$
\begin{array}{r}
12 \\
24 \\
326 \\
\times\ \ 48 \\
\hline
2608 \\
+\,1304 \\
\hline
15648
\end{array}
$$
  ← **Total product.**

The product is 15,648.

2. Multiply 5,369 by 842.

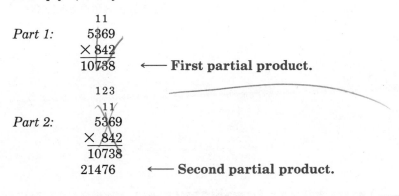

Part 1:
$$
\begin{array}{r}
11 \\
5369 \\
\times\ 842 \\
\hline
10738
\end{array}
$$
  ← **First partial product.**

Part 2:
$$
\begin{array}{r}
123 \\
11 \\
5369 \\
\times\ \ 842 \\
\hline
10738 \\
21476
\end{array}
$$
  ← **Second partial product.**

Part 3:
$$
\begin{array}{r}
257 \\
123 \\
11 \\
5369 \\
\times\ \ 842 \\
\hline
10738 \\
21476 \\
42952 \\
\hline
4520698
\end{array}
$$
  ← **Third partial product.**
  ← **Total product (Part 4).**

The product is 4,520,698.

3. Multiply 1,508 by 206.

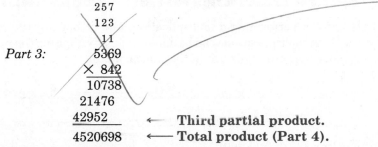

Part 1:
$$
\begin{array}{r}
3\ 4 \\
1508 \\
\times\ \ 206 \\
\hline
9048
\end{array}
$$
  ←— **First partial product (in first column from the right).**

Part 2:
$$
\begin{array}{r}
3\ 4 \\
1508 \\
\times\ \ 206 \\
\hline
9048
\end{array}
$$
     Since 0 times 1508 is 0, the partial product will not change the identity of the total product (which is obtained by addition).
Go to the next partial product.

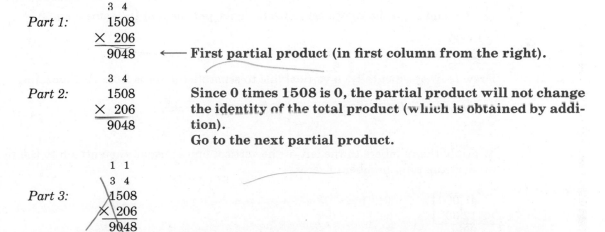

Part 3:
$$
\begin{array}{r}
1\ \ 1 \\
3\ 4 \\
1508 \\
\times\ \ 206 \\
\hline
9048 \\
3016 \\
\hline
310648
\end{array}
$$
  ← **Third partial product (in third column from the right).**
  ← **Total product (Part 4).**

The product is 310,648

★ **PRACTICE SET C**

1. Multiply 73 by 14.

2. Multiply 86 by 52.

3. Multiply 419 by 85.

4. Multiply 2,376 by 613.

5. Multiply 8,107 by 304.

6. Multiply 66,260 by 1,008.

7. Multiply 209 by 501.

8. Multiply 24 by 10.

9. Multiply 3,809 by 1,000.

10. Multiply 813 by 10,000.

☐ **MULTIPLICATIONS WITH NUMBERS ENDING IN ZERO**

Often, when performing a multiplication, one or both of the factors will end in zeros. Such multiplications can be done quickly by aligning the numbers so that the rightmost nonzero digits are in the same column.

☆ **SAMPLE SET D**

Perform the multiplication (49,000)(1,200).

$$
(49{,}000)(1{,}200) = \begin{array}{r} 49000 \\ \times\ 1200 \end{array}
$$

Since 9 and 2 are the rightmost nonzero digits, put them in the same column.

$$
\begin{array}{r} 49000 \\ \times 1200 \end{array}
$$

Draw (perhaps mentally) a vertical line to separate the zeros from the nonzeros.

$$
\begin{array}{r} 49|000 \\ \times 12|00 \end{array}
$$

Multiply the numbers to the left of the vertical line as usual, then attach to the right end of this product the total number of zeros.

$$
\begin{array}{r}
49|000 \\
\times\ 12|00 \\
\hline
98 \\
49 \\
\hline
588|00000
\end{array}
$$

Attach these 5 zeros to 588.

The product is 58,800,000.

★ **PRACTICE SET D**

1. Multiply 1,800 by 90.

2. Multiply 420,000 by 300.

3. Multiply 20,500,000 by 140,000.

☐ CALCULATORS

Most multiplications are performed using a calculator.

☆ **SAMPLE SET E**

---

1. Multiply 75,891 by 263.

		Display Reads
Type	75891	75891
Press	$\times$	75891
Type	263	263
Press	$=$	19959333

The product is 19,959,333.

2. Multiply 4,510,000,000,000 by 1,700.

		Display Reads
Type only	451	451
Press	$\times$	451
Type only	17	17
Press	$=$	7667

The display now reads 7667. We'll have to add the zeros ourselves. There are a total of 12 zeros. Attaching 12 zeros to 7667, we get 7,667,000,000,000,000.

The product is 7,667,000,000,000,000.

3. Multiply 57,847,298 by 38,976.

		Display Reads
Type	57847298	57847298
Press	$\times$	57847298
Type	38976	38976
Press	$=$	2.2546563 12

The display now reads 2.2546563 12. What kind of number is this? This is an example of a whole number written in **scientific notation**. We'll study this concept when we get to decimal numbers.

★ **PRACTICE SET E**

Use a calculator to perform each multiplication.

1. 52×27 **2.** $1,448 \times 6,155$ **3.** $8,940,000 \times 205,000$

Answers to Practice Sets are on p. 65.

Section 2.1 EXERCISES

For problems 1–55, perform the multiplications. You may check each product with a calculator.

1. 8
  $\times 3$

2. 3
  $\times 5$

3. 8
  $\times 6$

4. 5
  $\times 7$

5. 6×1

6. 4×5

7. 75×3

8. 35×5

9. 45
  $\times 6$

10. 31
  $\times 7$

11. 97
  $\times 6$

12. 75
  $\times 57$

13. 64
  $\times 15$

14. 73
  $\times 15$

15. 81
  $\times 95$

16. 31
  $\times 33$

17. 57×64

18. 76×42

19. 894×52

20. 684×38

21. 115
  $\times \ 22$

22. 706
  $\times \ 81$

23. 328
  $\times \ 21$

24. 550
  $\times \ 94$

25. 930×26

26. 318×63

27. 582
  $\times 127$

28. 247
  $\times 116$

29. 305
    ×225

30. 782
    ×547

47. 387
    ×190

48. 3,400
    × 70

31. 771
    ×663

32. 638
    ×516

49. 460,000
    × 14,000

50. 558,000,000
    ×   81,000

33. 1,905 × 710

34. 5,757 × 5,010

51. 37,000
    ×   120

52. 498,000
    ×   0

35. 3,106
    ×1,752

36. 9,300
    ×1,130

53. 4,585,000
    ×   140

54. 30,700,000
    ×   180

37. 7,057
    ×5,229

38. 8,051
    ×5,580

55. 8,000
    ×   10

39. 5,804
    ×4,300

40. 357
    × 16

56. Suppose a theater holds 426 people. If the theater charges $4 per ticket and sells every seat, how much money would they take in?

41. 724
    × 0

42. 2,649
    × 41

57. In an English class, a student is expected to read 12 novels during the semester and prepare a report on each one of them. If there are 32 students in the class, how many reports will be prepared?

43. 5,173
    ×   8

44. 1,999
    ×   0

45. 1,666
    ×   0

46. 51,730
    × 142

58. In a mathematics class, a final exam consists of 65 problems. If this exam is given to 28 people, how many problems must the instructor grade?

59. A business law instructor gives a 45 problem exam to two of her classes. If each class has 37 people in it, how many problems will the instructor have to grade?

60. An algebra instructor gives an exam that consists of 43 problems to four of his classes. If the classes have 25, 28, 31, and 35 students in them, how many problems will the instructor have to grade?

61. In statistics, the term "standard deviation" refers to a number that is calculated from certain data. If the data indicate that one standard deviation is 38 units, how many units is three standard deviations?

62. Soft drinks come in cases of 24 cans. If a supermarket sells 857 cases during one week, how many individual cans were sold?

63. There are 60 seconds in 1 minute and 60 minutes in 1 hour. How many seconds are there in 1 hour?

64. There are 60 seconds in 1 minute, 60 minutes in one hour, 24 hours in one day, and 365 days in one year. How many seconds are there in 1 year?

65. Light travels 186,000 miles in one second. How many miles does light travel in one year? (*Hint:* Can you use the result of the previous problem?)

66. An elementary school cafeteria sells 328 lunches every day. Each lunch costs $1. How much money does the cafeteria bring in in 2 weeks?

67. A computer company is selling stock for $23 a share. If 87 people each buy 55 shares, how much money would be brought in?

EXERCISES FOR REVIEW

(1.1) **68.** In the number 421,998, how may ten thousands are there?

(1.3) **69.** Round 448,062,187 to the nearest hundred thousand.

(1.4) **70.** Find the sum. 22,451 + 18,976.

(1.5) **71.** Subtract 2,289 from 3,001.

(1.6) **72.** Specify which property of addition justifies the fact that (a first whole number + a second whole number) = (the second whole number + the first whole number)

★ **Answers to Practice Sets (2.1)**

A. 1. 4×12. Multiplier is 4. Multiplicand is 12.   2. 8×36. Multiplier is 8. Multiplicand is 36.
3. 5×0. Multiplier is 5. Multiplicand is 0.
4. $12{,}000 \times 1{,}847$. Multiplier is 12,000. Multiplicand is 1,847.

B. 1. 185   2. 624   3. 3,752   4. 320,152   5. 904,797

C. 1. 1,022   2. 4,472   3. 35,615   4. 1,456,488   5. 2,464,528   6. 66,790,080   7. 104,709
8. 240   9. 3,809,000   10. 8,130,000

D. 1. 162,000   2. 126,000,000   3. 2,870,000,000,000

E. 1. 1,404   2. 8,912,440   3. 1,832,700,000,000

2.2 Concepts of Division of Whole Numbers

Section Overview

- ☑ **DIVISION**
- ☑ **DIVISION INTO ZERO** $\left(\text{ZERO AS A DIVIDEND: } \dfrac{0}{a}, \ a \neq 0\right)$
- ☑ **DIVISION BY ZERO** $\left(\text{(ZERO AS A DIVISOR: } \dfrac{a}{0}, \ a \neq 0\right)$
- ☑ **DIVISION BY AND INTO ZERO** $\Big(\text{ZERO AS A DIVIDEND AND}$
  DIVISOR: $\dfrac{0}{0}\Big)$
- ☑ **CALCULATORS**

☑ DIVISION

Division is a description of repeated subtraction.

In the process of division, the concern is how many times one number is contained in another number. For example, we might be interested in how many 5's are contained in 15. The word *times* is significant because it implies a relationship between division and multiplication.

There are several notations used to indicate division. Suppose Q records the number of times 5 is contained in 15. We can indicate this by writing

$$\begin{array}{c} Q \\ 5\overline{)15} \end{array} \qquad \frac{15}{5} = Q$$

5 into 15   15 divided by 5

$$\underbrace{15/5 = Q} \qquad \underbrace{15 \div 5 = Q}$$

15 divided by 5   15 divided by 5

Each of these division notations describes the *same* number, represented here by the symbol Q. Each notation also converts to the same multiplication form. It is

$$15 = 5 \times Q$$

In division,

Dividend
Divisor
Quotient

1. the number being divided into is called the **dividend.**
2. the number dividing into the dividend is the **divisor.**
3. the result of the division is called the **quotient.**

$$\text{divisor} \overline{)\text{dividend}}^{\text{quotient}}$$

$$\frac{\text{dividend}}{\text{divisor}} = \text{quotient}$$

$$\text{dividend/divisor} = \text{quotient} \qquad \text{dividend} \div \text{divisor} = \text{quotient}$$

☆ SAMPLE SET A

Find the following quotients using multiplication facts.

1. $18 \div 6$

Since $6 \times 3 = 18$,

$18 \div 6 = 3$

Notice also that

$$\left. \begin{array}{r} 18 \\ -\ 6 \\ \hline 12 \\ -\ 6 \\ \hline 6 \\ -\ 6 \\ \hline 0 \end{array} \right\} \text{Repeated subtraction}$$

Thus, 6 is contained in 18 three times.

2. $\dfrac{24}{3}$

Since $3 \times 8 = 24$,

$\dfrac{24}{3} = 8$

Notice also that 3 could be subtracted exactly 8 times from 24. This implies that 3 is contained in 24 eight times.

3. $\dfrac{36}{6}$

Since $6 \times 6 = 36$,

$\dfrac{36}{6} = 6$

Thus, there are 6 sixes in 36.

4. $9\overline{)72}$

Since $9 \times 8 = 72$,

$$9\overline{)72}^{\,8}$$

Thus, there are 8 nines in 72.

★ **PRACTICE SET A**

Use multiplication facts to determine the following quotients.

1. $32 \div 8$ **2.** $18 \div 9$ **3.** $\dfrac{25}{5}$ **4.** $\dfrac{48}{8}$ **5.** $\dfrac{28}{7}$ **6.** $4\overline{)36}$

☐ DIVISION INTO ZERO $\left(\text{ZERO AS A DIVIDEND: } \dfrac{0}{a}, a \neq 0\right)$

Let's look at what happens when the dividend (the number being divided into) is zero, and the divisor (the number doing the dividing) is any whole number except zero. The question is

What number, if any, is $\dfrac{0}{\text{any nonzero whole number}}$?

Let's represent this unknown quotient by Q. Then,

$$\frac{0}{\text{any nonzero whole number}} = Q$$

Converting this division problem to its corresponding multiplication problem, we get

$0 = Q \times$ (any nonzero whole number)

From our knowledge of multiplication, we can understand that if the product of two whole numbers is zero, then one or both of the whole numbers must be zero. Since any nonzero whole number is certainly not zero, Q must represent zero. Then,

$$\frac{0}{\text{any nonzero whole number}} = 0$$

Zero Divided By Any Nonzero Whole Number Is Zero

> Zero divided by any nonzero whole number is zero.

☒ DIVISON BY ZERO $\left(\text{ZERO AS A DIVISOR: } \dfrac{a}{0}, a \neq 0\right)$

Now we ask,

What number, if any, is $\dfrac{\text{any nonzero whole number}}{0}$?

Letting Q represent a possible quotient, we get

$$\frac{\text{any nonzero whole number}}{0} = Q$$

Converting to the corresponding multiplication form, we have

(any nonzero whole number) $= Q \times 0$

Since $Q \times 0 = 0$, (any nonzero whole number) $= 0$. But this is absurd. This would mean that $6 = 0$, or $37 = 0$. A nonzero whole number *cannot* equal 0! Thus,

$$\frac{\text{any nonzero whole number}}{0} \text{ } does \text{ } not \text{ name a number}$$

Division by Zero is Undefined

> Division by zero does not name a number. It is, therefore, undefined.

❑ DIVISION BY AND INTO ZERO
$\left(\text{ZERO AS A DIVIDEND AND DIVISOR: } \frac{0}{0}\right)$

We are now curious about zero divided by zero $\left(\frac{0}{0}\right)$. If we let Q represent a potential quotient, we get

$$\frac{0}{0} = Q$$

Converting to the multiplication form,

$$0 = Q \times 0$$

This results in

$$0 = 0$$

This is a statement that is true regardless of the number used in place of Q. For example,

$$\frac{0}{0} = 5, \text{ since } 0 = \quad 5 \times 0.$$

$$\frac{0}{0} = 31, \text{ since } 0 = \quad 31 \times 0.$$

$$\frac{0}{0} = 286, \text{ since } 0 = 286 \times 0.$$

A *unique* quotient cannot be determined.

Indeterminant

Since the result of the division is inconclusive, we say that $\frac{0}{0}$ is **indeterminant.**

$\frac{0}{0}$ is Indeterminant

> The division $\frac{0}{0}$ is indeterminant.

☆ SAMPLE SET B

Perform, if possible, each division.

1. $\frac{19}{0}$.   Since division by 0 does not name a whole number, no quotient exists, and we state

$\frac{19}{0}$ is undefined

2. $0\overline{)14}$.   Since division by 0 does not name a defined number, no quotient exists, and we state

$0\overline{)14}$ is undefined

3. $9\overline{)0}$.   Since division into 0 by any nonzero whole number results in 0, we have

$9\overline{)\overset{0}{0}}$

4. $\frac{0}{7}$.   Since division into 0 by any nonzero whole number results in 0, we have

$$\frac{0}{7} = 0$$

★ **PRACTICE SET B**

Perform, if possible, the following divisions.

1. $\dfrac{5}{0}$ **2.** $\dfrac{0}{4}$ **3.** $0\overline{)0}$ **4.** $0\overline{)8}$ **5.** $\dfrac{9}{0}$ **6.** $\dfrac{0}{1}$

❑ **CALCULATORS**

Divisions can also be performed using a calculator.

☆ **SAMPLE SET C**

1. Divide 24 by 3.

		Display Reads
Type	24	24
Press	$\div$	24
Type	3	3
Press	$=$	8

The display now reads 8, and we conclude that $24 \div 3 = 8$.

2. Divide 0 by 7.

		Display Reads
Type	0	0
Press	$\div$	0
Type	7	7
Press	$=$	0

The display now reads 0, and we conclude that $0 \div 7 = 0$.

3. Divide 7 by 0.

Since division by zero is undefined, the calculator should register some kind of error message.

		Display Reads
Type	7	7
Press	$\div$	7
Type	0	0
Press	$=$	Error

The error message indicates an undefined operation was attempted, in this case, division by zero.

★ PRACTICE SET C

Use a calculator to perform each division.

1. $35 \div 7$ **2.** $56 \div 8$ **3.** $0 \div 6$ **4.** $3 \div 0$ **5.** $0 \div 0$

Answers to Practice Sets are on p. 71.

Section 2.2 EXERCISES

For problems 1–25, determine the quotients (if possible). You may use a calculator to check the result.

1. $4\overline{)32}$ **2.** $7\overline{)42}$

3. $6\overline{)18}$ **4.** $2\overline{)14}$

5. $3\overline{)27}$ **6.** $1\overline{)6}$

7. $4\overline{)28}$ **8.** $\dfrac{30}{5}$

9. $\dfrac{16}{4}$ **10.** $24 \div 8$

11. $10 \div 2$ **12.** $21 \div 7$

13. $21 \div 3$ **14.** $0 \div 6$

15. $8 \div 0$ **16.** $12 \div 4$

17. $3\overline{)9}$ **18.** $0\overline{)0}$

19. $7\overline{)0}$ **20.** $6\overline{)48}$

21. $\dfrac{15}{3}$ **22.** $\dfrac{35}{0}$

23. $56 \div 7$ **24.** $\dfrac{0}{9}$

25. $72 \div 8$

26. Write $\dfrac{16}{2} = 8$ using three different notations.

27. Write $\dfrac{27}{9} = 3$ using three different notations.

28. In the statement $6\overline{)24}^{\,4}$
   6 is called the _____ ,
   24 is called the _____ ,
   4 is called the _____ .

29. In the statement $56 \div 8 = 7$,

   7 is called the _____ ,
   8 is called the _____ ,
   56 is called the _____ .

EXERCISES FOR REVIEW

(1.1) **30.** What is the largest digit?

(1.4) **31.** Find the sum.   8,006
                              $+4,118$

(1.5) **32.** Find the difference.   631
                                     -589

(1.6) **33.** Use the numbers 2, 3, and 7 to illustrate the associative property of addition.

(2.1) **34.** Find the product.   86
                                  $\times 12$

★ **ANSWERS TO PRACTICE SETS (2.2)**

A. **1.** 4   **2.** 2   **3.** 5   **4.** 6   **5.** 4   **6.** 9

B. **1.** undefined   **2.** 0   **3.** indeterminant   **4.** undefined   **5.** undefined   **6.** 0

C. **1.** 5   **2.** 7   **3.** 0

4. An error message tells us that this operation is undefined. The particular message depends on the calculator.

5. An error message tells us that this operation cannot be performed. Some calculators actually set $0 \div 0$ equal to 1. We know better! $0 \div 0$ is indeterminant.

2.3 Division of Whole Numbers

Section Overview

- ❑ **DIVISION WITH A SINGLE DIGIT DIVISOR**
- ❑ **DIVISION WITH A MULTIPLE DIGIT DIVISOR**
- ❑ **DIVISION WITH A REMAINDER**
- ❑ **CALCULATORS**

❑ DIVISION WITH A SINGLE DIGIT DIVISOR

Our experience with multiplication of whole numbers allows us to perform such divisions as $75 \div 5$. We perform the division by performing the corresponding multiplication, $5 \times Q = 75$. Each division we considered in Section 2.2 had a one-digit quotient. Now we will consider divisions in which the quotient may consist of two or more digits. For example, $75 \div 5$.

Let's examine the division $75 \div 5$. We are asked to determine how many 5's are contained in 75. We'll approach the problem in the following way.

1. Make an educated guess based on experience with multiplication.
2. Find how close the estimate is by multiplying the estimate by 5.
3. If the product obtained in step 2 is less than 75, find out how much less by subtracting it from 75.
4. If the product obtained in step 2 is greater than 75, decrease the estimate until the product is less than 75. Decreasing the estimate makes sense because we do not wish to exceed 75.

We can suggest from this discussion that the process of division consists of

The Four Steps in Division

1. an educated guess
2. a multiplication
3. a subtraction
4. bringing down the next digit (if necessary)

The educated guess can be made by determining how many times the divisor is contained in the dividend by using only one or two digits of the dividend.

☆ SAMPLE SET A

1. Find $75 \div 5$.

$5\overline{)75}$ **Rewrite the problem using a division bracket.**

$\overset{10}{5\overline{)75}}$ **Make an educated guess by noting that one 5 is contained in 75 at most 10 times.**
Since 7 is the tens digit, we estimate that 5 goes into 75 at most 10 times.

$$
\begin{array}{r}
10 \\
5\overline{)75} \\
-50 \\
\hline
25
\end{array}
$$
Now determine how close the estimate is.
10 fives is $10 \times 5 = 50$. Subtract 50 from 75.
Estimate the number of 5's in 25.
There are exactly 5 fives in 25.

$$
\begin{array}{r}
5 \\
10 \\
5\overline{)75} \\
-50 \\
\hline
25 \\
-25 \\
\hline
0
\end{array}
$$
10 fives + 5 fives = 15 fives.
There are 15 fives contained in 75.

Check: $75 \overset{?}{=} 15 \times 5$
    $75 \overset{\checkmark}{=} 75$

Thus, $75 \div 5 = 15$.

The notation in this division can be shortened by writing.

$$
\begin{array}{r}
15 \\
5\overline{)75} \\
5\downarrow \\
\hline
25 \\
25 \\
\hline
0
\end{array}
$$
$\left\{\begin{array}{ll} \textbf{Divide:} & \text{5 goes into 7 at most 1 time.} \\ \textbf{Multiply:} & 1 \times 5 = 5. \text{ Write 5 below 7.} \\ \textbf{Subtract:} & 7 - 5 = 2. \text{ Bring down 5.} \end{array}\right.$
$\left\{\begin{array}{ll} \textbf{Divide:} & \text{5 goes into 25 exactly 5 times.} \\ \textbf{Multiply:} & 5 \times 5 = 25. \text{ Write 25 below 25.} \\ \textbf{Subtract:} & 25 - 25 = 0. \end{array}\right.$

2. Find $4{,}944 \div 8$.

$8\overline{)4944}$ **Rewrite the problem using a division bracket.**

$$
\begin{array}{r}
600 \\
8\overline{)4944} \\
-4800 \\
\hline
144
\end{array}
$$
8 goes into 49 at most 6 times, and 9 is in the hundreds column. We'll guess 600.
Then, $8 \times 600 = 4800$.

```
      10
     600
  8) 4944
    -4800
      144
    -  80
       64
```

8 goes into 14 at most 1 time, and 4 is in the tens column. We'll guess 10.

```
       8
      10
     600
  8) 4944
    -4800
      144
    -  80
       64
    -  64
        0
```

8 goes into 64 exactly 8 times.

600 eights + 10 eights + 8 eights = 618 eights.

Check: $4944 \overset{?}{=} 8 \times 618$

$4944 \overset{\checkmark}{=} 4944$

Thus, $4,944 \div 8 = 618$.

As in the first problem, the notation in this division can be shortened by eliminating the subtraction signs and the zeros in each educated guess.

```
      618
  8) 4944
     48↓|
     14 |
      8↓
      64
      64
       0
```

$\begin{cases} \text{Divide:} & \text{8 goes into 49 at most 6 times.} \\ \text{Multiply:} & 6 \times 8 = 48. \text{ Write 48 below 49.} \\ \text{Subtract:} & 49 - 48 = 1. \text{ Bring down the 4.} \end{cases}$

$\begin{cases} \text{Divide:} & \text{8 goes into 14 at most 1 time.} \\ \text{Multiply:} & 1 \times 8 = 8. \text{ Write 8 below 14.} \\ \text{Subtract:} & 14 - 8 = 6. \text{ Bring down the 4.} \end{cases}$

$\begin{cases} \text{Divide:} & \text{8 goes into 64 exactly 8 times.} \\ \text{Multiply:} & 8 \times 8 = 64. \text{ Write 64 below 64.} \\ \text{Subtract:} & 64 - 64 = 0. \end{cases}$

NOTE: Not all divisions end in zero. We will examine such divisions in a subsequent subsection.

★ PRACTICE SET A

Perform the following divisions.

1. $126 \div 7$ **2.** $324 \div 4$ **3.** $2,559 \div 3$ **4.** $5,645 \div 5$ **5.** $757,125 \div 9$

☐ DIVISION WITH A MULTIPLE DIGIT DIVISOR

The process of division also works when the divisor consists of two or more digits. We now make educated guesses using the first digit of the divisor and one or two digits of the dividend.

☆ **SAMPLE SET B**

1. Find $2,232 \div 36$.

$$36\overline{)2232}$$

Use the first digit of the divisor and the first two digits of the dividend to make the educated guess.

3 goes into 22 at most 7 times.
Try 7: $7 \times 36 = 252$ which is greater than 223. Reduce the estimate.
Try 6: $6 \times 36 = 216$ which is less than 223.

$$\begin{array}{r} 6 \\ 36\overline{)\ 2232} \\ -216\downarrow \\ \hline 72 \end{array}$$

Multiply: $6 \times 36 = 216$. **Write 216 below 223.**
Subtract: $223 - 216 = 7$. **Bring down the 2.**

Divide 3 into 7 to estimate the number of times 36 goes into 72. The 3 goes into 7 at most 2 times.
Try 2: $2 \times 36 = 72$.

$$\begin{array}{r} 62 \\ 36\overline{)2232} \\ 216\downarrow \\ \hline 72 \\ -72 \\ \hline 0 \end{array}$$

Check: $2232 \overset{?}{=} 36 \times 62$
$2232 \overset{?}{=} 2232$

Thus, $2,232 \div 36 = 62$.

2. Find $2,417,228 \div 802$.

$$802\overline{)2417228}$$

First, the educated guess: $24 \div 8 = 3$. Then $3 \times 802 = 2406$, which is less than 2417. Use 3 as the guess. Since $3 \times 802 = 2406$, and 2406 has four digits, place the 3 above the fourth digit of the dividend.

$$\begin{array}{r} 3 \\ 802\overline{)\ 2417228} \\ -2406\downarrow \\ \hline 112 \end{array}$$

Subtract: $2417 - 2406 = 11$.
Bring down the 2.

The divisor 802 goes into 112 at most 0 times. Use 0.

$$\begin{array}{r} 30 \\ 802\overline{)\ 2417228} \\ -2406\downarrow \\ \hline 112 \\ -0\downarrow \\ \hline 1122 \end{array}$$

Multiply: $0 \times 802 = 0$.
Subtract: $112 - 0 = 112$.
Bring down the 2.

The 8 goes into 11 at most 1 time, and $1 \times 802 = 802$, which is less than 1122. Try 1.

$$\begin{array}{r} 301 \\ 802\overline{)\ 2417228} \\ -2406\downarrow|\,| \\ \hline 112| \\ -0\downarrow| \\ \hline 1122| \\ -802\downarrow \\ \hline 3208 \end{array}$$

Subtract $1122 - 802 = 320$.
Bring down the 8.

8 goes into 32 at most 4 times.
$4 \times 802 = 3208$.
Use 4.

$$
\begin{array}{r}
3014 \\
802 \overline{)\ 2417228} \\
-2406\downarrow|| \\
\hline
112\ | \\
-0\downarrow \\
\hline
1122\ | \\
-802\downarrow \\
\hline
3208 \\
-3208 \\
\hline
0
\end{array}
$$

Check: $2417228 \overset{?}{=} 3014 \times 802$
            $2417228 \overset{?}{=} 2417228$

Thus, $2{,}417{,}228 \div 802 = 3{,}014$.

★ **PRACTICE SET B**

Perform the following divisions.

1. $1{,}376 \div 32$ **2.** $6{,}160 \div 55$ **3.** $18{,}605 \div 61$ **4.** $144{,}768 \div 48$

❑ **DIVISION WITH A REMAINDER**

We might wonder how many times 4 is contained in 10.
Repeated subtraction yields

$$
\begin{array}{r}
10 \\
-\ 4 \\
\hline
6 \\
-\ 4 \\
\hline
2
\end{array}
$$

Since the remainder is less than 4, we stop the subtraction. Thus, 4 goes into 10 two times with 2 remaining. We can write this as a division as follows.

$$
\begin{array}{r}
2 \\
4 \overline{)\ 10} \\
-\ 8 \\
\hline
2
\end{array}
$$ **Divide:**   **4 goes into 10 at most 2 times.**
         Multiply: $2 \times 4 = 8$. **Write 8 below 0.**
         Subtract: $10 - 8 = 2.$

Since 4 does not divide into 2 (the remainder is less than the divisor) and there are no digits to bring down to continue the process, we are done. We write

$$
\begin{array}{r}
2\ \text{R2} \\
4 \overline{)\ 10} \\
-\ 8 \\
\hline
2
\end{array}
$$ or $10 \div 4 = \underbrace{2\ \text{R2}}_{2\ \text{with remainder 2}}$

☆ **SAMPLE SET C**

1. Find $85 \div 3$.

$$
\begin{array}{r}
28 \\
3{\overline{\smash{\big)}\,85}} \\
6\downarrow \\
\hline
25 \\
24 \\
\hline
1
\end{array}
$$

$\left\{\begin{array}{lll} \text{Divide:} & \text{3 goes into 8 at most 2 times.} \\ \text{Multiply:} & 2 \times 3 = 6. \text{ Write 6 below 8.} \\ \text{Subtract:} & 8 - 6 = 2. \text{ Bring down the 5.} \end{array}\right.$

$\left\{\begin{array}{lll} \text{Divide:} & \text{3 goes into 25 at most 8 times.} \\ \text{Multiply:} & 3 \times 8 = 24. \text{ Write 24 below 25.} \\ \text{Subtract:} & 25 - 24 = 1 \end{array}\right.$

There are no more digits to bring down to continue the process. We are done. One is the remainder.

Check: Multiply 28 and 3, then add 1.

$$
\begin{array}{r}
28 \\
\times\ 3 \\
\hline
84 \\
+\ 1 \\
\hline
85
\end{array}
$$

Thus, $85 \div 3 = 28$ R1.

2. Find $726 \div 23$.

$$
\begin{array}{r}
31 \\
23{\overline{\smash{\big)}\,726}} \\
69\downarrow \\
\hline
36 \\
23 \\
\hline
13
\end{array}
$$

Check: Multiply 31 by 23, then add 13.

$$
\begin{array}{r}
31 \\
\times\ 23 \\
\hline
93 \\
62 \\
\hline
713 \\
+\ 13 \\
\hline
726
\end{array}
$$

Thus, $726 \div 23 = 31$ R13.

★ **PRACTICE SET C**

Perform the following divisions.

1. $75 \div 4$ **2.** $346 \div 8$ **3.** $489 \div 21$ **4.** $5{,}016 \div 82$ **5.** $41{,}196 \div 67$

☐ **CALCULATORS**

The calculator can be useful for finding quotients with single and multiple digit divisors. If, however, the division should result in a remainder, the calculator is unable to provide us with the particular value of the remainder. Also, some calculators (most nonscientific) are unable to perform divisions in which one of the numbers has more than eight digits.

☆ SAMPLE SET D

Use a calculator to perform each division.

1. 328 ÷ 8

Type   328

Press $\boxed{\div}$

Type   8

Press $\boxed{=}$

The display now reads 41.

2. 53,136 ÷ 82

Type   53136

Press $\boxed{\div}$

Type   82

Press $\boxed{=}$

The display now reads 648.

3. 730,019,001 ÷ 326

We first try to enter 730,019,001 but find that we can only enter 73001900. If our calculator has only an eight-digit display (as most nonscientific calculators do), we will be unable to use the calculator to perform this division.

4. 3727 ÷ 49

Type   3727

Press $\boxed{\div}$

Type   49

Press $\boxed{=}$

The display now reads 76.061224.

This number is an example of a decimal number (see Chapter 6). When a decimal number results in a calculator division, we can conclude that the division produces a remainder.

★ PRACTICE SET D

Use a calculator to perform each division.

1. 3,330 ÷ 74   **2.** 63,365 ÷ 115   **3.** 21,996,385,287 ÷ 53   **4.** 4,558 ÷ 67

Answers to Practice Sets are on p. 81.

Section 2.3 EXERCISES

For problems 1–55, perform the divisions.

Problems 1–38 can be checked with a calculator by multiplying the divisor and quotient then adding the remainder.

1. $52 \div 4$

2. $776 \div 8$

3. $603 \div 9$

4. $240 \div 8$

5. $208 \div 4$

6. $576 \div 6$

7. $21 \div 7$

8. $0 \div 0$

9. $140 \div 2$

10. $528 \div 8$

11. $244 \div 4$

12. $0 \div 7$

13. $177 \div 3$

14. $96 \div 8$

15. $67 \div 1$

16. $896 \div 56$

17. $1,044 \div 12$

18. $988 \div 19$

19. $5,238 \div 97$

20. $2,530 \div 55$

21. $4,264 \div 82$

22. $637 \div 13$

23. $3,420 \div 90$

24. $5,655 \div 87$

25. $2,115 \div 47$

26. $9,328 \div 22$

27. $55,167 \div 71$

28. $68,356 \div 92$

29. $27,702 \div 81$

30. $6,510 \div 31$

31. $60,536 \div 94$ **32.** $31,844 \div 38$

33. $23,985 \div 45$ **34.** $60,606 \div 74$

35. $2,975,400 \div 285$ **36.** $1,389,660 \div 795$

37. $7,162,060 \div 879$ **38.** $7,561,060 \div 909$

39. $38 \div 9$ **40.** $97 \div 4$

41. $199 \div 3$ **42.** $573 \div 6$

43. $10,701 \div 13$ **44.** $13,521 \div 53$

45. $3,628 \div 90$ **46.** $10,592 \div 43$

47. $19,965 \div 30$ **48.** $8,320 \div 21$

49. $61,282 \div 64$ **50.** $1,030 \div 28$

51. $7,319 \div 11$ **52.** $3,628 \div 90$

53. $35,279 \div 77$ **54.** $52,196 \div 55$

55. $67,751 \div 68$

For problems 56–60, use a calculator to find the quotients.

56. $4,346 \div 53$ **57.** $3,234 \div 77$

58. $6,771 \div 37$ **59.** $4,272,320 \div 520$

60. $7,558,110 \div 651$

61. A mathematics instructor at a high school is paid $17,775 for 9 months. How much money does this instructor make each month?

62. A couple pays $4,380 a year for a one-bedroom apartment. How much does this couple pay each month for this apartment?

63. Thirty-six people invest a total of $17,460 in a particular stock. If they each invested the same amount, how much did each person invest?

64. Each of the 28 students in a mathematics class buys a textbook. If the bookstore sells $644 worth of books, what is the price of each book?

65. A certain brand of refrigerator has an automatic ice cube maker that makes 336 ice cubes in one day. If the ice machine makes ice cubes at a constant rate, how many ice cubes does it make each hour?

66. A beer manufacturer bottles 52,380 ounces of beer each hour. If each bottle contains the same number of ounces of beer, and the manufacturer fills 4,365 bottles per hour, how many ounces of beer does each bottle contain?

67. A computer program consists of 68,112 bits. 68,112 bits equals 8,514 bytes. How many bits in one byte?

68. A 26-story building in San Francisco has a total of 416 offices. If each floor has the same number of offices, how many floors does this building have?

69. A college has 67 classrooms and a total of 2,546 desks. How many desks are in each classroom if each classroom has the same number of desks?

EXERCISES FOR REVIEW

(1.1) **70.** What is the value of 4 in the number 124,621?

(1.3) **71.** Round 604,092 to the nearest hundred thousand.

(1.6) **72.** What whole number is the additive identity?

(2.1) **73.** Find the product. $6,256 \times 100$.

(2.2) **74.** Find the quotient. $0 \div 11$.

★ **Answers to Practice Sets (2.3)**

A. **1.** 18   **2.** 81   **3.** 853   **4.** 1,129   **5.** 84,125

B. **1.** 43   **2.** 112   **3.** 305   **4.** 3,016

C. **1.** 18 R3   **2.** 43 R2   **3.** 23 R6   **4.** 61 R14   **5.** 614 R58

D. **1.** 45   **2.** 551

3. Since the dividend has more than eight digits, this division cannot be performed on most nonscientific calculators. On others, the answer is 415,026,137.4.

4. This division results in 68.02985075, a decimal number, and therefore, we cannot, at this time, find the value of the remainder. Later, we will discuss decimal numbers.

2.4 Some Interesting Facts about Division

Section Overview	☐ **DIVISION BY 2, 3, 4, AND 5** ☐ **DIVISION BY 6, 8, 9, AND 10**

Quite often, we are able to determine if a whole number is divisible by another whole number just by observing some simple facts about the number. Some of these facts are listed in this section.

☐ DIVISION BY 2, 3, 4, AND 5

Division by 2

> A whole number is **divisible by 2** if its *last digit* is 0, 2, 4, 6, or 8.

The numbers 80, 112, 64, 326, and 1,008 are all divisible by 2 since the last digit of each is 0, 2, 4, 6, or 8, respectively.

The numbers 85 and 731 are *not* divisible by 2.

Division by 3

> A whole number is **divisible by 3** if the *sum of its digits* is divisible by 3.

The number 432 is divisible by 3 since $4 + 3 + 2 = 9$ and 9 is divisible by 3.

$432 \div 3 = 144$

The number 25 is *not* divisible by 3 since $2 + 5 = 7$, and 7 is not divisible by 3.

Division by 4

> A whole number is **divisible by 4** if its *last two digits* form a number that is divisible by 4.

The number 31,048 is divisible by 4 since the last two digits, 4 and 8, form a number, 48, that is divisible by 4.

$31048 \div 4 = 7262$

The number 137 is not divisible by 4 since 37 is not divisible by 4.

Division by 5

> A whole number is **divisible by 5** if its *last digit* is 0 or 5.

☆ **SAMPLE SET A**

The numbers 65, 110, 8,030, and 16,955 are each divisible by 5 since the last digit of each is 0 or 5.

★ **PRACTICE SET A**

State which of the following whole numbers are divisible by 2, 3, 4, or 5. A number may be divisible by more than one number.

1. 26   **2.** 81   **3.** 51   **4.** 385

5. 6,112   **6.** 470   **7.** 113,154

☐ DIVISION BY 6, 8, 9, AND 10

Division by 6

A number is **divisible by 6** if it is divisible by *both* 2 and 3.

The number 234 is divisible by 2 since its last digit is 4. It is also divisible by 3 since $2 + 3 + 4 = 9$ and 9 is divisible by 3. Therefore, 234 is divisible by 6.

The number 6,532 is *not* divisible by 6. Although its last digit is 2, making it divisible by 2, the sum of its digits, $6 + 5 + 3 + 2 = 16$, and 16 is not divisible by 3.

Division by 8

A whole number is **divisible by 8** if its *last three digits* form a number that is divisible by 8.

The number 4,000 is divisible by 8 since 000 is divisible by 8.
The number 13,128 is divisible by 8 since 128 is divisible by 8.
The number 1,170 is *not* divisible by 8 since 170 is not divisible by 8.

Division by 9

A whole number is **divisible by 9** is the *sum of its digits* is divisible by 9.

The number 702 is divisible by 9 since $7 + 0 + 2$ is divisible by 9.
The number 6588 is divisible by 9 since $6 + 5 + 8 + 8 = 27$ is divisible by 9.
The number 14,123 is *not* divisible by 9 since $1 + 4 + 1 + 2 + 3 = 11$ is not divisible by 9.

Division by 10

A whole number is **divisible by 10** if its *last digit* is 0.

☆ **SAMPLE SET B**

The numbers 30, 170, 16,240, and 865,000 are all divisible by 10.

★ **PRACTICE SET B**

State which of the following whole numbers are divisible 6, 8, 9, or 10. Some numbers may be divisible by more than one number.

1. 900   **2.** 6,402   **3.** 6,660   **4.** 55,116

Answers to Practice Sets are on p. 85.

Section 2.4 EXERCISES

For problems 1–30, specify if the whole number is divisible by 2, 3, 4, 5, 6, 8, 9, or 10. Write "none" if the number is not divisible by any digit other than 1. Some numbers may be divisible by more than one number.

1. 48 _____

2. 85 _____

3. 30 _____

4. 83 _____

5. 98 _____

6. 972 _____

7. 892 _____

8. 676 _____

9. 903 _____

10. 800 _____

11. 223 _____

12. 836 _____

13. 665 _____

14. 4,381 _____

15. 2,195 _____

16. 2,544 _____

17. 5,172 _____

18. 1,307 _____

19. 1,050 _____

20. 3,898 _____

21. 1,621 _____

22. 27,808 _____

23. 45,764 _____

24. 49,198 _____

25. 296,122 _____

26. 178,656 _____

27. 5,102,417 _____

28. 16,990,792 _____

29. 620,157,659 _____

30. 457,687,705 _____

EXERCISES FOR REVIEW

(1.1) **31.** In the number 412, how many tens are there?

(1.5) **32.** Subtract 613 from 810.

(1.6) **33.** Add 35, 16, and 7 in two different ways.

(2.2) **34.** Find the quotient $35 \div 0$, if it exists.

(2.3) **35.** Find the quotient. $3654 \div 42$.

★ **Answers to Practice Sets (2.4)**

A. **1.** 2   **2.** 3   **3.** 3   **4.** 5   **5.** 2, 4   **6.** 2, 5   **7.** 2, 3

B. **1.** 6, 9, 10   **2.** 6   **3.** 6, 9, 10   **4.** 6, 9

2.5 Properties of Multiplication

Section Overview

- ☐ THE COMMUTATIVE PROPERTY OF MULTIPLICATION
- ☐ THE ASSOCIATIVE PROPERTY OF MULTIPLICATION
- ☐ THE MULTIPLICATIVE IDENTITY

We will now examine three simple but very important properties of multiplication.

☐ THE COMMUTATIVE PROPERTY OF MULTIPLICATION

Commutative Property of Multiplication

The product of two whole numbers is the same regardless of the order of the factors.

☆ **SAMPLE SET A**

Multiply the two whole numbers.

$$6 \cdot 7 = 42$$
$$7 \cdot 6 = 42$$

The numbers 6 and 7 can be multiplied in any order. Regardless of the order they are multiplied, the product is 42.

★ **PRACTICE SET A**

Use the commutative property of multiplication to find the products in two ways.

1.
```
15
   6
```

2.
```
432
428
```

☐ THE ASSOCIATIVE PROPERTY OF MULTIPLICATION

Associative Property of Multiplication

If three whole numbers are multiplied, the product will be the same if the first two are multiplied first and then that product is multiplied by the third, or if the second two are multiplied first and that product is multiplied by the first. Note that the order of the factors is maintained.

It is a common mathematical practice to *use parentheses* to show which pair of numbers is to be combined first.

☆ **SAMPLE SET B**

Multiply the whole numbers.

```
8
   3
14
```

$(8 \cdot 3) \cdot 14 = 24 \cdot 14 = 336$

$8 \cdot (3 \cdot 14) = 8 \cdot 42 = 336$

★ **PRACTICE SET B**

Use the associative property of multiplication to find the products in two ways.

1.
```
7
   3
8
```

2.
```
73
   18
126
```

☐ THE MULTIPLICATIVE IDENTITY

The Multiplicative Identity is 1

The whole number 1 is called the **multiplicative identity,** since any whole number multiplied by 1 is not changed.

☆ SAMPLE SET C

Multiply the whole numbers.

12		$12 \cdot 1 = 12$
	1	$1 \cdot 12 = 12$

★ PRACTICE SET C

Multiply the whole numbers.

843	
	1

Answers to Practice Sets are on p. 88.

Section 2.5 EXERCISES

For problems 1–12, multiply the numbers.

1.
9	
	26

2.
18	
	41

7.
3	
	7
12	

8.
	40
16	
	5

3.
	42
96	

4.
	6
192	

9.
	22
10	
	97

10.
110	
	85
0	

5.
1000	
	326

6.
	1400
70	

11.
462	
1	
	18

12.
3,178	
	5
101	

For problems 13–16, show that the quantities yield the same products by performing the multiplications.

13. $(4 \cdot 8) \cdot 2$ and $4 \cdot (8 \cdot 2)$

14. $(100 \cdot 62) \cdot 4$ and $100 \cdot (62 \cdot 4)$

15. $23 \cdot (11 \cdot 106)$ and $(23 \cdot 11) \cdot 106$

16. $1 \cdot (5 \cdot 2)$ and $(1 \cdot 5) \cdot 2$

17. The fact that

(a first number · a second number) · a third number = a first number · (a second number · a third number)

is an example of the _____ property of multiplication.

18. The fact that

1 · any number = that particular number

is an example of the _____ property of multiplication.

19. Use the numbers 7 and 9 to illustrate the commutative property of multiplication.

20. Use the numbers 6, 4, and 7 to illustrate the associative property of multiplication.

EXERCISES FOR REVIEW

(1.1) **21.** In the number 84,526,098,441, how many millions are there?

(1.4) **22.** Replace the letter m with the whole number that makes the addition true.

$$\begin{array}{r} 85 \\ + m \\ \hline 97 \end{array}$$

(1.6) **23.** Use the numbers 4 and 15 to illustrate the commutative property of addition.

(2.2) **24.** Find the product. $8,000,000 \times 1,000$.

(2.4) **25.** Specify which of the digits 2, 3, 4, 5, 6, 8, 10 are divisors of the number 2,244.

★ **Answers to Practice Sets (2.5)**

A. **1.** $15.6 = 90$ and $6 \cdot 15 = 90$ **2.** $432 \cdot 428 = 184,896$ and $428 \cdot 432 = 184,896$

B. **1.** 168   **2.** 165,564

C. 843

Chapter 2   SUMMARY OF KEY CONCEPTS

Multiplication (2.1)

Multiplication is a description of repeated addition.

$$\underbrace{7 + 7 + 7 + 7}_{\text{7 appears 4 }\textit{times}}$$

This expression is described by writing 4×7.

Multiplicand/Multiplier/ Product (2.1)

In a multiplication of whole numbers, the repeated addend is called the *multiplicand,* and the number that records the number of times the multiplicand is used is the *multiplier.* The result of the multiplication is the *product.*

Factors (2.1)

In a multiplication, the numbers being multiplied are also called *factors.* Thus, the multiplicand and the multiplier can be called factors.

Division (2.2)

Division is a description of repeated subtraction.

Dividend/Divisor/Quotient (2.2)

In a division, the number divided into is called the *dividend,* and the number dividing into the dividend is called the *divisor.* The result of the division is called the *quotient.*

$$\text{divisor}\overline{)\text{dividend}}^{\text{quotient}}$$

Division into Zero (2.2)

Zero divided by any nonzero whole number is zero.

Division by Zero (2.2)

Division by zero does not name a whole number. It is, therefore, undefined. The quotient $\dfrac{0}{0}$ is indeterminant.

Division by 2, 3, 4, 5, 6, 8, 9, 10 (2.4)

Division by the whole numbers 2, 3, 4, 5, 6, 8, 9, and 10 can be determined by noting some certain properties of the particular whole number.

Commutative Property of Multiplication (2.5)

The product of two whole numbers is the same regardless of the order of the factors.

$$3 \times 5 = 5 \times 3$$

Associative Property of Multiplication (2.5)

If three whole numbers are to be multiplied, the product will be the same if the first two are multiplied first and then that product is multiplied by the third, or if the second two are multiplied first and then that product is multiplied by the first.

$$(3 \times 5) \times 2 = 3 \times (5 \times 2)$$

Note that the order of the factors is maintained.

Multiplicative Identity (2.5)

The whole number 1 is called the *multiplicative identity* since any whole number multiplied by 1 is not changed.

$$4 \times 1 = 4$$
$$1 \times 4 = 4$$

EXERCISE SUPPLEMENT

Section 2.1

1. In the multiplication $5 \times 9 = 45$, 5 and 9 are called _____ and 45 is called the _____.

2. In the multiplication $4 \times 8 = 32$, 4 and 8 are called _____ and 32 is called the _____.

Section 2.2

3. In the division $24 \div 6 = 4$, 6 is called the _____, and 4 is called the _____.

4. In the division $36 \div 2 = 18$, 2 is called the _____, and 18 is called the _____.

Section 2.4

5. A number is divisible by 2 only if its last digit is _____.

6. A number is divisible by 3 only if _____ of its digits is divisible by 3.

7. A number is divisible by 4 only if the rightmost two digits form a number that is _____.

Sections 2.1, 2.3

Find each product or quotient.

8. $\begin{array}{r} 24 \\ \times\ 3 \\ \hline \end{array}$

9. $\begin{array}{r} 14 \\ \times\ 8 \\ \hline \end{array}$

10. $21 \div 7$

11. $35 \div 5$

12. $\begin{array}{r} 36 \\ \times 22 \\ \hline \end{array}$

13. $\begin{array}{r} 87 \\ \times 35 \\ \hline \end{array}$

14. $\begin{array}{r} 117 \\ \times\ 42 \\ \hline \end{array}$

15. $208 \div 52$

16. $\begin{array}{r} 521 \\ \times\ 87 \\ \hline \end{array}$

17. $\begin{array}{r} 1005 \\ \times\ \ 15 \\ \hline \end{array}$

18. $1338 \div 446$

19. $2814 \div 201$

20. $\begin{array}{r} 5521 \\ \times\ \ 8 \\ \hline \end{array}$

21. $\begin{array}{r} 6016 \\ \times\ \ 7 \\ \hline \end{array}$

22. $576 \div 24$

23. $3969 \div 63$

24. $\begin{array}{r} 5482 \\ \times\ 322 \\ \hline \end{array}$

25. $\begin{array}{r} 9104 \\ \times\ 115 \\ \hline \end{array}$

26. $\begin{array}{r} 6102 \\ \times 1000 \\ \hline \end{array}$

27. $\begin{array}{r} 10101 \\ \times 10000 \\ \hline \end{array}$

28. $162{,}006 \div 31$

29. $0 \div 25$

30. $25 \div 0$

31. $4280 \div 10$

32. $2126000 \div 100$

33. $84 \div 15$

34. $126 \div 4$

35. $424 \div 0$

36. $1198 \div 46$

37. $995 \div 31$

38. $0 \div 18$

39. $\begin{array}{r} 2162 \\ \times 1421 \\ \hline \end{array}$

40. 0×0

41. 5×0

42. 64×1

43. 1×0

44. $0 \div 3$

45. $14 \div 0$

46. $35 \div 1$

47. $1 \div 1$

Section 2.5

48. Use the commutative property of multiplication to rewrite 36×128.

49. Use the commutative property of multiplication to rewrite 114×226.

50. Use the associative property of multiplication to rewrite $(5 \cdot 4) \cdot 8$.

51. Use the associative property of multiplication to rewrite $16 \cdot (14 \cdot 0)$.

Sections 2.1, 2.3

52. A computer store is selling diskettes for $4 each. At this price, how much would 15 diskettes cost?

53. Light travels 186,000 miles in one second. How far does light travel in 23 seconds?

54. A dinner bill for eight people comes to exactly $112. How much should each person pay if they all agree to split the bill equally?

55. Each of the 33 students in a math class buys a textbook. If the bookstore sells $1089 worth of books, what is the price of each book?

1. _____

1. **(2.1)** In the multiplication of $8 \times 7 = 56$, what are the names given to the 8 and 7 and the 56?

2. _____

2. **(2.1)** Multiplication is a description of what repeated process?

3. **(2.2)** In the division $12 \div 3 = 4$, what are the names given to the 3 and the 4?

3. _____

4. **(2.4)** Name the digits that a number must end in to be divisible by 2.

5. **(2.5)** Name the property of multiplication that states that the order of the factors in a multiplication can be changed without changing the product.

4. _____

6. **(2.5)** Which number is called the multiplicative identity?

5. _____

For problems 7–17, find the product or quotient.

6. _____

7. **(2.1)** 14×6

8. **(2.1)** 37×0

7. _____

9. **(2.1)** 352×1000

8. _____

9. _____

10. **(2.1)** 5986×70

10. _____

11. **(2.1)** 21×12

11. _____

12. **(2.2)** $856 \div 0$

12. _____

13. **(2.2)** $0 \div 8$

13. _____

14. _____

14. (2.3) $136 \div 8$

15. _____

15. (2.3) $432 \div 24$

16. _____

16. (2.3) $5286 \div 37$

17. _____

17. (2.5) 211×1

For problems 18–20, use the numbers 216, 1,005, and 640.

18. _____

18. (2.4) Which numbers are divisible by 3?

19. _____

19. (2.4) Which number is divisible by 4?

20. _____

20. (2.4) Which number(s) is divisible by 5?

3

Exponents, Roots, and Factorizations of Whole Numbers

After completing this chapter, you should

Section 3.1 Exponents and Roots
- understand and be able to read exponential notation
- understand the concept of root and be able to read root notation
- be able to use a calculator having the y^x key to determine a root

Section 3.2 Grouping Symbols and the Order of Operations
- understand the use of grouping symbols
- understand and be able to use the order of operations
- use the calculator to determine the value of a numerical expression

Section 3.3 Prime Factorization of Natural Numbers
- be able to determine the factors of a whole number
- be able to distinguish between prime and composite numbers
- be familiar with the fundamental principle of arithmetic
- be able to find the prime factorization of a whole number

Section 3.4 The Greatest Common Factor
- be able to find the greatest common factor of two or more whole numbers

Section 3.5 The Least Common Multiple
- be able to find the least common multiple of two or more whole numbers

3.1 Exponents and Roots

<table>
<tr><td>Section
Overview</td><td>☐ **EXPONENTIAL NOTATION**
☐ **READING EXPONENTIAL NOTATION**
☐ **ROOTS**
☐ **READING ROOT NOTATION**
☐ **CALCULATORS**</td></tr>
</table>

☐ EXPONENTIAL NOTATION

Exponential Notation

We have noted that multiplication is a description of repeated addition. **Exponential notation** is a description of repeated multiplication.

Suppose we have the repeated multiplication

$$8 \cdot 8 \cdot 8 \cdot 8 \cdot 8$$

The factor 8 is repeated 5 times. Exponential notation uses a *superscript* for the number of times the factor is repeated. The superscript is placed on the repeated factor, 8^5, in this case. The superscript is called an **exponent.**

Exponent

The Function of an Exponent

> An **exponent** records the number of identical factors that are repeated in a multiplication.

☆ SAMPLE SET A

Write the following multiplication using exponents.

1. $3 \cdot 3$.   Since the factor 3 appears 2 times, we record this as

3^2

2. $62 \cdot 62 \cdot 62 \cdot 62 \cdot 62 \cdot 62 \cdot 62 \cdot 62 \cdot 62$.   Since the factor 62 appears 9 times, we record this as

62^9

Expand (write without exponents) each number.

3. 12^4.   The exponent 4 is recording 4 factors of 12 in a multiplication. Thus,

$12^4 = 12 \cdot 12 \cdot 12 \cdot 12$

4. 706^3.   The exponent 3 is recording 3 factors of 706 in a multiplication. Thus,

$706^3 = 706 \cdot 706 \cdot 706$

★ PRACTICE SET A

Write the following using exponents.

1. $37 \cdot 37$ **2.** $16 \cdot 16 \cdot 16 \cdot 16 \cdot 16$ **3.** $9 \cdot 9 \cdot 9 \cdot 9 \cdot 9 \cdot 9 \cdot 9 \cdot 9 \cdot 9 \cdot 9$

Write each number without exponents.

4. 85^3 **5.** 4^7 **6.** $1,739^2$

❏ READING EXPONENTIAL NOTATION

In a number such as 8^5,

Base

8 is called the **base.**

Exponent, Power

5 is called the **exponent,** or **power.**

8^5 is read as "eight to the fifth power," or more simply as "eight to the fifth," or "the fifth power of eight."

Squared

When a whole number is raised to the second power, it is said to be **squared.** The number 5^2 can be read as

5 to the second power, or
5 to the second, or
5 squared.

Cubed

When a whole number is raised to the third power, it is said to be **cubed.** The number 5^3 can be read as

5 to the third power, or
5 to the third, or
5 cubed.

When a whole number is raised to the power of 4 or higher, we simply say that that number is raised to that particular power. The number 5^8 can be read as

5 to the eighth power, or just
5 to the eighth.

❏ ROOTS

In the English language, the word "root" can mean a source of something. In mathematical terms, the word "root" is used to indicate that one number is the source of another number through repeated multiplication.

We know that $49 = 7^2$, that is, $49 = 7 \cdot 7$. Through repeated multiplication, 7 is the source of 49. Thus, 7 is a root of 49. Since two 7's must be multiplied together to produce 49, the 7 is called the second or **square root** of 49.

Square Root

We know that $8 = 2^3$, that is, $8 = 2 \cdot 2 \cdot 2$. Through repeated multiplication, 2 is the source of 8. Thus, 2 is a root of 8. Since three 2's must be multiplied together to produce 8, 2 is called the third or **cube root** of 8.

Cube Root

We can continue this way to see such roots as fourth roots, fifth roots, sixth roots, and so on.

❏ READING ROOT NOTATION

There is a symbol used to indicate roots of a number. It is called the radical sign $\sqrt[n]{}$

The Radical Sign $\sqrt[n]{}$

The symbol $\sqrt[n]{}$ is called a **radical sign** and indicates the nth root of a number.

We discuss *particular roots* using the radical sign as follows:

Square Root

$\sqrt[2]{\text{number}}$ indicates the **square root** of the number under the radical sign. It is customary to drop the 2 in the radical sign when discussing square roots. The symbol $\sqrt{}$ is understood to be the square root radical sign.

$$\sqrt{49} = 7 \quad \text{since} \quad 7 \cdot 7 = 7^2 = 49$$

Cube Root

$\sqrt[3]{\text{number}}$ indicates the **cube root** of the number under the radical sign.

$$\sqrt[3]{8} = 2 \quad \text{since} \quad 2 \cdot 2 \cdot 2 = 2^3 = 8$$

Fourth Root $\sqrt[4]{\text{number}}$ indicates the **fourth root** of the number under the radical sign.

$\sqrt[4]{81} = 3$ since $3 \cdot 3 \cdot 3 \cdot 3 = 3^4 = 81$

In an expression such as $\sqrt[5]{32}$,

Radical Sign $\sqrt{}$ is called the **radical sign.**
Index   5 is called the **index.** (The index describes the indicated root.)
Radicand   32 is called the **radicand.**
Radical $\sqrt[5]{32}$ is called a **radical** (or radical expression).

☆ **SAMPLE SET B**

Find each root.

1. $\sqrt{25}$.   To determine the square root of 25, we ask, "What whole number squared equals 25?" From our experience with multiplication, we know this number to be 5. Thus,

$\sqrt{25} = 5$

Check: $5 \cdot 5 = 5^2 = 25$.

2. $\sqrt[5]{32}$.   To determine the fifth root of 32, we ask, "What whole number raised to the fifth power equals 32?" This number is 2.

$\sqrt[5]{32} = 2$

Check: $2 \cdot 2 \cdot 2 \cdot 2 \cdot 2 = 2^5 = 32$.

★ **PRACTICE SET B**

Find the following roots using only a knowledge of multiplication.

1. $\sqrt{64}$ 2. $\sqrt{100}$ 3. $\sqrt[3]{64}$ 4. $\sqrt[6]{64}$

⬜ CALCULATORS

Calculators with the $\boxed{\sqrt{x}}$, $\boxed{y^x}$, and $\boxed{1/x}$ keys can be used to find or approximate roots.

☆ **SAMPLE SET C**

1. Use the calculator to find $\sqrt{121}$.

		Display Reads
Type	121	121
Press	$\boxed{\sqrt{x}}$	11

2. Find $\sqrt[7]{2187}$.

		Display Reads
Type	2187	2187
Press	$\boxed{y^x}$	2187
Type	7	7
Press	$\boxed{1/x}$	.14285714
Press	$\boxed{=}$	3

$\sqrt[7]{2187} = 3$.   (Which means that $3^7 = 2187$.)

★ **PRACTICE SET C**

Use a calculator to find the following roots.

1. $\sqrt[3]{729}$ **2.** $\sqrt[4]{8503056}$ **3.** $\sqrt{53361}$ **4.** $\sqrt[12]{16777216}$

Answers to Practice Sets are on p. 101.

Section 3.1 EXERCISES

For problems 1–9, write the expressions using exponential notation.

1. $4 \cdot 4$

2. $12 \cdot 12$

3. $9 \cdot 9 \cdot 9 \cdot 9$

4. $10 \cdot 10 \cdot 10 \cdot 10 \cdot 10 \cdot 10 \cdot$

5. $826 \cdot 826 \cdot 826$

6. $3{,}021 \cdot 3{,}021 \cdot 3{,}021 \cdot 3{,}021 \cdot 3{,}021$

7. $\underbrace{6 \cdot 6 \cdot \cdots \cdot 6}_{\text{85 factors of 6}}$

8. $\underbrace{2 \cdot 2 \cdot \cdots \cdot 2}_{\text{112 factors of 2}}$

9. $\underbrace{1 \cdot 1 \cdot \cdots \cdot 1}_{\text{3,008 factors of 1}}$

For problems 10–15, expand the terms. (Do not find the actual value.)

10. 5^3 **11.** 7^4

12. 15^2 **13.** 117^5

14. 61^6 **15.** 30^2

For problems 16–45, determine the value of each of the powers. Use a calculator to check each result.

16. 3^2 **17.** 4^2

18. 1^2 **19.** 10^2

20. 11^2 **21.** 12^2

22. 13^2 **23.** 15^2

24. 1^4 **25.** 3^4

26. 7^3 **27.** 10^3

28. 100^2

29. 8^3

30. 5^5

31. 9^3

32. 6^2

33. 7^1

34. 1^{28}

35. 2^7

36. 0^5

37. 8^4

38. 5^8

39. 6^9

40. 25^3

41. 42^2

42. 31^3

43. 15^5

44. 2^{20}

45. 816^2

For problems 46–63, find the roots (using your knowledge of multiplication). Use a calculator to check each result.

46. $\sqrt{9}$

47. $\sqrt{16}$

48. $\sqrt{36}$

49. $\sqrt{64}$

50. $\sqrt{121}$

51. $\sqrt{144}$

52. $\sqrt{169}$

53. $\sqrt{225}$

54. $\sqrt[3]{27}$

55. $\sqrt[5]{32}$

56. $\sqrt[4]{256}$

57. $\sqrt[3]{216}$

58. $\sqrt[7]{1}$

59. $\sqrt{400}$

60. $\sqrt{900}$

61. $\sqrt{10,000}$

62. $\sqrt{324}$

63. $\sqrt{3,600}$

For problems 64–75, use a calculator with the keys $\boxed{\sqrt{x}}$, $\boxed{y^x}$, and $\boxed{1/x}$ to find each of the values.

64. $\sqrt{676}$

65. $\sqrt{1,156}$

66. $\sqrt{46,225}$

67. $\sqrt{17,288,964}$

68. $\sqrt[3]{3,375}$

69. $\sqrt[4]{331,776}$

70. $\sqrt[8]{5,764,801}$

71. $\sqrt[12]{16,777,216}$

72. $\sqrt[8]{16,777,216}$

73. $\sqrt[10]{9,765,625}$

74. $\sqrt[4]{160,000}$

75. $\sqrt[3]{531,441}$

EXERCISES FOR REVIEW

(1.6) **76.** Use the numbers 3, 8, and 9 to illustrate the associative property of addition.

(2.1) **77.** In the multiplication $8 \cdot 4 = 32$, specify the name given to the numbers 8 and 4.

(2.2) **78.** Does the quotient $15 \div 0$ exist? If so, what is it?

(2.2) **79.** Does the quotient $0 \div 15$ exist? If so, what is it?

(2.5) **80.** Use the numbers 4 and 7 to illustrate the commutative property of multiplication.

★ **ANSWERS TO PRACTICE SETS (3.1)**

A. **1.** 37^2 **2.** 16^5 **3.** 9^{10} **4.** $85 \cdot 85 \cdot 85$ **5.** $4 \cdot 4 \cdot 4 \cdot 4 \cdot 4 \cdot 4 \cdot 4$ **6.** $1739 \cdot 1739$

B. **1.** 8   **2.** 10   **3.** 4   **4.** 2

C. **1.** 9   **2.** 54   **3.** 231   **4.** 4

3.2 Grouping Symbols and the Order of Operations

Section Overview
- ❑ **GROUPING SYMBOLS**
- ❑ **MULTIPLE GROUPING SYMBOLS**
- ❑ **THE ORDER OF OPERATIONS**
- ❑ **CALCULATORS**

❑ GROUPING SYMBOLS

Grouping symbols are used to indicate that a particular collection of numbers and meaningful operations are to be grouped together and considered as one number. The grouping symbols commonly used in mathematics are the following:

( )
[ ]
{ }
———

Parentheses:	( )
Brackets:	[ ]
Braces:	{ }
Bar:	———

In a computation in which more than one operation is involved, grouping symbols indicate which operation to perform first. If possible, we perform operations *inside* grouping symbols first.

☆ **SAMPLE SET A**

If possible, determine the value of each of the following.

1. $9 + (3 \cdot 8)$

Since 3 and 8 are within parentheses, they are to be combined first.

$$9 + (3 \cdot 8) = 9 + 24$$
$$= 33$$

Continued

Thus,

$$9 + (3 \cdot 8) = 33$$

2. $(10 \div 0) \cdot 6$

Since $10 \div 0$ is undefined, this operation is meaningless, and we attach no value to it. We write, "undefined."

★ **PRACTICE SET A**

If possible, determine the value of each of the following.

1. $16 - (3 \cdot 2)$ **2.** $5 + (7 \cdot 9)$ **3.** $(4 + 8) \cdot 2$ **4.** $28 \div (18 - 11)$

5. $(33 \div 3) - 11$ **6.** $4 + (0 \div 0)$

❑ MULTIPLE GROUPING SYMBOLS

When a set of grouping symbols occurs *inside* another set of grouping symbols, we perform the operations within the innermost set first.

☆ **SAMPLE SET B**

Determine the value of each of the following.

1. $2 + (8 \cdot 3) - (5 + 6)$

Combine 8 and 3 first, then combine 5 and 6.

$2 + 24 - 11$ **Now combine left to right.**
$26 - 11$
15

2. $10 + [30 - (2 \cdot 9)]$

Combine 2 and 9 since they occur in the innermost set of parentheses.

$10 + [30 - 18]$ **Now combine 30 and 18.**
$10 + 12$
22

★ **PRACTICE SET B**

Determine the value of each of the following.

1. $(17 + 8) + (9 + 20)$ **2.** $(55 - 6) - (13 \cdot 2)$ **3.** $23 + (12 \div 4) - (11 \cdot 2)$

4. $86 + [14 \div (10 - 8)]$ **5.** $31 + \{9 + [1 + (35 - 2)]\}$ **6.** $\{6 - [24 \div (4 \cdot 2)]\}^3$

☐ THE ORDER OF OPERATIONS

Sometimes there are no grouping symbols indicating which operations to perform first. For example, suppose we wish to find the value of $3 + 5 \cdot 2$. We could do *either* of two things:

1. Add 3 and 5, then multiply this sum by 2.

$$3 + 5 \cdot 2 = 8 \cdot 2$$
$$= 16$$

2. Multiply 5 and 2, then add 3 to this product.

$$3 + 5 \cdot 2 = 3 + 10$$
$$= 13$$

We now have two values for one number. To determine the correct value, we must use the *accepted order of operations*.

Order of Operations

> 1. Perform all operations *inside* grouping symbols, beginning with the innermost set, in the order 2, 3, 4 described below.
> 2. Perform all exponential and root operations.
> 3. Perform all multiplications and divisions, moving left to right.
> 4. Perform all additions and subtractions, moving left to right.

☆ SAMPLE SET C

Determine the value of each of the following.

1. $21 + 3 \cdot 12$.   **Multiply first.**
    $21 + 36$ **Add.**
    57

2. $(15 - 8) + 5 \cdot (6 + 4)$.   **Simplify inside parentheses first.**
    $7 + 5 \cdot 10$ **Multiply.**
    $7 + 50$ **Add.**
    57

3. $63 - (4 + 6 \cdot 3) + 76 - 4$.   **Simplify first within the parentheses by multiplying, then adding.**

    $63 - (4 + 18) + 76 - 4$
    $63 - 22 + 76 - 4$ **Now perform the additions and subtractions, moving left to right.**
    $41 + 76 - 4$ **Add 41 and 76:** $41 + 76 = 117$.
    $117 - 4$ **Subtract 4 from 117:** $117 - 4 = 113$.
    113

Continued

4. $7 \cdot 6 - 4^2 + 1^5$.   **Evaluate the exponential forms, moving left to right.**
   $7 \cdot 6 - 16 + 1$ **Multiply 7 and 6: $7 \cdot 6 = 42$.**
   $42 - 16 + 1$ **Subtract 16 from 42: $42 - 16 = 26$.**
   $26 + 1$ **Add 26 and 1: $26 + 1 = 27$.**
   27

5. $6 \cdot (3^2 + 2^2) + 4^2$.   **Evaluate the exponential forms in the parentheses:**
   $3^2 = 9$ and $2^2 = 4$.

   $6 \cdot (9 + 4) + 4^2$ **Add the 9 and 4 in the parentheses: $9 + 4 = 13$.**
   $6 \cdot (13) + 4^2$ **Evaluate the exponential form: $4^2 = 16$.**
   $6 \cdot (13) + 16$ **Multiply 6 and 13: $6 \cdot 13 = 78$.**
   $78 + 16$ **Add 78 and 16: $78 + 16 = 94$.**
   94

6. $\dfrac{6^2 + 2^2}{4^2 + 6 \cdot 2^2} + \dfrac{1^3 + 8^2}{10^2 - 19 \cdot 5}$.   **Recall that the bar is a grouping symbol. The fraction**
   $\dfrac{6^2 + 2^2}{4^2 + 6 \cdot 2^2}$ **is equivalent to $(6^2 + 2^2) \div (4^2 + 6 \cdot 2^2)$.**

   $\dfrac{36 + 4}{16 + 6 \cdot 4} + \dfrac{1 + 64}{100 - 19 \cdot 5}$

   $\dfrac{36 + 4}{16 + 24} + \dfrac{1 + 64}{100 - 95}$

   $\dfrac{40}{40} + \dfrac{65}{5}$

   $1 + 13$
   14

★ PRACTICE SET C

Determine the value of each of the following.

1. $8 + (32 - 7)$ **2.** $(34 + 18 - 2 \cdot 3) + 11$ **3.** $8(10) + 4(2 + 3) - (20 + 3 \cdot 15 + 40 - 5)$

4. $5 \cdot 8 + 4^2 - 2^2$ **5.** $4(6^2 - 3^3) \div (4^2 - 4)$ **6.** $(8 + 9 \cdot 3) \div 7 + 5 \cdot (8 \div 4 + 7 + 3 \cdot 5)$

7. $\dfrac{3^3 + 2^3}{6^2 - 29} + 5\left(\dfrac{8^2 + 2^4}{7^2 - 3^2}\right) \div \dfrac{8 \cdot 3 + 1^8}{2^3 - 3}$

❑ CALCULATORS

Using a calculator is helpful for simplifying computations that involve large numbers.

☆ SAMPLE SET D

Use a calculator to determine each value.

1. $9,842 + 56 \cdot 85$

Perform the multiplication first.

	Key		Display Reads
Type	56		56
Press	$\times$		56
Type	85		85
Press	$+$		4760
Type	9842		9842
Press	$=$		14602

Now perform the addition.

The display now reads 14,602.

2. $42(27 + 18) + 105(810 \div 18)$

Operate inside the parentheses.

	Key		Display Reads
Type	27		27
Press	$+$		27
Type	18		18
Press	$-$		45
Press	$\times$		45
Type	42		42
Press	$=$		1890

Multiply by 42.

Place this result into memory by pressing the memory key.

Now operate in the other parentheses.

	Key		Display Reads
Type	810		810
Press	$\div$		810
Type	18		18
Press	$-$		45
Press	$\times$		45
Type	105		105
Press	$=$		4725
Press	$+$		4725
			1890
Press	$=$		6615

Now multiply by 105.

We are now ready to add these two quantities together. Press the memory recall key.

Thus, $42(27 + 18) + 105(810 \div 18) = 6,615$.

Continued

3. $16^4 + 37^3$

Nonscientific Calculators		
Key		**Display Reads**
Type	16	16
Press	$\times$	16
Type	16	16
Press	$\times$	256
Type	16	16
Press	$\times$	4096
Type	16	16
Press	$=$	65536
Press the memory key		
Type	37	37
Press	$\times$	37
Type	37	37
Press	$\times$	1369
Type	37	37
Press	$\times$	50653
Press	$+$	50653
Press memory recall key		65536
Press	$=$	116189

Calculators with y^x Key		
Key		**Display Reads**
Type	16	16
Press	y^x	16
Type	4	4
Press	$=$	4096
Press	$+$	4096
Type	37	37
Press	y^x	37
Type	3	3
Press	$=$	116189

Thus, $16^4 + 37^3 = 116{,}189$.

We can certainly see that the more powerful calculator simplifies computations.

4. Nonscientific calculators are unable to handle calculations involving very large numbers.

$85612 \cdot 21065$

Key		Display Reads
Type	85612	85612
Press	$\times$	85162
Type	21065	21065
Press	$=$	

This number is too big for the display of some calculators and we'll probably get some kind of error message. On some scientific calculators such large numbers are coped with by placing them in a form called "scientific notation." Others can do the multiplication directly. (1803416780)

★ **PRACTICE SET D**

Use a calculator to find each value.

1. $9{,}285 + 86(49)$ **2.** $55(84 - 26) + 120(512 - 488)$ **3.** $106^3 - 17^4$ **4.** $6{,}053^3$

Answers to Practice Sets are on p. 110.

Section 3.2 EXERCISES

For problems 1 – 43, find each value. Check each result with a calculator.

1. $2 + 3 \cdot (8)$

2. $18 + 7 \cdot (4 - 1)$

3. $3 + 8 \cdot (6 - 2) + 11$

4. $1 - 5 \cdot (8 - 8)$

5. $37 - 1 \cdot 6^2$

6. $98 \div 2 \div 7^2$

7. $(4^2 - 2 \cdot 4) - 2^3$

8. $\sqrt{9} + 14$

9. $\sqrt{100} + \sqrt{81} - 4^2$

10. $\sqrt[3]{8} + 8 - 2 \cdot 5$

11. $\sqrt[4]{16} - 1 + 5^2$

12. $61 - 22 + 4[3 \cdot (10) + 11]$

13. $121 - 4 \cdot [(4) \cdot (5) - 12] + \dfrac{16}{2}$

14. $\dfrac{(1 + 16) - 3}{7} + 5 \cdot (12)$

15. $\dfrac{8 \cdot (6 + 20)}{8} + \dfrac{3 \cdot (6 + 16)}{22}$

16. $10 \cdot [8 + 2 \cdot (6 + 7)]$

17. $21 \div 7 \div 3$

18. $10^2 \cdot 3 \div 5^2 \cdot 3 - 2 \cdot 3$

19. $85 \div 5 \cdot 5 - 85$

20. $\dfrac{51}{17} + 7 - 2 \cdot 5 \cdot \left(\dfrac{12}{3}\right)$

21. $2^2 \cdot 3 + 2^3 \cdot (6 - 2) - (3 + 17) + 11(6)$

22. $26 - 2 \cdot \left\{\dfrac{6 + 20}{13}\right\}$

23. $2 \cdot \{(7 + 7) + 6 \cdot [4 \cdot (8 + 2)]\}$

24. $0 + 10(0) + 15 \cdot \{4 \cdot 3 + 1\}$

25. $18 + \dfrac{7 + 2}{9}$

26. $(4 + 7) \cdot (8 - 3)$

27. $(6 + 8) \cdot (5 + 2 - 4)$

28. $(21 - 3) \cdot (6 - 1) \cdot (7) + 4(6 + 3)$

29. $(10 + 5) \cdot (10 + 5) - 4 \cdot (60 - 4)$

36. $\dfrac{(1 + 6)^2 + 2}{3 \cdot 6 + 1}$

30. $6 \cdot \{2 \cdot 8 + 3\} - (5) \cdot (2) + \dfrac{8}{4} +$
$(1 + 8) \cdot (1 + 11)$

37. $\dfrac{6^2 - 1}{2^3 - 3} + \dfrac{4^3 + 2 \cdot 3}{2 \cdot 5}$

31. $2^5 + 3 \cdot (8 + 1)$

38. $\dfrac{5(8^2 - 9 \cdot 6)}{2^5 - 7} + \dfrac{7^2 - 4^2}{2^4 - 5}$

32. $3^4 + 2^4 \cdot (1 + 5)$

39. $\dfrac{(2 + 1)^3 + 2^3 + 1^{10}}{6^2} - \dfrac{15^2 - [2 \cdot 5]^2}{5 \cdot 5^2}$

33. $1^6 + 0^8 + 5^2 \cdot (2 + 8)^3$

40. $\dfrac{6^3 - 2 \cdot 10^2}{2^2} + \dfrac{18(2^3 + 7^2)}{2(19) - 3^3}$

34. $(7) \cdot (16) - 3^4 + 2^2 \cdot (1^7 + 3^2)$

35. $\dfrac{2^3 - 7}{5^2}$

41. $2 \cdot \{6 + [10^2 - 6\sqrt{25}]\}$

42. $181 - 3 \cdot \{2\sqrt{36} + 3\sqrt[3]{64}\}$

43. $\dfrac{2 \cdot (\sqrt{81} - \sqrt[3]{125})}{4^2 - 10 + 2^2}$

EXERCISES FOR REVIEW

(1.6)   **44.** The fact that

0 + any whole number = that particular whole number

is an example of which property of addition?

(2.1)   **45.** Find the product. $4{,}271 \times 630$.

(2.2)   **46.** In the statement $27 \div 3 = 9$, what name is given to the result 9?

(2.6)   **47.** What number is the multiplicative identity?

(2.6)   **48.** Find the value of 2^4.

★ **ANSWERS TO PRACTICE SETS (3.2)**

A.   **1.** 10   **2.** 68   **3.** 24   **4.** 4   **5.** 0   **6.** not possible (indeterminant)

B.   **1.** 54   **2.** 23   **3.** 4   **4.** 93   **5.** 74   **6.** 27

C.   **1.** 33   **2.** 57   **3.** 0   **4.** 52   **5.** 3   **6.** 125   **7.** 7

D.   **1.** 13,499   **2.** 6,070   **3.** 1,107,495

4. This number is too big for a nonscientific calculator. A scientific calculator will probably give you $2.217747109 \times 10^{11}$.

3.3 Prime Factorization of Natural Numbers

Section Overview

☐ **FACTORS**
☐ **DETERMINING THE FACTORS OF A WHOLE NUMBER**
☐ **PRIME AND COMPOSITE NUMBERS**
☐ **THE FUNDAMENTAL PRINCIPLE OF ARITHMETIC**
☐ **THE PRIME FACTORIZATION OF A NATURAL NUMBER**

☐ FACTORS

From observations made in the process of multiplication, we have seen that

(factor) $\cdot$ (factor) = product

Factors
Product

The two numbers being multiplied are the **factors** and the result of the multiplication is the **product.** Now, using our knowledge of division, we can see that a first number is a factor of a second number if the first number divides into the second number a whole number of times (without a remainder).

One Number as a Factor of Another

> A first number is a **factor** of a second number if the first number divides into the second number a whole number of times (without a remainder).

We show this in the following examples:

3 is a factor of 27, since $27 \div 3 = 9$, or $3 \cdot 9 = 27$.
7 is a factor of 56, since $56 \div 7 = 8$, or $7 \cdot 8 = 56$.
4 is *not* a factor of 10, since $10 \div 4 = 2$ R2. (There is a remainder.)

❑ DETERMINING THE FACTORS OF A WHOLE NUMBER

We can use the tests for divisibility from Section 2.4 to determine *all* the factors of a whole number.

☆ SAMPLE SET A

Find all the factors of 24.

Try 1: $24 \div 1 = 24$ **1 and 24 are factors.**
Try 2: 24 is even, so 24 is divisible by 2.

$24 \div 2 = 12$ **2 and 12 are factors.**

Try 3: $2 + 4 = 6$ and 6 is divisible by 3, so 24 is divisible by 3.

$24 \div 3 = 8$ **3 and 8 are factors**

Try 4: $24 \div 4 = 6$ **4 and 6 are factors.**
Try 5: $24 \div 5 = 4$ R4   **5 is *not* a factor.**

The next number to try is 6, but we already have that 6 is a factor. Once we come upon a factor that we already have discovered, we can stop.
All the whole number factors of 24 are 1, 2, 3, 4, 6, 8, 12, and 24.

★ PRACTICE SET A

Find all the factors of each of the following numbers.

1. 6   **2.** 12   **3.** 18   **4.** 5   **5.** 10   **6.** 33   **7.** 19

❑ PRIME AND COMPOSITE NUMBERS

Notice that the only factors of 7 are 1 and 7 itself, and that the only factors of 3 are 1 and 3 itself. However, the number 8 has the factors 1, 2, 4, and 8, and the number 10 has the factors 1, 2, 5, and 10. Thus, we can see that a whole number can have only *two* factors (itself and 1) and another whole number can have *several* factors.

We can use this observation to make a useful classification for whole numbers: prime numbers and composite numbers.

PRIME NUMBERS

Prime Number

> A whole number (greater than one) whose only factors are itself and 1 is called a **prime number.**

The Number 1 *is Not* a Prime Number

The first seven prime numbers are 2, 3, 5, 7, 11, 13, and 17. Notice that the whole number 1 is *not* considered to be a prime number, and the whole number 2 is the *first prime* and the *only even prime* number.

COMPOSITE NUMBERS

Composite Number

> A whole number composed of factors other than itself and 1 is called a **composite number.** Composite numbers are not prime numbers.

Some composite numbers are 4, 6, 8, 9, 10, 12, and 15.

☆ **SAMPLE SET B**

Determine which whole numbers are prime and which are composite.

1. 39.   Since 3 divides into 39, the number 39 is composite: $39 \div 3 = 13$.
2. 47.   A few division trials will assure us that 47 is only divisible by 1 and 47. Therefore, 47 is prime.

★ **PRACTICE SET B**

Determine which of the following whole numbers are prime and which are composite.

1. 3   **2.** 16   **3.** 21   **4.** 35   **5.** 47   **6.** 29

7. 101   **8.** 51

❑ THE FUNDAMENTAL PRINCIPLE OF ARITHMETIC

Prime numbers are very useful in the study of mathematics. We will see how they are used in subsequent sections. We now state the Fundamental Principle of Arithmetic.

Fundamental Principle of Arithmetic

> Except for the order of the factors, every natural number other than 1 can be factored in one and only one way as a product of prime numbers.

Prime Factorization

> When a number is factored so that all its factors are prime numbers, the factorization is called the **prime factorization** of the number.

The technique of prime factorization is illustrated in the following three examples.

1. $10 = 2 \cdot 5$.   Both 2 and 5 are primes. Therefore, $2 \cdot 5$ is the prime factorization of 10.
2. 11.   The number 11 is a prime number. Prime factorization applies *only* to composite numbers. Thus, 11 has *no* prime factorization.
3. $60 = 2 \cdot 30$.   The number 30 is not prime: $30 = 2 \cdot 15$.

$$60 = 2 \cdot 2 \cdot 15$$

The number 15 is not prime: $15 = 3 \cdot 5$.

$$60 = 2 \cdot 2 \cdot 3 \cdot 5$$

We'll use exponents.

$$60 = 2^2 \cdot 3 \cdot 5$$

The numbers 2, 3, and 5 are each prime. Therefore, $2^2 \cdot 3 \cdot 5$ is the prime factorization of 60.

☐ THE PRIME FACTORIZATION OF A NATURAL NUMBER

The following method provides a way of finding the prime factorization of a natural number.

The Method of Finding the Prime Factorization of a Natural Number

1. Divide the number repeatedly by the smallest prime number that will divide into it a whole number of times (without a remainder).
2. When the prime number used in step 1 no longer divides into the given number without a remainder, repeat the division process with the next largest prime that divides the given number.
3. Continue this process until the quotient is smaller than the divisor.
4. The prime factorization of the given number is the *product* of all these prime divisors. If the number has no prime divisors, it is a prime number.

We may be able to use some of the tests for divisibility we studied in Section 2.4 to help find the primes that divide the given number.

☆ SAMPLE SET C

1. Find the prime factorization of 60.

   Since the last digit of 60 is 0, which is even, 60 is divisible by 2. We will repeatedly divide by 2 until we no longer can. We shall divide as follows:

   ```
   2 | 60
     2 | 30      30 is divisible by 2 again.
       3 | 15    15 is not divisible by 2, but it is divisible by 3, the next prime.
         5 | 5   5 is not divisible by 3, but it is divisible by 5, the next prime.
             1
   ```

   The quotient 1 is finally smaller than the divisor 5, and the prime factorization of 60 is the product of these prime divisors.

   $$60 = 2 \cdot 2 \cdot 3 \cdot 5$$

   We use exponents when possible.

   $$60 = 2^2 \cdot 3 \cdot 5$$

Continued

2. Find the prime factorization of 441.

441 is not divisible by 2 since its last digit is not divisible by 2.
441 is divisible by 3 since $4 + 4 + 1 = 9$ and 9 is divisible by 3.

$\begin{array}{r|l} 3 & 441 \\ 3 & 147 \\ 7 & 49 \\ 7 & 7 \\ & 1 \end{array}$ **147 is divisible by 3 ($1 + 4 + 7 = 12$).**
49 is not divisible by 3, nor is it divisible by 5. It is divisible by 7.

The quotient 1 is finally smaller than the divisor 7, and the prime factorization of 441 is the product of these prime divisors.

$441 = 3 \cdot 3 \cdot 7 \cdot 7$

Use exponents.

$441 = 3^2 \cdot 7^2$

3. Find the prime factorization of 31.

31 is not divisible by 2.	**Its last digit is not even.**
	$31 \div 2 = 15 \text{ R}1$
	The quotient, 15, is larger than the divisor, 2. Continue.
31 is not divisible by 3.	**The digits $3 + 1 = 4$, and 4 is not divisible by 3.**
	$31 \div 3 = 10 \text{ R}1$
	The quotient, 10, is larger than the divisor, 3. Continue.
31 is not divisible by 5.	**The last digit of 31 is not 0 or 5.**
	$31 \div 5 = 6 \text{ R}1$
	The quotient, 6, is larger than the divisor, 5. Continue.
31 is not divisible by 7.	**Divide by 7.**
	$31 \div 7 = 4 \text{ R}1$
	The quotient, 4, is smaller than the divisor, 7. We can stop the process and conclude that 31 is a prime number.

The number 31 is a prime number.

★ PRACTICE SET C

Find the prime factorization of each whole number.

1. 22   **2.** 40   **3.** 48   **4.** 63   **5.** 945   **6.** 1,617

7. 17   **8.** 61

Answers to Practice Sets are on p. 117.

Section 3.3 EXERCISES

For problems 1–10, determine the missing factor(s).

1. $14 = 7 \cdot$ _____

2. $20 = 4 \cdot$ _____

3. $36 = 9 \cdot$ _____

4. $42 = 21 \cdot$ _____

5. $44 = 4 \cdot$ _____

6. $38 = 2 \cdot$ _____

7. $18 = 3 \cdot$ _____ $\cdot$ _____

8. $28 = 2 \cdot$ _____ $\cdot$ _____

9. $300 = 2 \cdot 5 \cdot$ _____ $\cdot$ _____

10. $840 = 2 \cdot$ _____ $\cdot$ _____ $\cdot$ _____

For problems 11–20, find all the factors of each of the numbers.

11. 16   **12.** 22

13. 56   **14.** 105

15. 220   **16.** 15

17. 32   **18.** 80

19. 142   **20.** 218

For problems 21–40, determine which of the whole numbers are prime and which are composite.

21. 23   **22.** 25

23. 27

24. 2

39. 4,575

40. 119

25. 3

26. 5

For problems 41–50, find the prime factorization of each of the whole numbers.

41. 26

42. 38

27. 7

28. 9

29. 11

30. 34

43. 54

44. 62

31. 55

32. 63

45. 56

46. 176

33. 1,044

34. 924

47. 480

48. 819

35. 339

36. 103

37. 209

38. 667

49. 2,025

50. 148,225

EXERCISES FOR REVIEW

(1.3)   **51.** Round 26,584 to the nearest ten.

(1.5)   **52.** How much bigger is 106 than 79?

(2.2)   **53.** True or false? Zero divided by any nonzero whole number is zero.

(2.3)   **54.** Find the quotient. $10{,}584 \div 126$.

(3.2)   **55.** Find the value of $\sqrt{121} - \sqrt{81} + 6^2 \div 3$.

★ **ANSWERS TO PRACTICE SETS (3.3)**

A. **1.** 1, 2, 3, 6   **2.** 1, 2, 3, 4, 6, 12   **3.** 1, 2, 3, 6, 9, 18   **4.** 1, 5   **5.** 1, 2, 5, 10   **6.** 1, 3, 11, 33
7. 1, 19

B. **1.** prime   **2.** composite   **3.** composite   **4.** composite   **5.** prime   **6.** prime   **7.** prime
8. composite

C. **1.** $22 = 2 \cdot 11$ **2.** $40 = 2^3 \cdot 5$ **3.** $48 = 2^4 \cdot 3$ **4.** $63 = 3^2 \cdot 7$ **5.** $945 = 3^3 \cdot 5 \cdot 7$
6. $1617 = 3 \cdot 7^2 \cdot 11$ **7.** 17 is prime   **8.** 61 is prime

3.4 The Greatest Common Factor

Section
Overview

☐ **THE GREATEST COMMON FACTOR (GCF)**
☐ **A METHOD FOR DETERMINING THE GREATEST COMMON FACTOR**

☐ THE GREATEST COMMON FACTOR (GCF)

Using the method we studied in Section 3.3, we could obtain the prime factorizations of 30 and 42.

$$30 = 2 \cdot 3 \cdot 5$$
$$42 = 2 \cdot 3 \cdot 7$$

Common Factor

We notice that 2 appears as a factor in both numbers, that is, 2 is a **common factor** of 30 and 42. We also notice that 3 appears as a factor in both numbers. Three is also a common factor of 30 and 42.

Greatest Common Factor
(GCF)

When considering two or more numbers, it is often useful to know if there is a largest common factor of the numbers, and if so, what that number is. The largest common factor of two or more whole numbers is called the **greatest common factor,** and is abbreviated by **GCF.** The greatest common factor of a collection of whole numbers is useful in working with fractions (which we will do in Chapter 4).

☐ A METHOD FOR DETERMINING THE GREATEST COMMON FACTOR

A straightforward method for determining the GCF of two or more whole numbers makes use of both the prime factorization of the numbers and exponents.

Finding the GCF

To find the **greatest common factor (GCF)** of two or more whole numbers:

1. Write the prime factorization of each number, using exponents on repeated factors.
2. Write each base that is common to each of the numbers.
3. To each base listed in step 2, attach the *smallest exponent* that appears on it in either of the prime factorizations.
4. The GCF is the product of the numbers found in step 3.

Find the GCF of the following numbers.

1. 12 and 18

1. $12 = 2 \cdot 6 = 2 \cdot 2 \cdot 3 = 2^2 \cdot 3$
   $18 = 2 \cdot 9 = 2 \cdot 3 \cdot 3 = 2 \cdot 3^2$
2. The common bases are 2 and 3.
3. The *smallest exponents* appearing on 2 and 3 in the prime factorizations are, respectively, 1 and 1 (2^1 and 3^1), or 2 and 3.
4. The GCF is the product of these numbers.

   $2 \cdot 3 = 6$

The GCF of 30 and 42 is 6 because 6 is the largest number that divides both 30 and 42 without a remainder.

2. 18, 60, and 72

1. $18 = 2 \cdot 9 = 2 \cdot 3 \cdot 3 = 2 \cdot 3^2$
   $60 = 2 \cdot 30 = 2 \cdot 2 \cdot 15 = 2 \cdot 2 \cdot 3 \cdot 5 = 2^2 \cdot 3 \cdot 5$
   $72 = 2 \cdot 36 = 2 \cdot 2 \cdot 18 = 2 \cdot 2 \cdot 2 \cdot 9 = 2 \cdot 2 \cdot 2 \cdot 3 \cdot 3 = 2^3 \cdot 3^2$
2. The common bases are 2 and 3.
3. The smallest exponents appearing on 2 and 3 in the prime factorizations are, respectively, 1 and 1:

   2^1 from 18.
   3^1 from 60.

4. The GCF is the product of these numbers.

   GCF is $2 \cdot 3 = 6$

Thus, 6 is the largest number that divides 18, 60, and 72 without a remainder.

3. 700, 1,880, and 6,160

1. $\begin{aligned} 700 &= 2 \cdot 350 = 2 \cdot 2 \cdot 175 = 2 \cdot 2 \cdot 5 \cdot 35 \\ &= 2 \cdot 2 \cdot 5 \cdot 5 \cdot 7 \\ &= 2^2 \cdot 5^2 \cdot 7 \end{aligned}$

   $\begin{aligned} 1{,}880 &= 2 \cdot 940 = 2 \cdot 2 \cdot 470 = 2 \cdot 2 \cdot 2 \cdot 235 \\ &= 2 \cdot 2 \cdot 2 \cdot 5 \cdot 47 \\ &= 2^3 \cdot 5 \cdot 47 \end{aligned}$

   $\begin{aligned} 6{,}160 &= 2 \cdot 3{,}080 = 2 \cdot 2 \cdot 1{,}540 = 2 \cdot 2 \cdot 2 \cdot 770 \\ &= 2 \cdot 2 \cdot 2 \cdot 2 \cdot 385 \\ &= 2 \cdot 2 \cdot 2 \cdot 2 \cdot 5 \cdot 77 \\ &= 2 \cdot 2 \cdot 2 \cdot 2 \cdot 5 \cdot 7 \cdot 11 \\ &= 2^4 \cdot 5 \cdot 7 \cdot 11 \end{aligned}$

2. The common bases are 2 and 5
3. The smallest exponents appearing on 2 and 5 in the prime factorizations are, respectively, 2 and 1.

   2^2 from 700.
   5^1 from either 1,880 or 6,160.

4. The GCF is the product of these numbers.

   GCF is $2^2 \cdot 5 = 4 \cdot 5 = 20$

Thus, 20 is the largest number that divides 700, 1,880, and 6,160 without a remainder.

★ **PRACTICE SET A**

Find the GCF of the following numbers.

1. 24 and 36 **2.** 48 and 72 **3.** 50 and 140 **4.** 21 and 225

5. 450, 600, and 540

Answers to the Practice Set are on p. 121.

Section 3.4 EXERCISES

For problems 1–27, find the greatest common factor (GCF) of the numbers.

1. 6 and 8 **2.** 5 and 10

3. 8 and 12 **4.** 9 and 12

5. 20 and 24 **6.** 35 and 175

7. 25 and 45 **8.** 45 and 189

9. 66 and 165 **10.** 264 and 132

11. 99 and 135 **12.** 65 and 15

13. 33 and 77 **14.** 245 and 80

15. 351 and 165

16. 60, 140, and 100

17. 147, 343, and 231

18. 24, 30, and 45

19. 175, 225, and 400

20. 210, 630, and 182

21. 14, 44, and 616

22. 1,617, 735, and 429

23. 1,573, 4,862, and 3,553

24. 3,672, 68, and 920

25. 7, 2,401, 343, 16, and 807

26. 500, 77, and 39

27. 441, 275, and 221

EXERCISES FOR REVIEW

(2.1)   **28.** Find the product. $2{,}753 \times 4{,}006$.

(2.3)   **29.** Find the quotient. $954 \div 18$.

(2.4)   **30.** Specify which of the digits 2, 3, or 4 divide into 9,462.

(3.1)   **31.** Write $8 \times 8 \times 8 \times 8 \times 8 \times 8$ using exponents.

(3.3)   **32.** Find the prime factorization of 378.

3.5 The Least Common Multiple

Section Overview	
	☐ **MULTIPLES**
	☐ **COMMON MULTIPLES**
	☐ **THE LEAST COMMON MULTIPLE (LCM)**
	☐ **FINDING THE LEAST COMMON MULTIPLE**

☐ MULTIPLES

When a whole number is multiplied by other whole numbers, with the exception of zero, the resulting products are called **multiples** of the given whole number. Note that any whole number is a multiple of itself.

☆ **SAMPLE SET A**

MULTIPLES OF 2	MULTIPLES OF 3	MULTIPLES OF 8	MULTIPLES OF 10
$2 \times 1 = 2$	$3 \times 1 = 3$	$8 \times 1 = 8$	$10 \times 1 = 10$
$2 \times 2 = 4$	$3 \times 2 = 6$	$8 \times 2 = 16$	$10 \times 2 = 20$
$2 \times 3 = 6$	$3 \times 3 = 9$	$8 \times 3 = 24$	$10 \times 3 = 30$
$2 \times 4 = 8$	$3 \times 4 = 12$	$8 \times 4 = 32$	$10 \times 4 = 40$
$2 \times 5 = 10$	$3 \times 5 = 15$	$8 \times 5 = 40$	$10 \times 5 = 50$
$\vdots$	$\vdots$	$\vdots$	$\vdots$

★ **PRACTICE SET A**

Find the first five multiples of the following numbers.

1. 4   **2.** 5   **3.** 6   **4.** 7   **5.** 9

☐ COMMON MULTIPLES

There will be times when we are given two or more whole numbers and we will need to know if there are any multiples that are common to each of them. If there are, we will need to know what they are. For example, some of the multiples that are common to 2 and 3 are 6, 12, and 18.

☆ **SAMPLE SET B**

We can visualize common multiples using the number line.

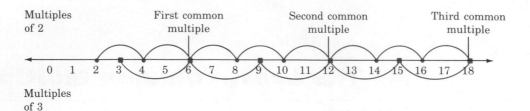

Notice that the common multiples can be divided by *both* whole numbers.

★ **PRACTICE SET B**

Find the first five common multiples of the following numbers.

1. 2 and 4 **2.** 3 and 4 **3.** 2 and 5 **4.** 3 and 6 **5.** 4 and 5

❑ THE LEAST COMMON MULTIPLE (LCM)

Notice that in our number line visualization of common multiples (above), the first common multiple is also the smallest, or **least common multiple,** abbreviated by **LCM.**

Least Common Multiple

> The **least common multiple,** LCM, of two or more whole numbers is the smallest whole number that each of the given numbers will divide into without a remainder.

The least common multiple will be extremely useful in working with fractions (Chapter 4).

❑ FINDING THE LEAST COMMON MULTIPLE

Finding the LCM

> To find the LCM of two or more numbers:
>
> 1. Write the prime factorization of each number, using exponents on repeated factors.
> 2. Write each base that appears in each of the prime factorizations.
> 3. To each base, attach the *largest exponent* that appears on it in the prime factorizations.
> 4. The LCM is the product of the numbers found in step 3.

There are some major differences between using the processes for obtaining the GCF and the LCM that we must note carefully:

The Difference Between the Processes for Obtaining the GCF and the LCM

1. Notice the difference between step 2 for the LCM and step 2 for the GCF. For the GCF, we use only the bases that are *common* in the prime factorizations, whereas for the LCM, we use *each* base that appears in the prime factorizations.
2. Notice the difference between step 3 for the LCM and step 3 for the GCF. For the GCF, we attach the *smallest* exponents to the common bases, whereas for the LCM, we attach the *largest* exponents to the bases.

☆ SAMPLE SET C

Find the LCM of the following numbers.

1. 9 and 12

1. $9 = 3 \cdot 3 = 3^2$
   $12 = 2 \cdot 6 = 2 \cdot 2 \cdot 3 = 2^2 \cdot 3$

2. The bases that appear in the prime factorizations are 2 and 3.
3. The *largest exponents* appearing on 2 and 3 in the prime factorizations are, respectively, 2 and 2:

   2^2 from 12.
   3^2 from 9.

4. The LCM is the product of these numbers.

   $LCM = 2^2 \cdot 3^2 = 4 \cdot 9 = 36$

Thus, 36 is the smallest number that both 9 and 12 divide into without remainders.

2. 90 and 630

1. $90 = 2 \cdot 45 = 2 \cdot 3 \cdot 15 = 2 \cdot 3 \cdot 3 \cdot 5 = 2 \cdot 3^2 \cdot 5$
   $630 = 2 \cdot 315 = 2 \cdot 3 \cdot 105 = 2 \cdot 3 \cdot 3 \cdot 35 = 2 \cdot 3 \cdot 3 \cdot 5 \cdot 7$
   $\qquad\qquad = 2 \cdot 3^2 \cdot 5 \cdot 7$

2. The bases that appear in the prime factorizations are 2, 3, 5, and 7.
3. The largest exponents that appear on 2, 3, 5, and 7 are, respectively, 1, 2, 1, and 1:

   2^1 from either 90 or 630.
   3^2 from either 90 or 630.
   5^1 from either 90 or 630.
   7^1 from 630.

4. The LCM is the product of these numbers.

   $LCM = 2 \cdot 3^2 \cdot 5 \cdot 7 = 2 \cdot 9 \cdot 5 \cdot 7 = 630$

Thus, 630 is the smallest number that both 90 and 630 divide into with no remainders.

3. 33, 110, and 484

1. $33 = 3 \cdot 11$
   $110 = 2 \cdot 55 = 2 \cdot 5 \cdot 11$
   $484 = 2 \cdot 242 = 2 \cdot 2 \cdot 121 = 2 \cdot 2 \cdot 11 \cdot 11 = 2^2 \cdot 11^2$.
2. The bases that appear in the prime factorizations are 2, 3, 5, and 11.
3. The largest exponents that appear on 2, 3, 5, and 11 are, respectively, 2, 1, 1, and 2:

   2^2 from 484.
   3^1 from 33.
   5^1 from 110
   11^2 from 484.

Continued

4. The LCM is the product of these numbers.

$$LCM = 2^2 \cdot 3 \cdot 5 \cdot 11^2$$
$$= 4 \cdot 3 \cdot 5 \cdot 121$$
$$= 7260$$

Thus, 7260 is the smallest number that 33, 110, and 484 divide into without remainders.

★ PRACTICE SET C

Find the LCM of the following numbers.

1. 20 and 54   **2.** 14 and 28   **3.** 6 and 63   **4.** 28, 40, and 98

5. 16, 27, 125, and 363

Answers to Practice Sets are on p. 126.

Section 3.5 EXERCISES

For problems 1–45, find the least common multiple of the numbers.

1. 8 and 12   **2.** 6 and 15

3. 8 and 10   **4.** 10 and 14

5. 4 and 6   **6.** 6 and 12

7. 9 and 18   **8.** 6 and 8

9. 5 and 6   **10.** 7 and 8

11. 3 and 4   **12.** 2 and 9

13. 7 and 9   **14.** 28 and 36   **29.** 18, 21, and 42

                                           30. 4, 5, and 21

15. 24 and 36   **16.** 28 and 42

                                           31. 45, 63, and 98

17. 240 and 360   **18.** 162 and 270

                                           32. 15, 25, and 40

19. 20 and 24   **20.** 25 and 30

                                           33. 12, 16, and 20

21. 24 and 54   **22.** 16 and 24

                                           34. 84 and 96

23. 36 and 48   **24.** 24 and 40

                                           35. 48 and 54

25. 15 and 21   **26.** 50 and 140

                                           36. 12, 16, and 24

27. 7, 11, and 33   **28.** 8, 10, and 15   **37.** 12, 16, 24, and 36

38. 6, 9, 12, and 18

42. 38, 92, 115, and 189

39. 8, 14, 28, and 32

43. 8 and 8

40. 18, 80, 108, and 490

44. 12, 12, and 12

41. 22, 27, 130, and 225

45. 3, 9, 12, and 3

EXERCISES FOR REVIEW

(1.3) **46.** Round 434,892 to the nearest ten thousand.

(1.5) **47.** How much bigger is 14,061 than 7,509?

(2.3) **48.** Find the quotient. 22,428 ÷ 14.

(3.1) **49.** Expand 84^3. Do not find the value.

(3.4) **50.** Find the greatest common factor of 48 and 72.

★ **ANSWERS TO PRACTICE SETS (3.5)**

A. **1.** 4, 8, 12, 16, 20   **2.** 5, 10, 15, 20, 25   **3.** 6, 12, 18, 24, 30   **4.** 7, 14, 21, 28, 35
5. 9, 18, 27, 36, 45

B. **1.** 4, 8, 12, 16, 20   **2.** 12, 24, 36, 48, 60   **3.** 10, 20, 30, 40, 50   **4.** 6, 12, 18, 24, 30
5. 20, 40, 60, 80, 100

C. **1.** 540   **2.** 28   **3.** 126   **4.** 1,960   **5.** 6,534,000

Exponential Notation (3.1)

Exponential notation is a description of repeated multiplication.

Exponent (3.1)

An *exponent* records the number of identical factors repeated in a multiplication.

In a number such as 7^3.

Base
Exponent
Power (3.1)

7 is called the *base*.
3 is called the *exponent,* or power.
7^3 is read "seven to the third power," or "seven cubed."

Squared
Cubed (3.1)

A number raised to the second power is often called *squared*. A number raised to the third power is often called *cubed*.

Root (3.1)

In mathematics, the word *root* is used to indicate that, through repeated multiplication, one number is the source of another number.

The Radical Sign √ (3.1)

The symbol $\sqrt{}$ is called a *radical sign* and indicates the square root of a number. The symbol $\sqrt[n]{}$ represents the *n*th root.

Radical
Index
Radicand (3.1)

An expression such as $\sqrt[4]{16}$ is called a *radical* and 4 is called the *index*. The number 16 is called the *radicand*.

Grouping Symbols (3.2)

Grouping symbols are used to indicate that a particular collection of numbers and meaningful operations are to be grouped together and considered as one number. The grouping symbols commonly used in mathematics are

Parentheses: (   )
Brackets: [   ]
Braces: {   }
Bar: ____

Order of Operations (3.2)

1. Perform all operations inside grouping symbols, beginning with the innermost set, in the order of 2, 3, and 4 below.
2. Perform all exponential and root operations, moving left to right.
3. Perform all multiplications and division, moving left to right.
4. Perform all additions and subtractions, moving left to right.

One Number as the Factor of Another (3.3)

A first number is a factor of a second number if the first number divides into the second number a whole number of times.

Prime Number (3.3)

A whole number greater than one whose only factors are itself and 1 is called a *prime number*. The whole number 1 is not a prime number. The whole number 2 is the first prime number and the only even prime number.

Composite Number (3.3)

A whole number greater than one that is composed of factors other than itself and 1 is called a *composite number*.

Fundamental Principle of Arithmetic (3.3)

Except for the order of factors, every whole number other than 1 can be written in one and only one way as a product of prime numbers.

Prime Factorization (3.3)

The prime factorization of 45 is $3 \cdot 3 \cdot 5$. The numbers that occur in this factorization of 45 are each prime.

Determining the Prime Factorization of a Whole Number (3.3)

There is a simple method, based on division by prime numbers, that produces the prime factorization of a whole number. For example, we determine the prime factorization of 132 as follows.

```
2 | 132
  2 | 66
    3 | 33
        11
```

The prime factorization of 132 is $2 \cdot 2 \cdot 3 \cdot 11 = 2^2 \cdot 3 \cdot 11$.

Common Factor **(3.4)**

A factor that occurs in each number of a group of numbers is called a *common factor*.

3 is a common factor to the group 18, 6, and 45

Greatest Common Factor (GCF) **(3.4)**

The largest common factor of a group of whole numbers is called the *greatest common factor*. For example, to find the greatest common factor of 12 and 20,

1. Write the prime factorization of each number.

$$12 = 2 \cdot 2 \cdot 3 = \quad 2^2 \cdot 3$$
$$60 = 2 \cdot 2 \cdot 3 \cdot 5 = 2^2 \cdot 3 \cdot 5$$

2. Write each base that is common to each of the numbers:

2 and 3

3. The smallest exponent appearing on 2 is 2.
   The smallest exponent appearing on 3 is 1.
4. The GCF of 12 and 60 is the product of the numbers 2^2 and 3.

$$2^2 \cdot 3 = 4 \cdot 3 = 12$$

Thus, 12 is the largest number that divides both 12 and 60 without a remainder.

Finding the GCF **(3.4)**

There is a simple method, based on prime factorization, that determines the GCF of a group of whole numbers.

Multiple **(3.5)**

When a whole number is multiplied by all other whole numbers, with the exception of zero, the resulting individual products are called *multiples* of that whole number. Some multiples of 7 are 7, 14, 21, and 28.

Common Multiples **(3.5)**

Multiples that are common to a group of whole numbers are called *common multiples*. Some common multiples of 6 and 9 are 18, 36, and 54.

The LCM **(3.5)**

The *least common multiple* (LCM) of a group of whole numbers is the smallest whole number that each of the given whole numbers divides into without a remainder. The least common multiple of 9 and 6 is 18.

Finding the LCM **(3.5)**

There is a simple method, based on prime factorization, that determines the LCM of a group of whole numbers. For example, the least common multiple of 28 and 72 is found in the following way.

1. Write the prime factorization of each number.

$$28 = 2 \cdot 2 \cdot 7 = 2^2 \cdot 7$$
$$72 = 2 \cdot 2 \cdot 2 \cdot 3 \cdot 3 = 2^3 \cdot 3^2$$

2. Write each base that appears in each of the prime factorizations, 2, 3, and 7.
3. To each of the bases listed in step 2, attach the *largest* exponent that appears on it in the prime factorization.

2^3, 3^2, and 7

4. The LCM is the product of the numbers found in step 3.

$$2^3 \cdot 3^2 \cdot 7 = 8 \cdot 9 \cdot 7 = 504$$

Thus, 504 is the smallest number that both 28 and 72 will divide into without a remainder.

The Difference Between the GCF and the LCM **(3.5)**

The GCF of two or more whole numbers is the largest number that divides into each of the given whole numbers.

The LCM of two or more whole numbers is the smallest whole number that each of the given numbers divides into without a remainder.

EXERCISE SUPPLEMENT

Section 3.1

For problems 1–25, determine the value of each power and root.

1. 3^3

2. 4^3

3. 0^5

4. 1^4

5. 12^2

6. 7^2

7. 8^2

8. 11^2

9. 2^5

10. 3^4

11. 15^2

12. 20^2

13. 25^2

14. $\sqrt{36}$

15. $\sqrt{225}$

16. $\sqrt[3]{64}$

17. $\sqrt[4]{16}$

18. $\sqrt{0}$

19. $\sqrt[3]{1}$

20. $\sqrt[3]{216}$

21. $\sqrt{144}$

22. $\sqrt{196}$

23. $\sqrt{1}$

24. $\sqrt[4]{0}$

25. $\sqrt[6]{64}$

Section 3.2

For problems 26–45, use the order of operations to determine each value.

26. $2^3 - 2 \cdot 4$

27. $5^2 - 10 \cdot 2 - 5$

28. $\sqrt{81} - 3^2 + 6 \cdot 2$

29. $15^2 + 5^2 \cdot 2^2$

30. $3 \cdot (2^2 + 3^2)$

31. $64 \cdot (3^2 - 2^3)$

32. $\dfrac{5^2 + 1}{13} + \dfrac{3^3 + 1}{14}$

33. $\dfrac{6^2 - 1}{5 \cdot 7} - \dfrac{49 + 7}{2 \cdot 7}$

34. $\dfrac{2 \cdot [3 + 5(2^2 + 1)]}{5 \cdot 2^3 - 3^2}$

35. $\dfrac{3^2 \cdot [2^5 - 1^4(2^3 + 25)]}{2 \cdot 5^2 + 5 + 2}$

36. $\dfrac{(5^2 - 2^3) - 2 \cdot 7}{2^2 - 1} + 5 \cdot \left[\dfrac{3^2 - 3}{2} + 1 \right]$

37. $(8 - 3)^2 + (2 + 3^2)^2$

38. $3^2 \cdot (4^2 + \sqrt{25}) + 2^3 \cdot (\sqrt{81} - 3^2)$

39. $\sqrt{16 + 9}$

40. $\sqrt{16} + \sqrt{9}$

41. Compare the results of problems 39 and 40. What might we conclude?

42. $\sqrt{18 \cdot 2}$

43. $\sqrt{6 \cdot 6}$

44. $\sqrt{7 \cdot 7}$

45. $\sqrt{8 \cdot 8}$

46. An _____ records the number of identical factors that are repeated in a multiplication.

Section 3.3

For problems 47–53, find all the factors of each number.

47. 18

48. 24

49. 11

50. 12

51. 51

52. 25

53. 2

54. What number is the smallest prime number?

For problems 55–64, write each number as a product of prime factors.

55. 55

56. 20

57. 80

58. 284

59. 700

60. 845

61. 1,614

62. 921

63. 29

64. 37

Section 3.4

For problems 65–75, find the greatest common factor of each collection of numbers.

65. 5 and 15

66. 6 and 14

67. 10 and 15

68. 6, 8, and 12

69. 18 and 24

70. 42 and 54

71. 40 and 60

72. 18, 48, and 72

73. 147, 189, and 315

74. 64, 72, and 108

75. 275, 297, and 539

Section 3.5

For problems 76–86, find the least common multiple of each collection of numbers.

76. 5 and 15

77. 6 and 14

78. 10 and 15

79. 36 and 90

80. 42 and 54

81. 8, 12, and 20

82. 40, 50, and 180

83. 135, 147, and 324

84. 108, 144, and 324

85. 5, 18, 25, and 30

86. 12, 15, 18, and 20

87. Find all divisors of 24.

88. Find all factors of 24.

89. Write all divisors of $2^3 \cdot 5^2 \cdot 7$.

90. Write all divisors of $6 \cdot 8^2 \cdot 10^3$.

91. Does 7 divide $5^3 \cdot 6^4 \cdot 7^2 \cdot 8^5$?

92. Does 13 divide $8^3 \cdot 10^2 \cdot 11^4 \cdot 13^2 \cdot 15$?

1. _____

1. (3.1) In the number 8^5, write the names used for the number 8 and the number 5.

2. _____

2. (3.1) Write using exponents.

$$12 \times 12 \times 12 \times 12 \times 12 \times 12 \times 12$$

3. _____

3. (3.1) Expand 9^4.

For problems 4–15, determine the value of each expression.

4. (3.2) 4^3

4. _____

5. (3.2) 1^5

5. _____

6. (3.2) 0^3

6. _____

7. (3.2) 2^6

7. _____

8. (3.2) $\sqrt{49}$

8. _____

9. (3.2) $\sqrt[3]{27}$

9. _____

10. _____

10. (3.2) $\sqrt[8]{1}$

11. _____

11. (3.2) $16 + 2 \cdot (8 - 6)$

12. (3.2) $5^3 - \sqrt{100} + 8 \cdot 2 - 20 \div 5$

12. _____

13. (3.2) $3 \cdot \dfrac{8^2 - 2 \cdot 3^2}{5^2 - 2} \cdot \dfrac{6^3 - 4 \cdot 5^2}{29}$

13. _____

14. (3.2) $\dfrac{20 + 2^4}{2^3 \cdot 2 - 5 \cdot 2} + \dfrac{5 \cdot 7 - \sqrt{81}}{7 + 3 \cdot 2}$

14. _____

15. (3.2) $[(8 - 3)^2 + (33 - 4\sqrt{49})] - 2[(10 - 3^2) + 9] - 5$

15. _____

For problems 16–20, find the prime factorization of each whole number. If the number is prime, write "prime."

16. (3.3) 18

16. _____

17. (3.3) 68

17. _____

132

18. _____

18. (3.3) 142

19. (3.3) 151

19. _____

20. (3.3) 468

20. _____

21. _____

For problems 21 and 22, find the greatest common factor.

21. (3.4) 200 and 36

22. _____

22. (3.4) 900 and 135

23. _____

23. (3.4) Write all the factors of 36.

24. _____

24. (3.4) Write all the divisors of 18.

25. _____

25. (3.4) Does 7 divide into $5^2 \cdot 6^3 \cdot 7^4 \cdot 8$? Explain.

133

26. _____

26. (3.4) Is 3 a factor of $2^6 \cdot 3^2 \cdot 5^3 \cdot 4^6$? Explain.

27. _____

27. (3.4) Does 13 divide into $11^3 \cdot 12^4 \cdot 15^2$? Explain.

For problems 28 and 29, find the least common multiple.

28. _____

28. (3.5) 432 and 180

29. _____

29. (3.5) 28, 40, and 95

4

Introduction to Fractions and Multiplication and Division of Fractions

After completing this chapter, you should

Section 4.1 Fractions of Whole Numbers
- understand the concept of fractions of whole numbers
- be able to recognize the parts of a fraction

Section 4.2 Proper Fractions, Improper Fractions, and Mixed Numbers
- be able to distinguish between proper fractions, improper fractions, and mixed numbers
- be able to convert an improper fraction to a mixed number
- be able to convert a mixed number to an improper fraction

Section 4.3 Equivalent Fractions, Reducing Fractions to Lowest Terms, and Raising Fractions to Higher Terms
- be able to recognize equivalent fractions
- be able to reduce a fraction to lowest terms
- be able to raise a fraction to higher terms

Section 4.4 Multiplication of Fractions
- understand the concept of multiplication of fractions
- be able to multiply one fraction by another
- be able to multiply mixed numbers
- be able to find powers and roots of various fractions

Section 4.5 Division of Fractions
- be able to determine the reciprocal of a number
- be able to divide one fraction by another

Section 4.6 Applications Involving Fractions
- be able to solve missing product statements
- be able to solve missing factor statements

4.1 Fractions of Whole Numbers

Section
Overview

❏ **MORE NUMBERS ON THE NUMBER LINE**
❏ **FRACTIONS OF WHOLE NUMBERS**
❏ **THE PARTS OF A FRACTION**
❏ **READING AND WRITING FRACTIONS**

❏ MORE NUMBERS ON THE NUMBER LINE

In Chapters 1, 2, and 3, we studied the whole numbers and methods of combining them. We noted that we could visually display the whole numbers by drawing a number line and placing closed circles at whole number locations.

By observing this number line, we can see that the whole numbers do not account for every point on the line. What numbers, if any, can be associated with these points? In this section we will see that many of the points on the number line, including the points already associated with whole numbers, can be associated with numbers called *fractions*.

❏ FRACTIONS OF WHOLE NUMBERS

The Nature of the Positive
Fractions

We can extend our collection of numbers, which now contains only the whole numbers, by including fractions of whole numbers. We can determine the nature of these fractions using the number line.

If we place a pencil at some whole number and proceed to travel to the right to the next whole number, we see that our journey can be *broken* into different types of equal parts as shown in the following examples.

(a) 1 part.

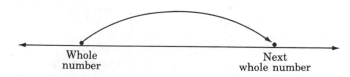

(b) 2 equal parts.

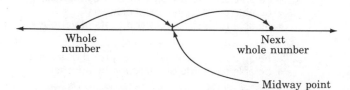

(c) 3 equal parts.

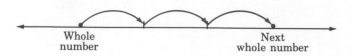

(d) 4 equal parts.

Notice that the number of parts, 2, 3, and 4, that we are breaking the original quantity into is always a *nonzero whole number*. The idea of breaking up a whole quantity gives us the word *fraction*. The word fraction comes from the Latin word "fractio" which means a breaking, or fracture.

The Latin Word Fractio

Suppose we break up the interval from some whole number to the next whole number into five equal parts.

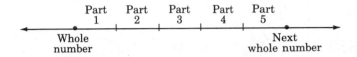

After starting to move from one whole number to the next, we decide to stop after covering only two parts. We have covered 2 parts of 5 equal parts. This situation is described by writing $\frac{2}{5}$.

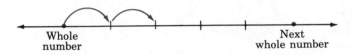

Positive Fraction

A number such as $\frac{2}{5}$ is called a **positive fraction,** or more simply, a **fraction.**

❑ THE PARTS OF A FRACTION

A fraction has *three parts.*

Fraction Bar

1. The fraction bar ——.

   The **fraction bar** serves as a grouping symbol. It separates a quantity into individual groups. These groups have names, as noted in 2 and 3 below.

2. The nonzero number below the fraction bar.

Denominator

   This number is called the **denominator** of the fraction, and it indicates the number of parts the whole quantity has been divided into. Notice that the denominator must be a nonzero whole number since the least number of parts any quantity can have is one.

3. The number above the fraction bar.

Numerator

   This number is called the **numerator** of the fraction, and it indicates how many of the specified parts are being considered. Notice that the numerator can be any whole number (including zero) since any number of the specified parts can be considered.

$$\frac{\text{whole number}}{\text{nonzero whole number}} \longleftrightarrow \frac{\text{numerator}}{\text{denominator}}$$

The diagrams in the following problems are illustrations of fractions.

1.

A whole
circle

The whole circle
divided into
3 equal parts

1 of the 3
equal parts

$\frac{1}{3}$ ← $\boxed{1}$ of $\boxed{3}$ equal parts

The fraction $\frac{1}{3}$ is read as "one third."

2.

A whole
rectangle

The whole
rectangle
divided into
5 equal parts

3 of the 5
equal parts

$\frac{3}{5}$ ← $\boxed{3}$ of $\boxed{5}$ equal parts

The fraction $\frac{3}{5}$ is read as "three fifths."

3.
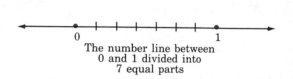

The number line
between 0 and 1

The number line between
0 and 1 divided into
7 equal parts

6 of the 7 equal parts

$\frac{6}{7}$ ← $\boxed{6}$ of the $\boxed{7}$ equal parts

The fraction $\frac{6}{7}$ is read as "six sevenths."

4.

A whole circle

The whole circle
divided into
4 equal parts

4 of the 4
equal parts

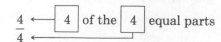

$\frac{4}{4}$ ← $\boxed{4}$ of the $\boxed{4}$ equal parts

When the numerator and denominator are equal, the fraction represents the entire quantity, and its value is 1.

$$\frac{\text{nonzero whole number}}{\text{same nonzero whole number}} = 1$$

★ **PRACTICE SET A**

Specify the numerator and denominator of the following fractions.

1. $\dfrac{4}{7}$ 2. $\dfrac{5}{8}$ 3. $\dfrac{10}{15}$ 4. $\dfrac{1}{9}$ 5. $\dfrac{0}{2}$

☐ READING AND WRITING FRACTIONS

In order to properly translate fractions from word form to number form, or from number form to word form, it is necessary to understand the use of the *hyphen*.

Use of the Hyphen

One of the main uses of the **hyphen** is to tell the reader that two words not ordinarily joined are to be taken in combination as a unit. Hyphens are *always* used for numbers between and including 21 and 99 (except those ending in zero).

☆ **SAMPLE SET B**

Write each fraction using whole numbers.

1. Fifty three-hundredths.   The hyphen joins the words three and hundredths and tells us to consider them as a single unit. Therefore,

   fifty three-hundredths translates as $\dfrac{50}{300}$

2. Fifty-three hundredths.   The hyphen joins the numbers fifty and three and tells us to consider them as a single unit. Therefore,

   fifty-three hundredths translates as $\dfrac{53}{100}$

3. Four hundred seven-thousandths.   The hyphen joins the words seven and thousandths and tells us to consider them as a single unit. Therefore,

   four hundred seven-thousandths translates as $\dfrac{400}{7,000}$

4. Four hundred seven thousandths.   The absence of hyphens indicates that the words *seven* and *thousandths* are to be considered individually.

   four hundred seven thousandths translates as $\dfrac{407}{1,000}$

Continued

Write each fraction using words.

5. $\frac{21}{85}$ translates as twenty-one eighty-fifths.

6. $\frac{200}{3,000}$ translates as two hundred three-thousandths.   A hyphen is needed between the words three and thousandths to tell the reader that these words are to be considered as a single unit.

7. $\frac{203}{1,000}$ translates as two hundred three thousandths.

★ PRACTICE SET B

Write the following fractions using whole numbers.

1. one tenth   **2.** eleven fourteenths

3. sixteen thirty-fifths   **4.** eight hundred seven-thousandths

Write the following using words.

5. $\frac{3}{8}$ **6.** $\frac{1}{10}$ **7.** $\frac{3}{250}$ **8.** $\frac{114}{3,190}$

Name the fraction that describes each shaded portion.

9.   **10.**

In problems 11 and 12, state the numerator and denominator, and write each fraction in words.

11. The number $\frac{5}{9}$ is used in converting from Fahrenheit to Celsius.

12. A dime is $\frac{1}{10}$ of a dollar.

Answers to Practice Sets are on p. 144.

Section 4.1 EXERCISES

For problems 1–10, specify the numerator and denominator in each fraction.

1. $\frac{3}{4}$

2. $\frac{9}{10}$

3. $\frac{1}{5}$

4. $\frac{5}{6}$

5. $\frac{7}{7}$

6. $\frac{4}{6}$

7. $\frac{0}{12}$

8. $\frac{25}{25}$

9. $\frac{18}{1}$

10. $\frac{0}{16}$

For problems 11–20, write the fractions using whole numbers.

11. four fifths

12. two ninths

13. fifteen twentieths

14. forty-seven eighty-thirds

15. ninety-one one hundred sevenths

16. twenty-two four hundred elevenths

17. six hundred five eight hundred thirty-fourths

18. three thousand three forty-four ten-thousandths

19. ninety-two one-millionths

20. one three-billionths

For problems 21–30, write the fractions using words.

21. $\frac{5}{9}$

22. $\frac{6}{10}$

23. $\frac{8}{15}$

24. $\frac{10}{13}$

25. $\frac{75}{100}$

26. $\frac{86}{135}$

27. $\frac{916}{1,014}$

28. $\frac{501}{10,001}$

29. $\frac{18}{31,608}$

30. $\frac{1}{500,000}$

For problems 31–34, name the fraction corresponding to the shaded portion.

31.

32.

33.

34.

For problems 35 – 38, shade the portion corresponding to the given fraction on the given figure.

35. $\dfrac{3}{5}$

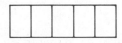

36. $\dfrac{1}{8}$

37. $\dfrac{6}{6}$

38. $\dfrac{0}{3}$

State the numerator and denominator and write in words each of the fractions appearing in the statements for problems 39 – 48.

39. A contractor is selling houses on $\dfrac{1}{4}$ acre lots.

40. The fraction $\dfrac{22}{7}$ is sometimes used as an approximation to the number π. (The symbol is read "pi.")

41. The fraction $\dfrac{4}{3}$ is used in finding the volume of a sphere.

42. One inch is $\dfrac{1}{12}$ of a foot.

43. About $\dfrac{2}{7}$ of the students in a college statistics class received a "B" in the course.

44. The probability of randomly selecting a club when drawing one card from a standard deck of 52 cards is $\dfrac{13}{52}$.

45. In a box that contains eight computer chips, five are known to be good and three are known to be defective. If three chips are selected at random, the probability that all three are defective is $\frac{1}{56}$.

46. In a room of 25 people, the probability that at least two people have the same birthdate (date and month, not year) is $\frac{569}{1,000}$.

47. The mean (average) of the numbers 21, 25, 43, and 36 is $\frac{125}{4}$.

48. If a rock falls from a height of 20 meters on Jupiter, the rock will be $\frac{32}{25}$ meters high after $\frac{6}{5}$ seconds.

EXERCISES FOR REVIEW

(1.6) **49.** Use the numbers 3 and 11 to illustrate the commutative property of addition.

(2.3) **50.** Find the quotient. $676 \div 26$.

(3.1) **51.** Write $7 \cdot 7 \cdot 7 \cdot 7 \cdot 7$ using exponents.

(3.2) **52.** Find the value of $\frac{8 \cdot (6 + 20)}{8} + \frac{3 \cdot (6 + 16)}{22}$.

(3.5) **53.** Find the least common multiple of 12, 16, and 18.

★ **Answers to Practice Sets (4.1)**

A. **1.** 4, 7   **2.** 5, 8   **3.** 10, 15   **4.** 1, 9   **5.** 0, 2

B. **1.** $\frac{1}{10}$ **2.** $\frac{11}{14}$ **3.** $\frac{16}{35}$ **4.** $\frac{800}{7,000}$

5. three eighths   **6.** one tenth   **7.** three two hundred fiftieths

8. one hundred fourteen three thousand one hundred ninetieths   **9.** $\frac{3}{8}$ **10.** $\frac{1}{16}$

11. 5, 9, five ninths   **12.** 1, 10, one tenth

4.2 Proper Fractions, Improper Fractions, and Mixed Numbers

Section Overview

☐ **POSITIVE PROPER FRACTIONS**
☐ **POSITIVE IMPROPER FRACTIONS**
☐ **POSITIVE MIXED NUMBERS**
☐ **RELATING POSITIVE IMPROPER FRACTIONS AND POSITIVE MIXED NUMBERS**
☐ **CONVERTING AN IMPROPER FRACTION TO A MIXED NUMBER**
☐ **CONVERTING A MIXED NUMBER TO AN IMPROPER FRACTION**

Now that we know what positive fractions are, we consider three types of positive fractions: proper fractions, improper fractions, and mixed numbers.

☐ POSITIVE PROPER FRACTIONS

Positive Proper Fractions

Fractions in which the whole number in the numerator is strictly less than the whole number in the denominator are called **positive proper fractions.** On the number line, proper fractions are located in the interval from 0 to 1. Positive proper fractions are always less than one.

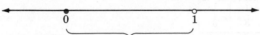

All proper fractions are located in this interval.

The closed circle at 0 indicates that 0 is included, while the open circle at 1 indicates that 1 is not included.

Some examples of positive proper fractions are

$$\frac{1}{2}, \quad \frac{3}{5}, \quad \frac{20}{27}, \quad \text{and} \quad \frac{106}{255}$$

Note that $1 < 2$, $3 < 5$, $20 < 27$, and $106 < 225$.

☐ POSITIVE IMPROPER FRACTIONS

Positive Improper Fractions

Fractions in which the whole number in the numerator is greater than or equal to the whole number in the denominator are called **positive improper fractions.** On the number line, improper fractions lie to the right of (and including) 1. Positive improper fractions are always greater than or equal to 1.

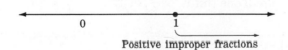

Positive improper fractions

Some examples of positive improper fractions are

$$\frac{3}{2}, \quad \frac{8}{5}, \quad \frac{4}{4}, \quad \text{and} \quad \frac{105}{16}$$

Note that $3 \geq 2$, $8 \geq 5$, $4 \geq 4$, and $105 \geq 16$.

☐ POSITIVE MIXED NUMBERS

A number of the form

nonzero whole number + proper fraction

Positive Mixed Number

is called a **positive mixed number.** For example, $2\dfrac{3}{5}$ is a mixed number. On the number line, mixed numbers are located in the interval to the right of (and including) 1. Mixed numbers are always greater than or equal to 1.

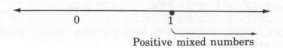

☐ RELATING POSITIVE IMPROPER FRACTIONS AND POSITIVE MIXED NUMBERS

A relationship between improper fractions and mixed numbers is suggested by two facts. The first is that improper fractions and mixed numbers are located in the same interval on the number line. The second fact, that mixed numbers are the sum of a natural number and a fraction, can be seen by making the following observations.

Divide a whole quantity into 3 equal parts.

| $\dfrac{1}{3}$ | $\dfrac{1}{3}$ | $\dfrac{1}{3}$ |

Now, consider the following examples by observing the respective shaded areas.

In the shaded region, there are 2 one thirds, or $\dfrac{2}{3}$.

$$2\left(\dfrac{1}{3}\right) = \dfrac{2}{3}$$

There are 3 one thirds, or $\dfrac{3}{3}$, or 1.

$$3\left(\dfrac{1}{3}\right) = \dfrac{3}{3} \qquad \text{or} \qquad 1$$

Thus,

$$\dfrac{3}{3} = 1$$

Improper fraction = whole number.

There are 4 one thirds, or $\dfrac{4}{3}$, or 1 and $\dfrac{1}{3}$.

$$4\left(\frac{1}{3}\right) = \frac{4}{3} \qquad \text{or} \qquad 1 \text{ and } \frac{1}{3}$$

The terms 1 and $\frac{1}{3}$ can be represented as $1 + \frac{1}{3}$ or $1\frac{1}{3}$.

Thus,

$$\frac{4}{3} = 1\frac{1}{3}.$$

Improper fraction = mixed number.

| $\frac{1}{3}$ | $\frac{1}{3}$ | $\frac{1}{3}$ | | $\frac{1}{3}$ | $\frac{1}{3}$ | $\frac{1}{3}$ |

There are 5 one thirds, or $\frac{5}{3}$, or 1 and $\frac{2}{3}$.

$$5\left(\frac{1}{3}\right) = \frac{5}{3} \qquad \text{or} \qquad 1 \text{ and } \frac{2}{3}$$

The terms 1 and $\frac{2}{3}$ can be represented as $1 + \frac{2}{3}$ or $1\frac{2}{3}$.

Thus,

$$\frac{5}{3} = 1\frac{2}{3}.$$

Improper fraction = mixed number.

| $\frac{1}{3}$ | $\frac{1}{3}$ | $\frac{1}{3}$ | | $\frac{1}{3}$ | $\frac{1}{3}$ | $\frac{1}{3}$ |

There are 6 one thirds, or $\frac{6}{3}$, or 2.

$$6\left(\frac{1}{3}\right) = \frac{6}{3} = 2$$

Thus,

$$\frac{6}{3} = 2$$

Improper fraction = whole number.

The following important fact is illustrated in the preceding examples.

Mixed Number = Natural Number + Proper Fraction

> **Mixed numbers** are the *sum* of a natural number and a proper fraction.
>
> Mixed number = (natural number) + (proper fraction)

For example $1\frac{1}{3}$ can be expressed as $1 + \frac{1}{3}$. The fraction $5\frac{7}{8}$ can be expressed as $5 + \frac{7}{8}$.

It is important to note that a number such as $5\frac{7}{8}$ does *not* indicate multiplication. To indicate multiplication, we would need to use a multiplication symbol (such as $\cdot$).

Note: $5\frac{7}{8}$ means $5 + \frac{7}{8}$ and *not* $5 \cdot \frac{7}{8}$, which means 5 times $\frac{7}{8}$ or 5 multiplied by $\frac{7}{8}$.

Thus, mixed numbers may be represented by improper fractions, and improper fractions may be represented by mixed numbers.

❑ CONVERTING IMPROPER FRACTIONS TO MIXED NUMBERS

To understand how we might convert an improper fraction to a mixed number, let's consider the fraction, $\frac{4}{3}$.

$$\underbrace{\boxed{\frac{1}{3}}\ \boxed{\frac{1}{3}}\ \boxed{\frac{1}{3}}}_{1}\quad +\quad \underbrace{\boxed{\frac{1}{3}}\ \boxed{\frac{1}{3}}\ \boxed{\frac{1}{3}}}_{\frac{1}{3}}$$

$$\frac{4}{3} = \underbrace{\frac{1}{3} + \frac{1}{3} + \frac{1}{3}}_{1} + \frac{1}{3}$$

$$= 1 + \frac{1}{3}$$

$$= 1\frac{1}{3}$$

Thus, $\frac{4}{3} = 1\frac{1}{3}$.

We can illustrate a procedure for converting an improper fraction to a mixed number using this example. However, the conversion is *more easily* accomplished by dividing the numerator by the denominator and using the result to write the mixed number.

Converting an Improper Fraction to a Mixed Number

> To convert an improper fraction to a mixed number, divide the numerator by the denominator.
>
> 1. The whole number part of the mixed numer is the quotient.
> 2. The fractional part of the mixed number is the remainder written over the divisor (the denominator of the improper fraction).

☆ SAMPLE SET A

Convert each improper fraction to its corresponding mixed number.

1. $\frac{5}{3}$.   Divide 5 by 3.

$$\begin{array}{r} 1 \\ 3\overline{)5} \\ \underline{3} \\ 2 \end{array}$$

← whole number part

← numerator of the fractional part

denominator of the fractional part

The improper fraction $\frac{5}{3} = 1\frac{2}{3}$.

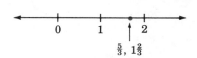

$\frac{5}{3}, 1\frac{2}{3}$

2. $\frac{46}{9}$.   Divide 46 by 9.

$$
\begin{array}{r}
5 \leftarrow \text{whole number part} \\
9\overline{)46} \\
45 \leftarrow \text{numerator of the fractional part} \\
\overline{1} \leftarrow \text{denominator of the fractional part}
\end{array}
$$

The improper fraction $\frac{46}{9} = 5\frac{1}{9}$.

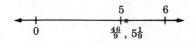

$\frac{46}{9}, 5\frac{1}{9}$

3. $\frac{83}{11}$.   Divide 83 by 11.

$$
\begin{array}{r}
7 \leftarrow \text{whole number part} \\
11\overline{)83} \\
77 \\
\overline{6} \leftarrow \text{numerator of the fractional part} \\
 \leftarrow \text{denominator of the fractional part}
\end{array}
$$

The improper fraction $\frac{83}{11} = 7\frac{6}{11}$.

$\frac{83}{11}, 7\frac{6}{11}$

4. $\frac{104}{4}$.   Divide 104 by 4.

$$
\begin{array}{r}
26 \leftarrow \text{whole number part} \\
4\overline{)104} \\
8 \\
\overline{24} \\
24 \\
\overline{0} \leftarrow \text{numerator of the fractional part} \\
 \leftarrow \text{denominator of the fractional part}
\end{array}
$$

$$\frac{104}{4} = 26\frac{0}{4} = 26$$

The improper fraction $\frac{104}{4} = 26$.

$\frac{104}{4}$

★ PRACTICE SET A

Convert each improper fraction to its corresponding mixed number.

1. $\dfrac{9}{2}$　　　**2.** $\dfrac{11}{3}$　　　**3.** $\dfrac{14}{11}$　　　**4.** $\dfrac{31}{13}$　　　**5.** $\dfrac{79}{4}$　　　**6.** $\dfrac{496}{8}$

☐ CONVERTING MIXED NUMBERS TO IMPROPER FRACTIONS

To understand how to convert a mixed number to an improper fraction, we'll recall

mixed number = (natural number) + (proper fraction)

and consider the following diagram.

$$\boxed{\tfrac{1}{3}}\ \boxed{\tfrac{1}{3}}\ \boxed{\tfrac{1}{3}} \qquad \boxed{\tfrac{1}{3}}\ \boxed{\tfrac{1}{3}}\ \boxed{\tfrac{1}{3}}$$

$$\underbrace{}_{1} \quad + \quad \underbrace{}_{\tfrac{2}{3}}$$

$$1\frac{2}{3} = 1 + \frac{2}{3}$$

$$= \underbrace{\frac{1}{3} + \frac{1}{3} + \frac{1}{3}}_{} + \underbrace{\frac{1}{3} + \frac{1}{3}}_{}$$

$$5 \cdot \frac{1}{3} = \frac{5}{3}$$

Recall that multiplication describes repeated addition.

Notice that $\dfrac{5}{3}$ can be obtained from $1\dfrac{2}{3}$ using multiplication in the following way.

Multiply: $3 \cdot 1 = 3$

$$1\frac{2}{3}$$

Add:　$3 + 2 = 5$. Place the 5 over the 3:　$\dfrac{5}{3}$.

The procedure for converting a mixed number to an improper fraction is illustrated in this example.

Converting a Mixed Number
to an Improper Fraction

To convert a mixed number to an improper fraction,

1. Multiply the denominator of the fractional part of the mixed number by the whole number part.
2. To this product, add the numerator of the fractional part.
3. Place this result over the denominator of the fractional part.

☆ **SAMPLE SET B**

Convert each mixed number to an improper fraction.

1. $5\frac{7}{8}$

    1. Multiply: $8 \cdot 5 = 40$.
    2. Add: $40 + 7 = 47$.

    3. Place 47 over 8: $\frac{47}{8}$.

    Thus, $5\frac{7}{8} = \frac{47}{8}$.

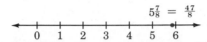

2. $16\frac{2}{3}$

    1. Multiply: $3 \cdot 16 = 48$.
    2. Add: $48 + 2 = 50$.

    3. Place 50 over 3: $\frac{50}{3}$.

    Thus, $16\frac{2}{3} = \frac{50}{3}$

★ **PRACTICE SET B**

Convert each mixed number to its corresponding improper fraction.

1. $8\frac{1}{4}$ **2.** $5\frac{3}{5}$ **3.** $1\frac{4}{15}$ **4.** $12\frac{2}{7}$

Answers to Practice Sets are on p. 154.

Section 4.2 EXERCISES

For problems 1–15, identify each expression as a proper fraction, an improper fraction, or a mixed number.

1. $\frac{3}{2}$ **2.** $\frac{4}{9}$

3. $\frac{5}{7}$ **4.** $\frac{1}{8}$

5. $6\frac{1}{4}$ **6.** $\frac{11}{8}$

7. $\dfrac{1,001}{12}$

8. $191\dfrac{4}{5}$

22. $\dfrac{121}{11}$

23. $\dfrac{165}{12}$

9. $1\dfrac{9}{13}$

10. $31\dfrac{6}{7}$

24. $\dfrac{346}{15}$

25. $\dfrac{5,000}{9}$

11. $3\dfrac{1}{40}$

12. $\dfrac{55}{12}$

26. $\dfrac{23}{5}$

27. $\dfrac{73}{2}$

13. $\dfrac{0}{9}$

14. $\dfrac{8}{9}$

28. $\dfrac{19}{2}$

29. $\dfrac{316}{41}$

15. $101\dfrac{1}{11}$

For problems 16–30, convert each of the improper fractions to its corresponding mixed number.

16. $\dfrac{11}{6}$

17. $\dfrac{14}{3}$

30. $\dfrac{800}{3}$

For problems 31–45, convert each of the mixed numbers to its corresponding improper fraction.

18. $\dfrac{25}{4}$

19. $\dfrac{35}{4}$

31. $4\dfrac{1}{8}$

32. $1\dfrac{5}{12}$

20. $\dfrac{71}{8}$

21. $\dfrac{63}{7}$

33. $6\dfrac{7}{9}$

34. $15\dfrac{1}{4}$

35. $10\frac{5}{11}$

36. $15\frac{3}{10}$

46. Why does $0\frac{4}{7}$ not qualify as a mixed number?

(*Hint:* See the definition of a mixed number.)

37. $8\frac{2}{3}$

38. $4\frac{3}{4}$

47. Why does 5 qualify as a mixed number? (*Hint:* See the definition of a mixed number.)

39. $21\frac{2}{5}$

40. $17\frac{9}{10}$

Calculator Problems

For problems 48–55, use a calculator to convert each mixed number to its corresponding improper fraction.

41. $9\frac{20}{21}$

42. $5\frac{1}{16}$

48. $35\frac{11}{12}$

49. $27\frac{5}{61}$

50. $83\frac{40}{41}$

51. $105\frac{21}{23}$

43. $90\frac{1}{100}$

44. $300\frac{43}{1,000}$

52. $72\frac{605}{606}$

53. $816\frac{19}{25}$

45. $19\frac{7}{8}$

54. $708\frac{42}{51}$

55. $6,012\frac{4,216}{8,117}$

EXERCISES FOR REVIEW

(1.3) **56.** Round 2,614,000 to the nearest thousand.

(2.1) **57.** Find the product. $1,004 \cdot 1,005$.

(2.4) **58.** Determine if 41,826 is divisible by 2 and 3.

(3.5) **59.** Find the least common multiple of 28 and 36.

(4.1) **60.** Specify the numerator and denominator of the fraction $\frac{12}{19}$.

★ **Answers to Practice Sets (4.2)**

A. **1.** $4\frac{1}{2}$ **2.** $3\frac{2}{3}$ **3.** $1\frac{3}{11}$ **4.** $2\frac{5}{13}$ **5.** $19\frac{3}{4}$ **6.** 62

B. **1.** $\frac{33}{4}$ **2.** $\frac{28}{5}$ **3.** $\frac{19}{15}$ **4.** $\frac{86}{7}$

4.3 Equivalent Fractions, Reducing Fractions to Lowest Terms, and Raising Fractions to Higher Terms

Section Overview

- ☐ **EQUIVALENT FRACTIONS**
- ☐ **REDUCING FRACTIONS TO LOWEST TERMS**
- ☐ **RAISING FRACTIONS TO HIGHER TERMS**

☐ EQUIVALENT FRACTIONS

Let's examine the following two diagrams.

| $\frac{1}{3}$ | $\frac{1}{3}$ | $\frac{1}{3}$ |

$\frac{2}{3}$ of the whole is shaded.

| $\frac{1}{6}$ | $\frac{1}{6}$ | $\frac{1}{6}$ | $\frac{1}{6}$ | $\frac{1}{6}$ | $\frac{1}{6}$ |

$\frac{4}{6}$ of the whole is shaded.

Notice that both $\frac{2}{3}$ and $\frac{4}{6}$ represent the *same part* of the whole, that is, they represent the same number.

Equivalent Fractions

> Fractions that have the same value are called **equivalent fractions.** Equivalent fractions may look different, but they are still the same point on the number line.

There is an interesting property that equivalent fractions satisfy.

$$\frac{2}{3} \diagdown\!\!\!\!\diagup \frac{4}{6}$$

A Test for Equivalent
Fractions Using the Cross
Product

These pairs of products are called **cross products.**

$$2 \cdot 6 \overset{?}{=} 3 \cdot 4$$
$$12 \overset{?}{=} 12$$

If the cross products are equal, the fractions are equivalent. If the cross products are not equal, the fractions are not equivalent.

Thus, $\frac{2}{3}$ and $\frac{4}{6}$ are equivalent, that is, $\frac{2}{3} = \frac{4}{6}$.

☆ SAMPLE SET A

Determine if the following pairs of fractions are equivalent.

1. $\frac{3}{4}$ and $\frac{6}{8}$.   **Test for equality of the cross products.**

$$\frac{3}{4} \times \frac{6}{8}$$

$3 \cdot 8 \overset{?}{=} 6 \cdot 4$
$24 \overset{?}{=} 24$ **The cross products are equal.**

The fractions $\frac{3}{4}$ and $\frac{6}{8}$ are equivalent, so $\frac{3}{4} = \frac{6}{8}$.

2. $\frac{3}{8}$ and $\frac{9}{16}$.   **Test for equality of the cross products.**

$$\frac{3}{8} \times \frac{9}{16}$$

$3 \cdot 16 \overset{?}{=} 9 \cdot 8$
$48 \neq 72$ **The cross products are *not* equal.**

The fractions $\frac{3}{8}$ and $\frac{9}{16}$ are not equivalent.

★ PRACTICE SET A

Determine if the pairs of fractions are equivalent.

1. $\frac{1}{2}, \frac{3}{6}$ **2.** $\frac{4}{5}, \frac{12}{15}$ **3.** $\frac{2}{3}, \frac{8}{15}$ **4.** $\frac{1}{8}, \frac{5}{40}$ **5.** $\frac{3}{12}, \frac{1}{4}$

❑ REDUCING FRACTIONS TO LOWEST TERMS

It is often very useful to *convert* one fraction to an equivalent fraction that has reduced values in the numerator and denominator. We can suggest a method for doing so by considering the equivalent fractions $\frac{9}{15}$ and $\frac{3}{5}$. First, divide both the numerator and denominator of $\frac{9}{15}$ by 3. The fractions $\frac{9}{15}$ and $\frac{3}{5}$ are equivalent.

(Can you prove this?) So, $\frac{9}{15} = \frac{3}{5}$. We wish to convert $\frac{9}{15}$ to $\frac{3}{5}$. Now divide the numerator and denominator of $\frac{9}{15}$ by 3, and see what happens.

$$\frac{9 \div 3}{15 \div 2} = \frac{3}{5}$$

The fraction $\frac{9}{15}$ is converted to $\frac{3}{5}$.

A natural question is "Why did we choose to divide by 3?" Notice that

$$\frac{9}{15} = \frac{3 \cdot 3}{5 \cdot 3}$$

We can see that the *factor* 3 is common to both the numerator and denominator.

From these observations we can suggest the following method for converting one fraction to an equivalent fraction that has reduced values in the numerator and denominator. The method is called **reducing a fraction.**

Reducing a Fraction

A fraction can be **reduced** by dividing *both* the numerator and denominator by the *same* nonzero whole number.

$$\frac{9}{12} = \frac{9 \div 3}{12 \div 3} = \frac{3}{4} \qquad \frac{16}{30} = \frac{16 \div 2}{30 \div 2} = \frac{8}{15}$$

$$\left(\text{Notice that } \frac{3}{3} = 1 \text{ and } \frac{2}{2} = 1 \right)$$

Consider the collection of equivalent fractions

$$\frac{5}{20}, \qquad \frac{4}{16}, \qquad \frac{3}{12}, \qquad \frac{2}{8}, \qquad \frac{1}{4}$$

Notice that each of the first four fractions can be *reduced* to the last fraction, $\frac{1}{4}$, by dividing both the numerator and denominator by, respectively, 5, 4, 3, and 2. When a fraction is converted to the fraction that has the smallest numerator and denominator in its collection of equivalent fractions, it is said to be **reduced to lowest**

Reduced to Lowest Terms

terms. The fractions $\frac{1}{4}, \frac{3}{8}, \frac{2}{5}$, and $\frac{7}{10}$ are all reduced to lowest terms.

Observe a very important property of a fraction that has been reduced to lowest terms. The *only* whole number that divides *both* the numerator and denominator without a remainder is the number 1. When 1 is the only whole number that divides two whole numbers, the two whole numbers are said to be **relatively prime.**

Relatively Prime

A fraction is reduced to lowest terms if its numerator and denominator are relatively prime.

METHODS OF REDUCING FRACTIONS TO LOWEST TERMS

Method 1:   DIVIDING OUT COMMON PRIMES

1. Write the numerator and denominator as a product of primes.
2. Divide the numerator and denominator by each of the common prime factors. We often indicate this division by drawing a slanted line through each divided out factor. This process is also called **cancelling common factors.**

Dividing Out (Cancelling) Common Factors

3. The product of the remaining factors in the numerator and the product of remaining factors of the denominator are relatively prime, and this fraction is reduced to lowest terms.

☆ SAMPLE SET B

Reduce each fraction to lowest terms.

1. $\dfrac{6}{18} = \dfrac{\overset{1}{\cancel{2}} \cdot \overset{1}{\cancel{3}}}{\underset{1}{\cancel{2}} \cdot \underset{1}{\cancel{3}} \cdot 3} = \dfrac{1}{3}$.　　　　　1 and 3 are relatively prime.

2. $\dfrac{16}{20} = \dfrac{\overset{1}{\cancel{2}} \cdot \overset{1}{\cancel{2}} \cdot 2 \cdot 2}{\underset{1}{\cancel{2}} \cdot \underset{1}{\cancel{2}} \cdot 5} = \dfrac{4}{5}$.　　　　　4 and 5 are relatively prime.

3. $\dfrac{56}{104} = \dfrac{\overset{1}{\cancel{2}} \cdot \overset{1}{\cancel{2}} \cdot \overset{1}{\cancel{2}} \cdot 7}{\underset{1}{\cancel{2}} \cdot \underset{1}{\cancel{2}} \cdot \underset{1}{\cancel{2}} \cdot 13} = \dfrac{7}{13}$.　　　7 and 13 are relatively prime (and also truly prime).

4. $\dfrac{315}{336} = \dfrac{\overset{1}{\cancel{3}} \cdot 3 \cdot 5 \cdot \overset{1}{\cancel{7}}}{2 \cdot 2 \cdot 2 \cdot 2 \cdot \underset{1}{\cancel{3}} \cdot \underset{1}{\cancel{7}}} = \dfrac{15}{16}$.　　　15 and 16 are relatively prime.

5. $\dfrac{8}{15} = \dfrac{2 \cdot 2 \cdot 2}{3 \cdot 5}$.　　　　　No common prime factors, so 8 and 15 are relatively prime.

The fraction $\dfrac{8}{15}$ is reduced to lowest terms.

★ PRACTICE SET B

Reduce each fraction to lowest terms.

1. $\dfrac{4}{8}$　　　2. $\dfrac{6}{15}$　　　3. $\dfrac{6}{48}$　　　4. $\dfrac{21}{48}$　　　5. $\dfrac{72}{42}$　　　6. $\dfrac{135}{243}$

Method 2:　DIVIDING OUT COMMON FACTORS

1. Mentally divide the numerator and the denominator by a factor that is common to each. Write the quotient above the original number.
2. Continue this process until the numerator and denominator are relatively prime.

☆ **SAMPLE SET C**

Reduce each fraction to lowest terms.

1. $\dfrac{25}{30}$.   5 divides into both 25 and 30.

$\dfrac{\overset{5}{\cancel{25}}}{\underset{6}{\cancel{30}}} = \dfrac{5}{6}$ 5 and 6 are relatively prime.

2. $\dfrac{18}{24}$.   Both numbers are even so we can divide by 2.

$\dfrac{\overset{9}{\cancel{18}}}{\underset{12}{\cancel{24}}}$ Now, both 9 and 12 are divisible by 3.

$\dfrac{\overset{\overset{3}{\cancel{9}}}{\cancel{18}}}{\underset{\underset{4}{\cancel{12}}}{\cancel{24}}} = \dfrac{3}{4}$ 3 and 4 are relatively prime.

3. $\dfrac{\overset{\overset{7}{\cancel{21}}}{\cancel{210}}}{\underset{\underset{5}{\cancel{15}}}{\cancel{150}}} = \dfrac{7}{5}$.   7 and 5 are relatively prime.

4. $\dfrac{36}{96} = \dfrac{18}{48} = \dfrac{9}{24} = \dfrac{3}{8}$.   3 and 8 are relatively prime.

★ **PRACTICE SET C**

Reduce each fraction to lowest terms.

1. $\dfrac{12}{16}$ 2. $\dfrac{9}{24}$ 3. $\dfrac{21}{84}$ 4. $\dfrac{48}{64}$ 5. $\dfrac{63}{81}$ 6. $\dfrac{150}{240}$

▢ RAISING FRACTIONS TO HIGHER TERMS

Equally as important as reducing fractions is raising fractions to higher terms. Raising a fraction to higher terms is the process of constructing an equivalent fraction that has higher values in the numerator and denominator than the original fraction.

The fractions $\dfrac{3}{5}$ and $\dfrac{9}{15}$ are equivalent, that is, $\dfrac{3}{5} = \dfrac{9}{15}$. Notice also,

$$\dfrac{3 \cdot 3}{5 \cdot 3} = \dfrac{9}{15}$$

Notice that $\frac{3}{3} = 1$ and that $\frac{3}{5} \cdot 1 = \frac{3}{5}$. We are not changing the value of $\frac{3}{5}$.

From these observations we can suggest the following method for converting one fraction to an equivalent fraction that has higher values in the numerator and denominator. This method is called **raising a fraction to higher terms.**

Raising a Fraction to Higher Terms

A fraction can be raised to an equivalent fraction that has higher terms in the numerator and denominator by multiplying both the numerator and denominator by the same nonzero whole number.

The fraction $\frac{3}{4}$ can be raised to $\frac{24}{32}$ by multiplying both the numerator and denominator by 8.

$$\frac{3}{4} = \frac{3 \cdot 8}{4 \cdot 8} = \frac{24}{32}$$

Notice that $\frac{8}{8} = 1$.

Most often, we will want to convert a given fraction to an equivalent fraction with a higher specified denominator. For example, we may wish to convert $\frac{5}{8}$ to an equivalent fraction that has denominator 32, that is,

$$\frac{5}{8} = \frac{?}{32}$$

This is possible to do because we know the process. We must multiply *both* the numerator and denominator of $\frac{5}{8}$ by the *same* nonzero whole number in order to obtain an equivalent fraction.

We have some information. The denominator 8 was raised to 32 by multiplying it by some nonzero whole number. Division will give us the proper factor. Divide the original denominator into the new denominator.

$$32 \div 8 = 4$$

Now, multiply the numerator 5 by 4.

$$5 \cdot 4 = 20$$

Thus,

$$\frac{5}{8} = \frac{5 \cdot 4}{8 \cdot 4} = \frac{20}{32}$$

So,

$$\frac{5}{8} = \frac{20}{32}$$

☆ **SAMPLE SET D**

Determine the missing numerator or denominator.

1. $\frac{3}{7} = \frac{?}{35}$.   Divide the original denominator into the new denominator.

$35 \div 7 = 5$ The quotient is 5. Multiply the original numerator by 5.

$\frac{3}{7} = \frac{3 \cdot 5}{7 \cdot 5} = \frac{15}{35}$ The missing numerator is 15.

Continued

2. $\dfrac{5}{6} = \dfrac{45}{?}$.   Divide the original numerator into the new numerator.

$45 \div 5 = 9$ The quotient is 9. Multiply the original denominator by 9.

$\dfrac{5}{6} = \dfrac{5 \cdot 9}{6 \cdot 9} = \dfrac{45}{54}$ The missing denominator is 45.

★ **PRACTICE SET D**

Determine the missing numerator or denominator.

1. $\dfrac{4}{5} = \dfrac{?}{40}$ 2. $\dfrac{3}{7} = \dfrac{?}{28}$ 3. $\dfrac{1}{6} = \dfrac{?}{24}$ 4. $\dfrac{3}{10} = \dfrac{45}{?}$ 5. $\dfrac{8}{15} = \dfrac{?}{165}$

Answers to Practice Sets are on p. 164.

Section 4.3 EXERCISES

For problems 1–15, determine if the pairs of fractions are equivalent.

1. $\dfrac{1}{2}, \dfrac{5}{10}$

2. $\dfrac{2}{3}, \dfrac{8}{12}$

3. $\dfrac{5}{12}, \dfrac{10}{24}$

4. $\dfrac{1}{2}, \dfrac{3}{6}$

5. $\dfrac{3}{5}, \dfrac{12}{15}$

6. $\dfrac{1}{6}, \dfrac{7}{42}$

7. $\dfrac{16}{25}, \dfrac{49}{75}$

8. $\dfrac{5}{28}, \dfrac{20}{112}$

9. $\dfrac{3}{10}, \dfrac{36}{110}$

10. $\dfrac{6}{10}, \dfrac{18}{32}$

11. $\dfrac{5}{8}, \dfrac{15}{24}$

20. $\dfrac{5}{6} = \dfrac{?}{18}$

12. $\dfrac{10}{16}, \dfrac{15}{24}$

21. $\dfrac{4}{5} = \dfrac{?}{25}$

13. $\dfrac{4}{5}, \dfrac{3}{4}$

22. $\dfrac{1}{2} = \dfrac{4}{?}$

14. $\dfrac{5}{7}, \dfrac{15}{21}$

23. $\dfrac{9}{25} = \dfrac{27}{?}$

15. $\dfrac{9}{11}, \dfrac{11}{9}$

24. $\dfrac{3}{2} = \dfrac{18}{?}$

For problems 16–35, determine the missing numerator or denominator.

25. $\dfrac{5}{3} = \dfrac{80}{?}$

16. $\dfrac{1}{3} = \dfrac{?}{12}$

26. $\dfrac{1}{8} = \dfrac{3}{?}$

17. $\dfrac{1}{5} = \dfrac{?}{30}$

18. $\dfrac{2}{3} = \dfrac{?}{9}$

27. $\dfrac{4}{5} = \dfrac{?}{100}$

19. $\dfrac{3}{4} = \dfrac{?}{16}$

28. $\dfrac{1}{2} = \dfrac{25}{?}$

29. $\dfrac{3}{16} = \dfrac{?}{96}$

30. $\dfrac{15}{16} = \dfrac{225}{?}$

31. $\dfrac{11}{12} = \dfrac{?}{168}$

32. $\dfrac{9}{13} = \dfrac{?}{286}$

33. $\dfrac{32}{33} = \dfrac{?}{1518}$

34. $\dfrac{19}{20} = \dfrac{1045}{?}$

35. $\dfrac{37}{50} = \dfrac{1369}{?}$

For problems 36–85, reduce, if possible, each of the fractions to lowest terms.

36. $\dfrac{6}{8}$

37. $\dfrac{8}{10}$

38. $\dfrac{5}{10}$

39. $\dfrac{6}{14}$

40. $\dfrac{3}{12}$

41. $\dfrac{4}{14}$

42. $\dfrac{1}{6}$

43. $\dfrac{4}{6}$

44. $\dfrac{18}{14}$

45. $\dfrac{20}{8}$

46. $\dfrac{4}{6}$

47. $\dfrac{10}{6}$

48. $\dfrac{6}{14}$

49. $\dfrac{14}{6}$

50. $\dfrac{10}{12}$

51. $\dfrac{16}{70}$

52. $\dfrac{40}{60}$

53. $\dfrac{20}{12}$

54. $\dfrac{32}{28}$

55. $\dfrac{36}{10}$

56. $\dfrac{36}{60}$

57. $\dfrac{12}{18}$

58. $\dfrac{18}{27}$

59. $\dfrac{18}{24}$

76. $\dfrac{30}{105}$

77. $\dfrac{46}{60}$

60. $\dfrac{32}{40}$

61. $\dfrac{11}{22}$

78. $\dfrac{75}{45}$

79. $\dfrac{40}{18}$

62. $\dfrac{27}{81}$

63. $\dfrac{17}{51}$

80. $\dfrac{108}{76}$

81. $\dfrac{7}{21}$

64. $\dfrac{16}{42}$

65. $\dfrac{39}{13}$

82. $\dfrac{6}{51}$

83. $\dfrac{51}{12}$

66. $\dfrac{44}{11}$

67. $\dfrac{66}{33}$

84. $\dfrac{8}{100}$

85. $\dfrac{51}{54}$

68. $\dfrac{15}{1}$

69. $\dfrac{15}{16}$

86. A ream of paper contains 500 sheets. What fraction of a ream of paper is 200 sheets? Be sure to reduce.

70. $\dfrac{15}{40}$

71. $\dfrac{36}{100}$

87. There are 24 hours in a day. What fraction of a day is 14 hours?

72. $\dfrac{45}{32}$

73. $\dfrac{30}{75}$

88. A full box contains 80 calculators. How many calculators are in $\dfrac{1}{4}$ of a box?

74. $\dfrac{121}{132}$

75. $\dfrac{72}{64}$

89. There are 48 plants per flat. How many plants are there in $\dfrac{1}{3}$ of a flat?

90. A person making $18,000 per year must pay $3,960 in income tax. What fraction of this person's yearly salary goes to the IRS?

93. $\dfrac{7}{15} = \dfrac{\cancel{7}}{\cancel{7}+8} = \dfrac{1}{8}$

For problems 91–95, find the mistake.

91. $\dfrac{3}{24} = \dfrac{\cancel{3}}{\cancel{3} \cdot 8} = \dfrac{0}{8} = 0$

94. $\dfrac{6}{7} = \dfrac{\cancel{5}+1}{\cancel{5}+2} = \dfrac{1}{2}$

92. $\dfrac{8}{10} = \dfrac{\cancel{2}+6}{\cancel{2}+8} = \dfrac{6}{8} = \dfrac{3}{4}$

95. $\dfrac{\cancel{9}}{\cancel{9}} = \dfrac{0}{0} = 0$

EXERCISES FOR REVIEW

(1.3)   **96.** Round 816 to the nearest thousand.

(2.2)   **97.** Perform the division: $0 \div 6$.

(3.3)   **98.** Find all the factors of 24.

(3.4)   **99.** Find the greatest common factor of 12 and 18.

(4.2)   **100.** Convert $\dfrac{15}{8}$ to a mixed number.

★ **Answers to Practice Sets (4.3)**

A.   **1.** $6 \overset{?}{=} 6$, yes   **2.** $60 \overset{?}{=} 60$, yes   **3.** $30 \neq 24$, no   **4.** $40 \overset{?}{=} 40$, yes   **5.** $12 \overset{?}{=} 12$, yes

B.   **1.** $\dfrac{1}{2}$ **2.** $\dfrac{2}{5}$ **3.** $\dfrac{1}{8}$ **4.** $\dfrac{7}{16}$ **5.** $\dfrac{12}{7}$ **6.** $\dfrac{5}{9}$

C.   **1.** $\dfrac{3}{4}$ **2.** $\dfrac{3}{8}$ **3.** $\dfrac{1}{4}$ **4.** $\dfrac{3}{4}$ **5.** $\dfrac{7}{9}$ **6.** $\dfrac{5}{8}$

D.   **1.** 32   **2.** 12   **3.** 4   **4.** 150   **5.** 88

4.4 Multiplication of Fractions

Section Overview

❑ **FRACTIONS OF FRACTIONS**
❑ **MULTIPLICATION OF FRACTIONS**
❑ **MULTIPLICATION OF FRACTIONS BY DIVIDING OUT COMMON FACTORS**
❑ **MULTIPLICATION OF MIXED NUMBERS**
❑ **POWERS AND ROOTS OF FRACTIONS**

❑ FRACTIONS OF FRACTIONS

We know that a fraction represents a part of a whole quantity. For example, two fifths of one unit can be represented by

$\frac{2}{5}$ of the whole is shaded.

A natural question is, what is a fractional part of a fractional quantity, or, what is a fraction of a fraction? For example, what is $\frac{2}{3}$ of $\frac{1}{2}$?

We can suggest an answer to this question by using a picture to examine $\frac{2}{3}$ of $\frac{1}{2}$.

First, let's represent $\frac{1}{2}$.

$\frac{1}{2}$ of the whole is shaded

Then divide each of the $\frac{1}{2}$ parts into 3 equal parts.

Each part is $\frac{1}{6}$ of the whole.

Now we'll take $\frac{2}{3}$ of the $\frac{1}{2}$ unit.

$\frac{2}{3}$ of $\frac{1}{2}$ is $\frac{2}{6}$, which reduces to $\frac{1}{3}$.

❏ MULTIPLICATION OF FRACTIONS

Now we ask, what arithmetic operation $(+, -, \times, \div)$ will produce $\dfrac{2}{6}$ from $\dfrac{2}{3}$ of $\dfrac{1}{2}$?

Notice that, if in the fractions $\dfrac{2}{3}$ and $\dfrac{1}{2}$, we multiply the numerators together and the denominators together, we get precisely $\dfrac{2}{6}$.

$$\frac{2 \cdot 1}{3 \cdot 2} = \frac{2}{6}$$

This reduces to $\dfrac{1}{3}$ as before.

Using this observation, we can suggest the following:

The Word "OF" Indicates Multiplication

The Method of Multiplying Fractions

1. The word "of" translates to the arithmetic operation "times."
2. To multiply two or more fractions, multiply the numerators together and then multiply the denominators together. Reduce if necessary.

$$\frac{\text{numerator 1}}{\text{denominator 1}} \cdot \frac{\text{numerator 2}}{\text{denominator 2}} = \frac{\text{numerator 1} \cdot \text{numerator 2}}{\text{denominator 1} \cdot \text{denominator 2}}$$

☆ SAMPLE SET A

Perform the following multiplications.

1. $\dfrac{3}{4} \cdot \dfrac{1}{6} = \dfrac{3 \cdot 1}{4 \cdot 6} = \dfrac{3}{24}.$ **Now, reduce.**

$$= \frac{\overset{1}{\cancel{3}}}{\underset{8}{\cancel{24}}} = \frac{1}{8}$$

Thus,

$$\frac{3}{4} \cdot \frac{1}{6} = \frac{1}{8}$$

This means that $\dfrac{3}{4}$ of $\dfrac{1}{6}$ is $\dfrac{1}{8}$, that is, $\dfrac{3}{4}$ of $\dfrac{1}{6}$ of a unit is $\dfrac{1}{8}$ of the original unit.

2. $\dfrac{3}{8} \cdot 4.$ **Write 4 as a fraction by writing $\dfrac{4}{1}$.**

$$\frac{3}{8} \cdot \frac{4}{1} = \frac{3 \cdot 4}{8 \cdot 1} = \frac{12}{8} = \frac{\overset{3}{\cancel{12}}}{\underset{2}{\cancel{8}}} = \frac{3}{2}$$

$$\frac{3}{8} \cdot 4 = \frac{3}{2}$$

This means that $\dfrac{3}{8}$ of 4 whole units is $\dfrac{3}{2}$ of one whole unit.

3. $\dfrac{2}{5} \cdot \dfrac{5}{8} \cdot \dfrac{1}{4} = \dfrac{2 \cdot 5 \cdot 1}{5 \cdot 8 \cdot 4} = \dfrac{\overset{1}{\cancel{10}}}{\underset{16}{\cancel{160}}} = \dfrac{1}{16}$

This means that $\dfrac{2}{5}$ of $\dfrac{5}{8}$ of $\dfrac{1}{4}$ of a whole unit is $\dfrac{1}{16}$ of the original unit.

★ **PRACTICE SET A**

Perform the following multiplications.

1. $\dfrac{2}{5} \cdot \dfrac{1}{6}$ **2.** $\dfrac{1}{4} \cdot \dfrac{8}{9}$ **3.** $\dfrac{4}{9} \cdot \dfrac{15}{16}$ **4.** $\left(\dfrac{2}{3}\right)\left(\dfrac{2}{3}\right)$ **5.** $\left(\dfrac{7}{4}\right)\left(\dfrac{8}{5}\right)$

6. $\dfrac{5}{6} \cdot \dfrac{7}{8}$ **7.** $\dfrac{2}{3} \cdot 5$ **8.** $\left(\dfrac{3}{4}\right)(10)$ **9.** $\dfrac{3}{4} \cdot \dfrac{8}{9} \cdot \dfrac{5}{12}$

☐ MULTIPLYING FRACTIONS BY DIVIDING OUT COMMON FACTORS

We have seen that to multiply two fractions together, we multiply numerators together, then denominators together, then reduce to lowest terms, if necessary. The reduction can be tedious if the numbers in the fractions are large. For example,

$$\dfrac{9}{16} \cdot \dfrac{10}{21} = \dfrac{9 \cdot 10}{16 \cdot 21} = \dfrac{90}{336} = \dfrac{45}{168} = \dfrac{15}{28}$$

We avoid the process of reducing if we divide out common factors *before* we multiply.

$$\dfrac{9}{16} \cdot \dfrac{10}{21} = \dfrac{\overset{3}{\cancel{9}}}{\underset{8}{\cancel{16}}} \cdot \dfrac{\overset{5}{\cancel{10}}}{\underset{7}{\cancel{21}}} = \dfrac{3 \cdot 5}{8 \cdot 7} = \dfrac{15}{56}$$

Divide 3 into 9 and 21, and divide 2 into 10 and 16. The product is a fraction that is reduced to lowest terms.

The Process of Multiplication by Dividing Out Common Factors

> To multiply fractions by dividing out common factors, divide out factors that are common to both a numerator and a denominator. The factor being divided out can appear in any numerator and any denominator.

☆ SAMPLE SET B

Perform the following multiplications.

1. $\dfrac{4}{5} \cdot \dfrac{5}{6}$

$$\dfrac{\overset{2}{\cancel{4}}}{\underset{1}{\cancel{5}}} \cdot \dfrac{\overset{1}{\cancel{5}}}{\underset{3}{\cancel{6}}} = \dfrac{2 \cdot 1}{1 \cdot 3} = \dfrac{2}{3}$$

Divide 4 and 6 by 2.
Divide 5 and 5 by 5.

2. $\dfrac{8}{12} \cdot \dfrac{8}{10}$

$$\dfrac{\overset{4}{\cancel{8}}}{\underset{3}{\cancel{12}}} \cdot \dfrac{\overset{2}{\cancel{8}}}{\underset{5}{\cancel{10}}} = \dfrac{4 \cdot 2}{3 \cdot 5} = \dfrac{8}{15}$$

Divide 8 and 10 by 2.
Divide 8 and 12 by 4.

3. $8 \cdot \dfrac{5}{12} = \dfrac{\overset{2}{\cancel{8}}}{1} \cdot \dfrac{5}{\underset{3}{\cancel{12}}} = \dfrac{2 \cdot 5}{1 \cdot 3} = \dfrac{10}{3}$

4. $\dfrac{35}{18} \cdot \dfrac{63}{105}$

$$\dfrac{\overset{1}{\cancel{\overset{7}{\cancel{35}}}}}{\underset{2}{\cancel{18}}} \cdot \dfrac{\overset{7}{\cancel{63}}}{\underset{\underset{3}{\cancel{21}}}{\cancel{105}}} = \dfrac{1 \cdot 7}{2 \cdot 3} = \dfrac{7}{6}$$

5. $\dfrac{13}{9} \cdot \dfrac{6}{39} \cdot \dfrac{1}{12}$

$$\dfrac{\overset{1}{\cancel{13}}}{9} \cdot \dfrac{\overset{\overset{1}{\cancel{2}}}{\cancel{6}}}{\underset{\underset{1}{\cancel{3}}}{\cancel{39}}} \cdot \dfrac{1}{\underset{6}{\cancel{12}}} = \dfrac{1 \cdot 1 \cdot 1}{9 \cdot 1 \cdot 6} = \dfrac{1}{54}$$

★ PRACTICE SET B

Perform the following multiplications.

1. $\dfrac{2}{3} \cdot \dfrac{7}{8}$ **2.** $\dfrac{25}{12} \cdot \dfrac{10}{45}$ **3.** $\dfrac{40}{48} \cdot \dfrac{72}{90}$ **4.** $7 \cdot \dfrac{2}{49}$ **5.** $12 \cdot \dfrac{3}{8}$

6. $\left(\dfrac{13}{7}\right)\left(\dfrac{14}{26}\right)$ **7.** $\dfrac{16}{10} \cdot \dfrac{22}{6} \cdot \dfrac{21}{44}$

❑ MULTIPLICATION OF MIXED NUMBERS

Multiplying Mixed Numbers

To perform a multiplication in which there are mixed numbers, it is convenient to first convert each mixed number to an improper fraction, then multiply.

☆ SAMPLE SET C

Perform the following multiplications. Convert improper fractions to mixed numbers.

1. $1\frac{1}{8} \cdot 4\frac{2}{3}$.

Convert each mixed number to an improper fraction.

$$1\frac{1}{8} = \frac{8 \cdot 1 + 1}{8} = \frac{9}{8}.$$

$$4\frac{2}{3} = \frac{4 \cdot 3 + 2}{3} = \frac{14}{3}.$$

$$\overset{3}{\underset{4}{\cancel{\frac{9}{8}}}} \cdot \overset{7}{\underset{1}{\cancel{\frac{14}{3}}}} = \frac{3 \cdot 7}{4 \cdot 1} = \frac{21}{4} = 5\frac{1}{4}$$

2. $16 \cdot 8\frac{1}{5}$.

Convert $8\frac{1}{5}$ to an improper fraction.

$$8\frac{1}{5} = \frac{5 \cdot 8 + 1}{5} = \frac{41}{5}.$$

$$\frac{16}{1} \cdot \frac{41}{5}.$$

There are no common factors to divide out.

$$\frac{16}{1} \cdot \frac{41}{5} = \frac{16 \cdot 41}{1 \cdot 5} = \frac{656}{5} = 131\frac{1}{5}$$

3. $9\frac{1}{6} \cdot 12\frac{3}{5}$.

Convert to improper fractions.

$$9\frac{1}{6} = \frac{6 \cdot 9 + 1}{6} = \frac{55}{6}.$$

$$12\frac{3}{5} = \frac{5 \cdot 12 + 3}{5} = \frac{63}{5}.$$

$$\overset{11}{\underset{2}{\cancel{\frac{55}{6}}}} \cdot \overset{21}{\underset{1}{\cancel{\frac{63}{5}}}} = \frac{11 \cdot 21}{2 \cdot 1} = \frac{231}{2} = 115\frac{1}{2}$$

4. $\frac{11}{8} \cdot 4\frac{1}{2} \cdot 3\frac{1}{3} = \frac{11}{8} \cdot \overset{3}{\underset{1}{\cancel{\frac{9}{2}}}} \cdot \overset{5}{\underset{1}{\cancel{\frac{10}{3}}}}$

$$= \frac{11 \cdot 3 \cdot 5}{8 \cdot 1 \cdot 1} = \frac{165}{8} = 20\frac{5}{8}$$

★ **PRACTICE SET C**

Perform the following multiplications. Convert improper fractions to mixed numbers.

1. $2\frac{2}{3} \cdot 2\frac{1}{4}$ **2.** $6\frac{2}{3} \cdot 3\frac{3}{10}$ **3.** $7\frac{1}{8} \cdot 12$ **4.** $2\frac{2}{5} \cdot 3\frac{3}{4} \cdot 3\frac{1}{3}$

◻ POWERS AND ROOTS OF FRACTIONS

☆ **SAMPLE SET D**

Find the value of each of the following.

1. $\left(\frac{1}{6}\right)^2 = \frac{1}{6} \cdot \frac{1}{6} = \frac{1 \cdot 1}{6 \cdot 6} = \frac{1}{36}$.

2. $\sqrt{\frac{9}{100}}$.   We're looking for a number, call it ?, such that when it is squared, $\frac{9}{100}$ is produced.

$(?)^2 = \frac{9}{100}$

We know that

$3^2 = 9$ and $10^2 = 100$

We'll try $\frac{3}{10}$. Since

$\left(\frac{3}{10}\right)^2 = \frac{3}{10} \cdot \frac{3}{10} = \frac{3 \cdot 3}{10 \cdot 10} = \frac{9}{100}$

$\sqrt{\frac{9}{100}} = \frac{3}{10}$

3. $4\frac{2}{5} \cdot \sqrt{\frac{100}{121}}$

$\frac{\overset{2}{\cancel{22}}}{\underset{1}{\cancel{5}}} \cdot \frac{\overset{2}{\cancel{10}}}{\underset{1}{\cancel{11}}} = \frac{2 \cdot 2}{1 \cdot 1} = \frac{4}{1} = 4$

$4\frac{2}{5} \cdot \sqrt{\frac{100}{121}} = 4$

★ PRACTICE SET D

Find the value of each of the following.

1. $\left(\dfrac{1}{8}\right)^2$ **2.** $\left(\dfrac{3}{10}\right)^2$ **3.** $\sqrt{\dfrac{4}{9}}$ **4.** $\sqrt{\dfrac{1}{4}}$ **5.** $\dfrac{3}{8} \cdot \sqrt{\dfrac{1}{9}}$ **6.** $9\dfrac{1}{3} \cdot \sqrt{\dfrac{81}{100}}$

7. $2\dfrac{8}{13} \cdot \sqrt{\dfrac{169}{16}}$

Answers to Practice Sets are on p. 175.

Section 4.4 EXERCISES

For problems 1–6, use the diagrams to find each of the following parts. Use multiplication to verify your result.

1. $\dfrac{3}{4}$ of $\dfrac{1}{3}$ **2.** $\dfrac{2}{3}$ of $\dfrac{3}{5}$

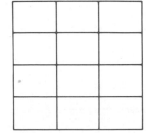   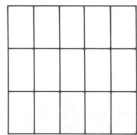

3. $\dfrac{2}{7}$ of $\dfrac{7}{8}$ **4.** $\dfrac{5}{6}$ of $\dfrac{3}{4}$

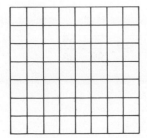   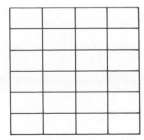

5. $\dfrac{1}{8}$ of $\dfrac{1}{8}$ **6.** $\dfrac{7}{12}$ of $\dfrac{6}{7}$

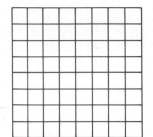   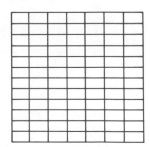

For problems 7–25, find each part without using a diagram.

7. $\dfrac{1}{2}$ of $\dfrac{4}{5}$.   **8.** $\dfrac{3}{5}$ of $\dfrac{5}{12}$

9. $\dfrac{1}{4}$ of $\dfrac{8}{9}$ **10.** $\dfrac{3}{16}$ of $\dfrac{12}{15}$

11. $\frac{2}{9}$ of $\frac{6}{5}$

12. $\frac{1}{8}$ of $\frac{3}{8}$

25. $2\frac{4}{5}$ of $5\frac{5}{6}$ of $7\frac{5}{7}$

For problems 26–70, find the products. Be sure to reduce.

13. $\frac{2}{3}$ of $\frac{9}{10}$

14. $\frac{18}{19}$ of $\frac{38}{54}$

26. $\frac{1}{3} \cdot \frac{2}{3}$

27. $\frac{1}{2} \cdot \frac{1}{2}$

15. $\frac{5}{6}$ of $2\frac{2}{5}$

16. $\frac{3}{4}$ of $3\frac{3}{5}$

28. $\frac{3}{4} \cdot \frac{3}{8}$

29. $\frac{2}{5} \cdot \frac{5}{6}$

17. $\frac{3}{2}$ of $2\frac{2}{9}$

18. $\frac{15}{4}$ of $4\frac{4}{5}$

30. $\frac{3}{8} \cdot \frac{8}{9}$

31. $\frac{5}{6} \cdot \frac{14}{15}$

19. $5\frac{1}{3}$ of $9\frac{3}{4}$

20. $1\frac{13}{15}$ of $8\frac{3}{4}$

32. $\frac{4}{7} \cdot \frac{7}{4}$

33. $\frac{3}{11} \cdot \frac{11}{3}$

21. $\frac{8}{9}$ of $\frac{3}{4}$ of $\frac{2}{3}$

22. $\frac{1}{6}$ of $\frac{12}{13}$ of $\frac{26}{36}$

34. $\frac{9}{16} \cdot \frac{20}{27}$

35. $\frac{35}{36} \cdot \frac{48}{55}$

23. $\frac{1}{2}$ of $\frac{1}{3}$ of $\frac{1}{4}$

24. $1\frac{3}{7}$ of $5\frac{1}{5}$ of $8\frac{1}{3}$

36. $\frac{21}{25} \cdot \frac{15}{14}$

37. $\frac{76}{99} \cdot \frac{66}{38}$

38. $\dfrac{3}{7} \cdot \dfrac{14}{18} \cdot \dfrac{6}{2}$

39. $\dfrac{4}{15} \cdot \dfrac{10}{3} \cdot \dfrac{27}{2}$

52. $\dfrac{3}{8} \cdot 24 \cdot \dfrac{2}{3}$

53. $\dfrac{5}{18} \cdot 10 \cdot \dfrac{2}{5}$

40. $\dfrac{14}{15} \cdot \dfrac{21}{28} \cdot \dfrac{45}{7}$

41. $\dfrac{8}{3} \cdot \dfrac{15}{4} \cdot \dfrac{16}{21}$

54. $\dfrac{16}{15} \cdot 50 \cdot \dfrac{3}{10}$

55. $5\dfrac{1}{3} \cdot \dfrac{27}{32}$

42. $\dfrac{18}{14} \cdot \dfrac{21}{35} \cdot \dfrac{36}{7}$

43. $\dfrac{3}{5} \cdot 20$

56. $2\dfrac{6}{7} \cdot 5\dfrac{3}{5}$

57. $6\dfrac{1}{4} \cdot 2\dfrac{4}{15}$

44. $\dfrac{8}{9} \cdot 18$

45. $\dfrac{6}{11} \cdot 33$

58. $9\dfrac{1}{3} \cdot \dfrac{9}{16} \cdot 1\dfrac{1}{3}$

59. $3\dfrac{5}{9} \cdot 1\dfrac{13}{14} \cdot 10\dfrac{1}{2}$

46. $\dfrac{18}{19} \cdot 38$

47. $\dfrac{5}{6} \cdot 10$

60. $20\dfrac{1}{4} \cdot 8\dfrac{2}{3} \cdot 16\dfrac{4}{5}$

61. $\left(\dfrac{2}{3}\right)^2$

48. $\dfrac{1}{9} \cdot 3$

49. $5 \cdot \dfrac{3}{8}$

62. $\left(\dfrac{3}{8}\right)^2$

63. $\left(\dfrac{2}{11}\right)^2$

50. $16 \cdot \dfrac{1}{4}$

51. $\dfrac{2}{3} \cdot 12 \cdot \dfrac{3}{4}$

64. $\left(\dfrac{8}{9}\right)^2$

65. $\left(\dfrac{1}{2}\right)^2$

66. $\left(\dfrac{3}{5}\right)^2 \cdot \dfrac{20}{3}$

67. $\left(\dfrac{1}{4}\right)^2 \cdot \dfrac{16}{15}$

73. $\sqrt{\dfrac{81}{121}}$

74. $\sqrt{\dfrac{36}{49}}$

75. $\sqrt{\dfrac{144}{25}}$

76. $\dfrac{2}{3} \cdot \sqrt{\dfrac{9}{16}}$

68. $\left(\dfrac{1}{2}\right)^2 \cdot \dfrac{8}{9}$

69. $\left(\dfrac{1}{2}\right)^2 \cdot \left(\dfrac{2}{5}\right)^2$

77. $\dfrac{3}{5} \cdot \sqrt{\dfrac{25}{81}}$

78. $\left(\dfrac{8}{5}\right)^2 \cdot \sqrt{\dfrac{25}{64}}$

70. $\left(\dfrac{3}{7}\right)^2 \cdot \left(\dfrac{1}{9}\right)^2$

79. $\left(1\dfrac{3}{4}\right)^2 \cdot \sqrt{\dfrac{4}{49}}$

For problems 71–80, find each value. Reduce answers to lowest terms or convert to mixed numbers.

71. $\sqrt{\dfrac{4}{9}}$

72. $\sqrt{\dfrac{16}{25}}$

80. $\left(2\dfrac{2}{3}\right)^2 \cdot \sqrt{\dfrac{36}{49}} \cdot \sqrt{\dfrac{64}{81}}$

EXERCISES FOR REVIEW

(1.1)   **81.** How many thousands in 342,810?

(1.4)   **82.** Find the sum of 22, 42, and 101.

(2.4)   **83.** Is 634,281 divisible by 3?

(3.3)   **84.** Is the whole number 51 prime or composite?

(4.3)   **85.** Reduce $\dfrac{36}{150}$ to lowest terms.

★ **Answers to Practice Sets (4.4)**

A. 1. $\frac{1}{15}$ 2. $\frac{2}{9}$ 3. $\frac{5}{12}$ 4. $\frac{4}{9}$ 5. $\frac{14}{5}$ 6. $\frac{35}{48}$ 7. $\frac{10}{3}$ 8. $\frac{15}{2}$ 9. $\frac{5}{18}$

B. 1. $\frac{7}{12}$ 2. $\frac{25}{54}$ 3. $\frac{2}{3}$ 4. $\frac{2}{7}$ 5. $\frac{9}{2}$ 6. 1   7. $\frac{14}{5}$

C. 1. 6   2. 22   3. $85\frac{1}{2}$ 4. 30

D. 1. $\frac{1}{64}$ 2. $\frac{9}{100}$ 3. $\frac{2}{3}$ 4. $\frac{1}{2}$ 5. $\frac{1}{8}$ 6. $8\frac{2}{5}$ 7. $8\frac{1}{2}$

4.5 Division of Fractions

Section Overview	☐ **RECIPROCALS** ☐ **DIVIDING FRACTIONS**

☐ **RECIPROCALS**

Reciprocals

> Two numbers whose product is 1 are called **reciprocals** of each other.

☆ **SAMPLE SET A**

The following pairs of numbers are reciprocals.

$\frac{3}{4}$ and $\frac{4}{3}$ $\frac{7}{16}$ and $\frac{16}{7}$ $\frac{1}{6}$ and $\frac{6}{1}$

$\frac{3}{4} \cdot \frac{4}{3} = 1$ $\frac{7}{16} \cdot \frac{16}{7} = 1$ $\frac{1}{6} \cdot \frac{6}{1} = 1$

Notice that we can find the reciprocal of a nonzero number in fractional form by inverting it (exchanging positions of the numerator and denominator).

★ **PRACTICE SET A**

Find the reciprocal of each number.

1. $\frac{3}{10}$ 2. $\frac{2}{3}$ 3. $\frac{7}{8}$ 4. $\frac{1}{5}$

5. $2\frac{2}{7}$ (*Hint:* Write this number as an improper fraction first.)   6. $5\frac{1}{4}$ 7. $10\frac{3}{16}$

❑ DIVIDING FRACTIONS

Our concept of division is that it indicates *how many times* one quantity is contained in another quantity. For example, using the diagram we can see that there are 6 one-thirds in 2.

 There are 6 one-thirds in 2.

Since 2 contains six $\frac{1}{3}$'s, we express this as

$$2 \div \boxed{\frac{1}{3}} = 6$$

Note also that $2 \cdot \boxed{3} = 6$

$\frac{1}{3}$ and 3 are reciprocals

Using these observations, we can suggest the following method for dividing a number by a fraction.

Dividing One Fraction by Another Fraction

> To divide a first fraction by a second, nonzero fraction, multiply the first fraction by the reciprocal of the second fraction.

Invert and Multiply

This method is commonly referred to as **"invert the divisor and multiply."**

☆ SAMPLE SET B

Perform the following divisions.

1. $\frac{1}{3} \div \frac{3}{4}$.   The divisor is $\frac{3}{4}$. Its reciprocal is $\frac{4}{3}$. Multiply $\frac{1}{3}$ by $\frac{4}{3}$.

$$\frac{1}{3} \cdot \frac{4}{3} = \frac{1 \cdot 4}{3 \cdot 3} = \frac{4}{9}$$

$$\frac{1}{3} \div \frac{3}{4} = \frac{4}{9}$$

2. $\frac{3}{8} \div \frac{5}{4}$.   The divisor is $\frac{5}{4}$. Its reciprocal is $\frac{4}{5}$. Multiply $\frac{3}{8}$ by $\frac{4}{5}$.

$$\frac{3}{\overset{}{\underset{2}{8}}} \cdot \frac{\overset{1}{4}}{5} = \frac{3 \cdot 1}{2 \cdot 5} = \frac{3}{10}$$

$$\frac{3}{8} \div \frac{5}{4} = \frac{3}{10}$$

3. $\dfrac{5}{6} \div \dfrac{5}{12}$.

The divisor is $\dfrac{5}{12}$. Its reciprocal is $\dfrac{12}{5}$. Multiply $\dfrac{5}{6}$ by $\dfrac{12}{5}$.

$$\dfrac{\overset{1}{\cancel{5}}}{\underset{1}{\cancel{6}}} \cdot \dfrac{\overset{2}{\cancel{12}}}{\underset{1}{\cancel{5}}} = \dfrac{1 \cdot 2}{1 \cdot 1} = \dfrac{2}{1} = 2$$

$$\dfrac{5}{6} \div \dfrac{5}{12} = 2$$

4. $2\dfrac{2}{9} \div 3\dfrac{1}{3}$.

Convert each mixed number to an improper fraction.

$$2\dfrac{2}{9} = \dfrac{9 \cdot 2 + 2}{9} = \dfrac{20}{9}.$$

$$3\dfrac{1}{3} = \dfrac{3 \cdot 3 + 1}{3} = \dfrac{10}{3}.$$

$$\dfrac{20}{9} \div \dfrac{10}{3}$$

The divisor is $\dfrac{10}{3}$. Its reciprocal is $\dfrac{3}{10}$. Multiply $\dfrac{20}{9}$ by $\dfrac{3}{10}$.

$$\dfrac{\overset{2}{\cancel{20}}}{\underset{3}{\cancel{9}}} \cdot \dfrac{\overset{1}{\cancel{3}}}{\underset{1}{\cancel{10}}} = \dfrac{2 \cdot 1}{3 \cdot 1} = \dfrac{2}{3}$$

$$2\dfrac{2}{9} \div 3\dfrac{1}{3} = \dfrac{2}{3}$$

5. $\dfrac{12}{11} \div 8$.

First conveniently write 8 as $\dfrac{8}{1}$.

$$\dfrac{12}{11} \div \dfrac{8}{1}$$

The divisor is $\dfrac{8}{1}$. Its reciprocal is $\dfrac{1}{8}$. Multiply $\dfrac{12}{11}$ by $\dfrac{1}{8}$.

$$\dfrac{\overset{3}{\cancel{12}}}{11} \cdot \dfrac{1}{\underset{2}{\cancel{8}}} = \dfrac{3 \cdot 1}{11 \cdot 2} = \dfrac{3}{22}$$

$$\dfrac{12}{11} \div 8 = \dfrac{3}{22}$$

6. $\dfrac{7}{8} \div \dfrac{21}{20} \cdot \dfrac{3}{35}$.

The divisor is $\dfrac{21}{20}$. Its reciprocal is $\dfrac{20}{21}$.

$$\dfrac{\overset{1}{\cancel{7}}}{\underset{2}{\cancel{8}}} \cdot \dfrac{\overset{\overset{1}{\cancel{5}}}{\cancel{20}}}{\underset{\underset{1}{\cancel{3}}}{\cancel{21}}} \cdot \dfrac{\overset{1}{\cancel{3}}}{\underset{7}{\cancel{35}}} = \dfrac{1 \cdot 1 \cdot 1}{2 \cdot 1 \cdot 7} = \dfrac{1}{14}$$

$$\dfrac{7}{8} \div \dfrac{21}{20} \cdot \dfrac{3}{25} = \dfrac{1}{14}$$

Continued

7. How many $2\frac{3}{8}$-inch-wide packages can be placed in a box 19 inches wide?

The problem is to determine how many two and three eighths are contained in 19, that is, what is $19 \div 2\frac{3}{8}$?

$2\frac{3}{8} = \frac{19}{8}$ Convert the divisor $2\frac{3}{8}$ to an improper fraction.

$19 = \frac{19}{1}$ Write the dividend 19 as $\frac{19}{1}$.

$\frac{19}{1} \div \frac{19}{8}$ The divisor is $\frac{19}{8}$. Its reciprocal is $\frac{8}{19}$.

$$\frac{\overset{1}{\cancel{19}}}{1} \cdot \frac{8}{\underset{1}{\cancel{19}}} = \frac{1 \cdot 8}{1 \cdot 1} = \frac{8}{1} = 8$$

Thus, 8 packages will fit into the box.

★ PRACTICE SET B

Perform the following divisions.

1. $\dfrac{1}{2} \div \dfrac{9}{8}$ **2.** $\dfrac{3}{8} \div \dfrac{9}{24}$ **3.** $\dfrac{7}{15} \div \dfrac{14}{15}$ **4.** $8 \div \dfrac{8}{15}$ **5.** $6\dfrac{1}{4} \div \dfrac{5}{12}$

6. $3\dfrac{1}{3} \div 1\dfrac{2}{3}$ **7.** $\dfrac{5}{6} \div \dfrac{2}{3} \cdot \dfrac{8}{25}$

8. A container will hold 106 ounces of grape juice. How many $6\frac{5}{8}$-ounce glasses of grape juice can be served from this container?

Determine each of the following quotients and then write a rule for this type of division.

9. $1 \div \dfrac{2}{3}$ **10.** $1 \div \dfrac{3}{8}$ **11.** $1 \div \dfrac{3}{4}$ **12.** $1 \div \dfrac{5}{2}$

13. When dividing 1 by a fraction, the quotient is the _____.

Answers to Practice Sets are on p. 181.

Section 4.5 EXERCISES

For problems 1–10, find the reciprocal of each number.

1. $\dfrac{4}{5}$

2. $\dfrac{8}{11}$

3. $\dfrac{2}{9}$

4. $\dfrac{1}{5}$

5. $3\dfrac{1}{4}$

6. $8\dfrac{1}{4}$

7. $3\dfrac{2}{7}$

8. $5\dfrac{3}{4}$

9. 1

10. 4

For problems 11–45, find each value.

11. $\dfrac{3}{8} \div \dfrac{3}{5}$

12. $\dfrac{5}{9} \div \dfrac{5}{6}$

13. $\dfrac{9}{16} \div \dfrac{15}{8}$

14. $\dfrac{4}{9} \div \dfrac{6}{15}$

15. $\dfrac{25}{49} \div \dfrac{4}{9}$

16. $\dfrac{15}{4} \div \dfrac{27}{8}$

17. $\dfrac{24}{75} \div \dfrac{8}{15}$

18. $\dfrac{5}{7} \div 0$

19. $\dfrac{7}{8} \div \dfrac{7}{8}$

20. $0 \div \dfrac{3}{5}$

21. $\dfrac{4}{11} \div \dfrac{4}{11}$

22. $\dfrac{2}{3} \div \dfrac{2}{3}$

23. $\dfrac{7}{10} \div \dfrac{10}{7}$

24. $\dfrac{3}{4} \div 6$

25. $\dfrac{9}{5} \div 3$

26. $4\dfrac{1}{6} \div 3\dfrac{1}{3}$

27. $7\dfrac{1}{7} \div 8\dfrac{1}{3}$

28. $1\dfrac{1}{2} \div 1\dfrac{1}{5}$

29. $3\dfrac{2}{5} \div \dfrac{6}{25}$

30. $5\dfrac{1}{6} \div \dfrac{31}{6}$

39. $4\dfrac{3}{25} \div 2\dfrac{56}{75}$

40. $\dfrac{1}{1000} \div \dfrac{1}{100}$

31. $\dfrac{35}{6} \div 3\dfrac{3}{4}$

32. $5\dfrac{1}{9} \div \dfrac{1}{18}$

41. $\dfrac{3}{8} \div \dfrac{9}{16} \cdot \dfrac{6}{5}$

42. $\dfrac{3}{16} \cdot \dfrac{9}{8} \cdot \dfrac{6}{5}$

33. $8\dfrac{3}{4} \div \dfrac{7}{8}$

34. $\dfrac{12}{8} \div 1\dfrac{1}{2}$

43. $\dfrac{4}{15} \div \dfrac{2}{25} \cdot \dfrac{9}{10}$

44. $\dfrac{21}{30} \cdot 1\dfrac{1}{4} \div \dfrac{9}{10}$

35. $3\dfrac{1}{8} \div \dfrac{15}{16}$

36. $11\dfrac{11}{12} \div 9\dfrac{5}{8}$

37. $2\dfrac{2}{9} \div 11\dfrac{2}{3}$

38. $\dfrac{16}{3} \div 6\dfrac{2}{5}$

45. $8\dfrac{1}{3} \cdot \dfrac{36}{75} \div 4$

EXERCISES FOR REVIEW

(1.1) **46.** What is the value of 5 in the number 504,216?

(2.1) **47.** Find the product of 2,010 and 160.

(2.5) **48.** Use the numbers 8 and 5 to illustrate the commutative property of multiplication.

(3.5) **49.** Find the least common multiple of 6, 16, and 72.

(4.4) **50.** Find $\dfrac{8}{9}$ of $6\dfrac{3}{4}$.

★ **Answers to Practice Sets (4.5)**

A. 1. $\dfrac{10}{3}$ 2. $\dfrac{3}{2}$ 3. $\dfrac{8}{7}$ 4. 5   5. $\dfrac{7}{16}$ 6. $\dfrac{4}{21}$ 7. $\dfrac{16}{163}$

B. 1. $\dfrac{4}{9}$ 2. 1   3. $\dfrac{1}{2}$ 4. 15   5. 15   6. 2   7. $\dfrac{2}{5}$ 8. 16 glasses   9. $\dfrac{3}{2}$ 10. $\dfrac{8}{3}$

11. $\dfrac{4}{3}$ 12. $\dfrac{2}{5}$ 13. is the reciprocal of the fraction.

4.6 Applications Involving Fractions

Section Overview

- ☐ **MULTIPLICATION STATEMENTS**
- ☐ **MISSING PRODUCT STATEMENTS**
- ☐ **MISSING FACTOR STATEMENTS**

☐ MULTIPLICATION STATEMENTS

Statement

A *statement* is a sentence that is either true or false. A mathematical statement of the form

product = (factor 1) · (factor 2)

Multiplication Statement

is a **multiplication statement.** Depending on the numbers that are used, it to can be either true or false.

Omitting exactly one of the three numbers in the statement will produce exactly one of the following three problems. For convenience, we'll represent the omitted (or missing) number with the letter M (M for Missing).

1. $M = $ (factor 1) · (factor 2)   Missing *product* statement.
2. $M \cdot$ (factor 2) = product   Missing *factor* statement.
3. (factor 1) · $M = $ product   Missing *factor* statement.

We are interested in developing and working with methods to determine the missing number that makes the statement true. Fundamental to these methods is the ability to translate two words to mathematical symbols. The word

of translates to *times*
is translates to *equals*

☐ MISSING PRODUCT STATEMENTS

The equation $M - 8 \cdot 4$ is a *missing product* statement. We can find the value of M that makes this statement true by *multiplying* the known factors.

Missing product statements can be used to determine the answer to a question such as, "What number is fraction 1 of fraction 2?

☆ **SAMPLE SET A**

1. Find $\dfrac{3}{4}$ of $\dfrac{8}{9}$.   We are being asked the question, "What number is $\dfrac{3}{4}$ of $\dfrac{8}{9}$?" We must translate from words to mathematical symbols.

Continued

$\underbrace{\text{What number}}$ is $\dfrac{3}{4}$ of $\dfrac{8}{9}$ becomes

$\downarrow$ $\downarrow$ $\downarrow$ $\downarrow$

M $\quad = \quad \dfrac{3}{4} \quad \cdot \quad \dfrac{8}{9}$ Multiply.

missing   known   known
product   factor   factor

$$M = \dfrac{\overset{1}{\cancel{3}}}{\underset{1}{\cancel{4}}} \cdot \dfrac{\overset{2}{\cancel{8}}}{\underset{3}{\cancel{9}}} = \dfrac{1 \cdot 2}{1 \cdot 3} = \dfrac{2}{3}$$

Thus, $\dfrac{3}{4}$ of $\dfrac{8}{9}$ is $\dfrac{2}{3}$.

2. $\underbrace{\text{What number}}$ is $\dfrac{3}{4}$ of 24

$\downarrow$ $\downarrow$ $\downarrow$ $\downarrow$

$M \quad = \quad \dfrac{3}{4} \quad \cdot \quad 24$

missing   known   known
product   factor   factor

$$M = \dfrac{3}{\underset{1}{\cancel{4}}} \cdot \dfrac{\overset{6}{\cancel{24}}}{1} = \dfrac{3 \cdot 6}{1 \cdot 1} = \dfrac{18}{1} = 18$$

Thus, 18 is $\dfrac{3}{4}$ of 24.

★ **PRACTICE SET A**

1. Find $\dfrac{3}{8}$ of $\dfrac{16}{15}$.

2. What number is $\dfrac{9}{10}$ of $\dfrac{5}{6}$?

3. $\dfrac{11}{16}$ of $\dfrac{8}{33}$ is what number?

❑ MISSING FACTOR STATEMENTS

The equation $8 \cdot M = 32$ is a *missing factor* statement. We can find the value of M that makes this statement true by *dividing* (since we know that $32 \div 8 = 4$).

$8 \cdot M = 32$ means that $M \quad = \quad 32 \quad \div \quad 8$

$\downarrow \qquad \downarrow \qquad \downarrow \qquad \downarrow$

missing $=$ product $\div$ known
factor   factor

Finding the Missing Factor

To find the missing factor in a missing factor statement, divide the product by the known factor.

missing factor = (product) ÷ (known factor)

Missing factor statements can be used to answer such questions as

1. $\dfrac{3}{8}$ of what number is $\dfrac{9}{4}$?

2. What part of $1\dfrac{2}{7}$ is $1\dfrac{13}{14}$?

☆ SAMPLE SET B

1. $\dfrac{3}{8}$ of what number is $\dfrac{9}{4}$?

$$\underset{\substack{\text{known} \\ \text{factor}}}{\dfrac{3}{8}} \cdot \underset{\substack{\text{missing} \\ \text{factor}}}{M} = \underset{\text{product}}{\dfrac{9}{4}}$$

Now, using

missing factor = (product) $\div$ (known factor)

We get

$$M = \dfrac{9}{4} \div \dfrac{3}{8} = \dfrac{9}{4} \cdot \dfrac{8}{3} = \dfrac{\overset{3}{\cancel{9}}}{\underset{1}{\cancel{4}}} \cdot \dfrac{\overset{2}{\cancel{8}}}{\underset{1}{\cancel{3}}}$$

$$= \dfrac{3 \cdot 2}{1 \cdot 1}$$

$$= 6$$

Check: $\dfrac{3}{8} \cdot 6 \overset{?}{=} \dfrac{9}{4}$

$$\dfrac{3}{\underset{4}{\cancel{8}}} \cdot \dfrac{\overset{3}{\cancel{6}}}{1} \overset{?}{=} \dfrac{9}{4}$$

$$\dfrac{3 \cdot 3}{4 \cdot 1} \overset{?}{=} \dfrac{9}{4}$$

$$\dfrac{9}{4} \overset{\checkmark}{=} \dfrac{9}{4}$$

Thus, $\dfrac{3}{8}$ of 6 is $\dfrac{9}{4}$.

2. What part of $1\dfrac{2}{7}$ is $1\dfrac{13}{14}$?

$$\underset{\substack{\text{missing} \\ \text{factor}}}{M} \cdot \underset{\substack{\text{known} \\ \text{factor}}}{1\dfrac{2}{7}} = \underset{\text{product}}{1\dfrac{13}{14}}$$

For convenience, let's convert the mixed numbers to improper fractions.

$$M \cdot \dfrac{9}{7} = \dfrac{27}{14}$$

Continued

Now, using

missing factor = (product) ÷ (known factor)

we get

$$M = \frac{27}{14} \div \frac{9}{7} = \frac{27}{14} \cdot \frac{7}{9} = \frac{\overset{3}{\cancel{27}}}{\underset{2}{\cancel{14}}} \cdot \frac{\overset{1}{\cancel{7}}}{\underset{1}{\cancel{9}}}$$

$$= \frac{3 \cdot 1}{2 \cdot 1}$$

$$= \frac{3}{2}$$

Check: $\frac{3}{2} \cdot \frac{9}{7} \overset{?}{=} \frac{27}{14}$

$\frac{3 \cdot 9}{2 \cdot 7} \overset{?}{=} \frac{27}{14}$

$\frac{27}{14} \overset{\checkmark}{=} \frac{27}{14}$

Thus, $\frac{3}{2}$ of $1\frac{2}{7}$ is $1\frac{13}{14}$.

★ **PRACTICE SET B**

1. $\frac{3}{5}$ of what number is $\frac{9}{20}$?   **2.** $3\frac{3}{4}$ of what number is $2\frac{2}{9}$?

3. What part of $\frac{3}{5}$ is $\frac{9}{10}$?   **4.** What part of $1\frac{1}{4}$ is $1\frac{7}{8}$?

Answers to Practice Sets are on p. 188.

Section 4.6 EXERCISES

1. Find $\frac{2}{3}$ of $\frac{3}{4}$.   **2.** Find $\frac{5}{8}$ of $\frac{1}{10}$.

3. Find $\dfrac{12}{13}$ of $\dfrac{13}{36}$.

10. $\dfrac{1}{10}$ of $\dfrac{1}{100}$ is what number?

4. Find $\dfrac{1}{4}$ of $\dfrac{4}{7}$.

11. $\dfrac{1}{100}$ of $\dfrac{1}{10}$ is what number?

5. $\dfrac{3}{10}$ of $\dfrac{15}{4}$ is what number?

12. $1\dfrac{5}{9}$ of $2\dfrac{4}{7}$ is what number?

6. $\dfrac{14}{15}$ of $\dfrac{20}{21}$ is what number?

13. $1\dfrac{7}{18}$ of $\dfrac{4}{15}$ is what number?

7. $\dfrac{3}{44}$ of $\dfrac{11}{12}$ is what number?

14. $1\dfrac{1}{8}$ of $1\dfrac{11}{16}$ is what number?

8. $\dfrac{1}{3}$ of 2 is what number?

15. Find $\dfrac{2}{3}$ of $\dfrac{1}{6}$ of $\dfrac{9}{2}$.

9. $\dfrac{1}{4}$ of 3 is what number?

16. Find $\dfrac{5}{8}$ of $\dfrac{9}{20}$ of $\dfrac{4}{9}$.

17. $\frac{5}{12}$ of what number is $\frac{5}{6}$?

18. $\frac{3}{14}$ of what number is $\frac{6}{7}$?

19. $\frac{10}{3}$ of what number is $\frac{5}{9}$?

20. $\frac{15}{7}$ of what number is $\frac{20}{21}$?

21. $\frac{8}{3}$ of what number is $1\frac{7}{9}$?

22. $\frac{1}{3}$ of what number is $\frac{1}{3}$?

23. $\frac{1}{6}$ of what number is $\frac{1}{6}$?

24. $\frac{3}{4}$ of what number is $\frac{3}{4}$?

25. $\frac{8}{11}$ of what number is $\frac{8}{11}$?

26. $\frac{3}{8}$ of what number is 0?

27. $\frac{2}{3}$ of what number is 1?

28. $3\frac{1}{5}$ of what number is 1?

29. $1\frac{9}{12}$ of what number is $5\frac{1}{4}$?

30. $3\frac{1}{25}$ of what number is $2\frac{8}{15}$?

31. What part of $\frac{2}{3}$ is $1\frac{1}{9}$?

32. What part of $\frac{9}{10}$ is $3\frac{3}{5}$?

33. What part of $\frac{8}{9}$ is $\frac{3}{5}$?

34. What part of $\frac{14}{15}$ is $\frac{7}{30}$?

35. What part of 3 is $\frac{1}{5}$?

36. What part of 8 is $\frac{2}{3}$?

37. What part of 24 is 9?

38. What part of 42 is 26?

39. Find $\frac{12}{13}$ of $\frac{39}{40}$.

40. $\frac{14}{15}$ of $\frac{12}{21}$ is what number?

41. $\frac{8}{15}$ of what number is $2\frac{2}{5}$?

42. $\frac{11}{15}$ of what number is $\frac{22}{35}$?

43. $\frac{11}{16}$ of what number is 1?

44. What part of $\frac{23}{40}$ is $3\frac{9}{20}$?

45. $\frac{4}{35}$ of $3\frac{9}{22}$ is what number?

EXERCISES FOR REVIEW

(1.6) 46. Use the numbers 2 and 7 to illustrate the commutative property of addition.

(2.2) 47. Is 4 divisible by 0?

(3.4) 48. Expand 3^7. Do not find the actual value.

(4.2) 49. Convert $3\frac{5}{12}$ to an improper fraction.

(4.5) 50. Find the value of $\frac{3}{8} \div \frac{9}{16} \cdot \frac{6}{5}$.

★ **Answers to Practice Sets (4.6)**

A. 1. $\frac{2}{5}$ 2. $\frac{3}{4}$ 3. $\frac{1}{6}$

B. 1. $\frac{3}{4}$ 2. $\frac{16}{27}$ 3. $1\frac{1}{2}$ 4. $1\frac{1}{2}$

Chapter 4   SUMMARY OF KEY CONCEPTS

Fraction (4.1)

The idea of breaking up a whole quantity into equal parts gives us the word *fraction*.

Fraction Bar
Denominator
Numerator (4.1)

A fraction has three parts:

1. The fraction bar ——— .
2. The nonzero whole number below the fraction bar is the *denominator*.
3. The whole number above the fraction bar is the *numerator*.

$$\frac{4 \;\leftarrow \text{numerator}}{5 \;\leftarrow \text{denominator}} \quad \leftarrow \text{fraction bar}$$

Proper Fraction (4.2)

Proper fractions are fractions in which the numerator is strictly less than the denominator.

$\frac{4}{5}$ is a proper fraction

Improper Fraction (4.2)

Improper fractions are fractions in which the numerator is greater than or equal to the denominator. Also, any nonzero number placed over 1 is an improper fraction.

$\frac{5}{4}, \frac{5}{5},$ and $\frac{5}{1}$ are improper fractions

Mixed Number (4.2)

A *mixed number* is a number that is the sum of a whole number and a proper fraction.

$1\frac{1}{5}$ is a mixed number $\left(1\frac{1}{5} = 1 + \frac{1}{5}\right)$

Correspondence Between Improper Fractions and Mixed Numbers (4.2)

Each improper fraction corresponds to a particular mixed number, and each mixed number corresponds to a particular improper fraction.

Converting an Improper Fraction to a Mixed Number (4.2)

A method, based on division, converts an improper fraction to an equivalent mixed number.

$\frac{5}{4}$ can be converted to $1\frac{1}{4}$

Converting a Mixed Number to an Improper Fraction (4.2)

A method, based on multiplication, converts a mixed number to an equivalent improper fraction.

$5\frac{7}{8}$ can be converted to $\frac{47}{8}$

Equivalent Fractions (4.3)

Fractions that represent the same quantity are *equivalent fractions*.

$\frac{3}{4}$ and $\frac{6}{8}$ are equivalent fractions

Test for Equivalent Fractions (4.3)

If the *cross products* of two fractions are equal, then the two fractions are equivalent.

$$\frac{3}{4} \bowtie \frac{6}{8}$$

$3 \cdot 8 \overset{?}{=} 4 \cdot 6$
$24 = 24$

Thus, $\frac{3}{4}$ and $\frac{6}{8}$ are equivalent.

Relatively Prime **(4.3)**	Two whole numbers are *relatively prime* when 1 is the only number that divides both of them.
	3 and 4 are relatively prime
Reduced to Lowest Terms **(4.3)**	A fraction is *reduced to lowest terms* if its numerator and denominator are relatively prime.
	The number $\dfrac{3}{4}$ is reduced to lowest terms, since 3 and 4 are relatively prime.
	The number $\dfrac{6}{8}$ is *not* reduced to lowest terms since 6 and 8 are not relatively prime.
Reducing Fractions to Lowest Terms **(4.3)**	Two methods, one based on dividing out common primes and one based on dividing out any common factors, are available for reducing a fraction to lowest terms.
Raising Fractions to Higher Terms **(4.3)**	A fraction can be raised to higher terms by multiplying both the numerator and denominator by the same nonzero number.
	$$\frac{3}{4} = \frac{3 \cdot 2}{4 \cdot 2} = \frac{6}{8}$$
The Word "OF" Means Multiplication **(4.4)**	In many mathematical applications, the word "of" means multiplication.
Multiplication of Fractions **(4.4)**	To multiply two or more fractions, multiply the numerators together and multiply the denominators together. Reduce if possible.
	$$\frac{5}{8} \cdot \frac{4}{15} = \frac{5 \cdot 4}{8 \cdot 15} = \frac{20}{120} = \frac{1}{6}$$
Multiplying Fractions by Dividing Out Common Factors **(4.4)**	Two or more fractions can be multiplied by first dividing out common factors and then using the rule for multiplying fractions.
	$$\frac{\cancel{5}^{\,1}}{\cancel{8}_{\,2}} \cdot \frac{\cancel{4}^{\,1}}{\cancel{15}_{\,3}} = \frac{1 \cdot 1}{2 \cdot 3} = \frac{1}{6}$$
Multiplication of Mixed Numbers **(4.4)**	To perform a multiplication in which there are mixed numbers, first convert each mixed number to an improper fraction, then multiply. This idea also applies to division of mixed numbers.
Reciprocals **(4.5)**	Two numbers whose product is 1 are reciprocals.
	7 and $\dfrac{1}{7}$ are reciprocals
Division of Fractions **(4.5)**	To divide one fraction by another fraction, multiply the dividend by the reciprocal of the divisor.
	$$\frac{4}{5} \div \frac{2}{15} = \frac{4}{5} \cdot \frac{15}{2}$$
Dividing 1 by a Fraction **(4.5)**	When dividing 1 by a fraction, the quotient is the reciprocal of the fraction.
	$$\frac{1}{\frac{3}{7}} = \frac{7}{3}$$
Multiplication Statements **(4.6)**	A mathematical statement of the form
	product = (factor 1) · (factor 2)
	is a multiplication statement.

By omitting one of the three numbers, one of three following problems result:

1. $M =$ (factor 1) · (factor 2)   Missing product statement.
2. product = (factor 1) · M Missing factor statement.
3. product = M · (factor 2)   Missing factor statement.

Missing products are determined by simply multiplying the known factors. *Missing factors* are determined by

missing factor = (product) ÷ (known factor)

EXERCISE SUPPLEMENT

Section 4.1

For problems 1 and 2, name the suggested fraction.

1.

2.

For problems 3–5, specify the numerator and denominator.

3. $\dfrac{4}{5}$

4. $\dfrac{5}{12}$

5. $\dfrac{1}{3}$

For problems 6–10, write each fraction using digits.

6. Three fifths

7. Eight elevenths

8. Sixty-one forty-firsts

9. Two hundred six-thousandths

10. Zero tenths

For problems 11–15, write each fraction using words.

11. $\dfrac{10}{17}$

12. $\dfrac{21}{38}$

13. $\dfrac{606}{1431}$

14. $\dfrac{0}{8}$

15. $\dfrac{1}{16}$

For problems 16–18, state each numerator and denominator and write each fraction using digits.

16. One minute is one sixtieth of an hour.

17. In a box that contains forty-five electronic components, eight are known to be defective. If three components are chosen at random from the box, the probability that all three are defective is fifty-six fourteen thousand one hundred ninetieths.

18. About three fifths of the students in a college algebra class received a "B" in the course.

For problems 19 and 20, shade the region corresponding to the given fraction.

19. $\dfrac{1}{4}$

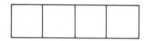

20. $\dfrac{3}{7}$

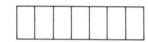

Section 4.2

For problems 21–29, convert each improper fraction to a mixed number.

21. $\dfrac{11}{4}$

22. $\dfrac{15}{2}$

23. $\dfrac{51}{8}$

24. $\dfrac{121}{15}$

25. $\dfrac{356}{3}$

26. $\dfrac{3}{2}$

27. $\dfrac{5}{4}$

28. $\dfrac{20}{5}$

29. $\dfrac{9}{3}$

For problems 30–40, convert each mixed number to an improper fraction.

30. $5\dfrac{2}{3}$

31. $16\dfrac{1}{8}$

32. $18\dfrac{1}{3}$

33. $3\dfrac{1}{5}$

34. $2\dfrac{9}{16}$

35. $17\dfrac{20}{21}$

36. $1\dfrac{7}{8}$

37. $1\dfrac{1}{2}$

38. $2\dfrac{1}{2}$

39. $8\dfrac{6}{7}$

40. $2\dfrac{9}{2}$

41. Why does $0\dfrac{1}{12}$ not qualify as a mixed number?

42. Why does 8 qualify as a mixed number?

Section 4.3

For problems 43–47, determine if the pairs of fractions are equivalent.

43. $\dfrac{1}{2}, \dfrac{15}{30}$

44. $\dfrac{8}{9}, \dfrac{32}{36}$

45. $\dfrac{3}{14}, \dfrac{24}{110}$

46. $2\dfrac{3}{8}, \dfrac{38}{16}$

47. $\dfrac{108}{77}, 1\dfrac{5}{13}$

For problems 48–60, reduce, if possible, each fraction.

48. $\dfrac{10}{25}$

49. $\dfrac{32}{44}$

50. $\dfrac{102}{266}$

51. $\dfrac{15}{33}$

52. $\dfrac{18}{25}$

53. $\dfrac{21}{35}$

54. $\dfrac{9}{16}$

55. $\dfrac{45}{85}$

56. $\dfrac{24}{42}$

57. $\dfrac{70}{136}$

58. $\dfrac{182}{580}$

59. $\dfrac{325}{810}$

60. $\dfrac{250}{1000}$

For problems 61–72, determine the missing numerator or denominator.

61. $\dfrac{3}{7} = \dfrac{?}{35}$

62. $\dfrac{4}{11} = \dfrac{?}{99}$

63. $\dfrac{1}{12} = \dfrac{?}{72}$

64. $\dfrac{5}{8} = \dfrac{25}{?}$

65. $\dfrac{11}{9} = \dfrac{33}{?}$

66. $\dfrac{4}{15} = \dfrac{24}{?}$

67. $\dfrac{14}{15} = \dfrac{?}{45}$

68. $\dfrac{0}{5} = \dfrac{?}{20}$

69. $\dfrac{12}{21} = \dfrac{96}{?}$

70. $\dfrac{14}{23} = \dfrac{?}{253}$

71. $\dfrac{15}{16} = \dfrac{180}{?}$

72. $\dfrac{21}{22} = \dfrac{336}{?}$

Sections 4.4 and 4.5

For problems 73–95, perform each multiplication and division.

73. $\dfrac{4}{5} \cdot \dfrac{15}{16}$

74. $\dfrac{8}{9} \cdot \dfrac{3}{24}$

75. $\dfrac{1}{10} \cdot \dfrac{5}{12}$

76. $\dfrac{14}{15} \div \dfrac{7}{5}$

77. $\dfrac{5}{6} \cdot \dfrac{13}{22} \cdot \dfrac{11}{39}$

78. $\dfrac{2}{3} \div \dfrac{15}{7} \cdot \dfrac{5}{6}$

79. $3\dfrac{1}{2} \div \dfrac{7}{2}$

80. $2\dfrac{4}{9} \div \dfrac{11}{45}$

81. $\dfrac{8}{15} \cdot \dfrac{3}{16} \cdot \dfrac{5}{24}$

82. $\dfrac{8}{15} \div 3\dfrac{3}{5} \cdot \dfrac{9}{16}$

83. $\dfrac{14}{15} \div 3\dfrac{8}{9} \cdot \dfrac{10}{21}$

84. $18 \cdot 5\dfrac{3}{4}$

85. $3\dfrac{3}{7} \cdot 2\dfrac{1}{12}$

86. $4\dfrac{1}{2} \div 2\dfrac{4}{7}$

87. $6\dfrac{1}{2} \div 3\dfrac{1}{4}$

88. $3\dfrac{5}{16} \div 2\dfrac{7}{18}$

89. $7 \div 2\dfrac{1}{3}$

90. $17 \div 4\dfrac{1}{4}$

91. $\dfrac{5}{8} \div 1\dfrac{1}{4}$

92. $2\dfrac{2}{3} \cdot 3\dfrac{3}{4}$

93. $20 \cdot \dfrac{18}{4}$

94. $0 \div 4\dfrac{1}{8}$

95. $1 \div 6\dfrac{1}{4} \cdot \dfrac{25}{4}$

Section 4.6

96. Find $\dfrac{8}{9}$ of $\dfrac{27}{2}$.

97. What part of $\dfrac{3}{8}$ is $\dfrac{21}{16}$?

98. What part of $3\frac{1}{5}$ is $1\frac{7}{25}$?

99. Find $6\frac{2}{3}$ of $\frac{9}{15}$.

100. $\frac{7}{20}$ of what number is $\frac{14}{35}$?

101. What part of $4\frac{1}{16}$ is $3\frac{3}{4}$?

102. Find $8\frac{3}{10}$ of $16\frac{2}{3}$.

103. $\frac{3}{20}$ of what number is $\frac{18}{30}$?

104. Find $\frac{1}{3}$ of 0.

105. Find $\frac{11}{12}$ of 1.

1. _____

1. **(4.1)** Shade a portion that corresponds to the fraction $\frac{5}{8}$.

2. _____

2. **(4.1)** Specify the numerator and denominator of the fraction $\frac{5}{9}$.

3. **(4.1)** Write the fraction five elevenths.

3. _____

4. **(4.1)** Write, in words, $\frac{4}{5}$.

4. _____

5. **(4.2)** Which of the fractions is a proper fraction?

$$4\frac{1}{12}, \frac{5}{12}, \frac{12}{5}$$

5. _____

6. **(4.2)** Convert $3\frac{4}{7}$ to an improper fraction.

6. _____

7. **(4.2)** Convert $\frac{16}{5}$ to a mixed number.

7. _____

8. **(4.3)** Determine if $\frac{5}{12}$ and $\frac{20}{48}$ are equivalent fractions.

8. _____

For problems 9–11, reduce, if possible, each fraction to lowest terms.

9. _____

9. **(4.3)** $\frac{21}{35}$

10. _____

10. **(4.3)** $\dfrac{15}{51}$

11. **(4.3)** $\dfrac{104}{480}$

11. _____

For problems 12 and 13, determine the missing numerator or denominator.

12. _____

12. **(4.3)** $\dfrac{5}{9} = \dfrac{?}{36}$

13. **(4.3)** $\dfrac{4}{3} = \dfrac{32}{?}$

13. _____

For problems 14–25, find each value.

14. _____

14. **(4.4)** $\dfrac{15}{16} \cdot \dfrac{4}{25}$

15. _____

15. **(4.4)** $3\dfrac{3}{4} \cdot 2\dfrac{2}{9} \cdot 6\dfrac{3}{5}$

16. _____

16. **(4.4)** $\sqrt{\dfrac{25}{36}}$

17. _____

17. **(4.4)** $\sqrt{\dfrac{4}{9}} \cdot \sqrt{\dfrac{81}{64}}$

198

18. _____

18. **(4.4)** $\dfrac{11}{30} \cdot \sqrt{\dfrac{225}{121}}$

19. _____

19. **(4.5)** $\dfrac{4}{15} \div 8$

20. _____

20. **(4.5)** $\dfrac{8}{15} \cdot \dfrac{5}{12} \div 2\dfrac{4}{9}$

21. _____

21. **(4.5)** $\left(\dfrac{6}{5}\right)^3 \div \sqrt{1\dfrac{11}{25}}$

22. _____

22. **(4.6)** Find $\dfrac{5}{12}$ of $\dfrac{24}{25}$.

23. _____

23. **(4.6)** $\dfrac{2}{9}$ of what number is $\dfrac{1}{18}$?

24. _____

24. **(4.6)** $1\dfrac{5}{7}$ of $\dfrac{21}{20}$ is what number?

25. _____

25. **(4.6)** What part of $\dfrac{9}{14}$ is $\dfrac{6}{7}$?

199

5

Addition and Subtraction of Fractions, Comparing Fractions, and Complex Fractions

After completing this chapter, you should

Section 5.1 Addition and Subtraction of Fractions with Like Denominators
- be able to add and subtract fractions with like denominators

Section 5.2 Addition and Subtraction of Fractions with Unlike Denominators
- be able to add and subtract fractions with unlike denominators

Section 5.3 Addition and Subtraction of Mixed Numbers
- be able to add and subtract mixed numbers

Section 5.4 Comparing Fractions
- understand ordering of numbers and be familiar with grouping symbols
- be able to compare two or more fractions

Section 5.5 Complex Fractions
- be able to distinguish between simple and complex fractions
- be able to convert a complex fraction to a simple fraction

Section 5.6 Combinations of Operations with Fractions
- gain a further understanding of the order of operations

5.1 Addition and Subtraction of Fractions with Like Denominators

Section
Overview

☐ **ADDITION OF FRACTIONS WITH LIKE DENOMINATORS**
☐ **SUBTRACTION OF FRACTIONS WITH LIKE DENOMINATORS**

☐ ADDITION OF FRACTIONS WITH LIKE DENOMINATORS

Let's examine the following diagram.

| $\frac{1}{5}$ | $\frac{1}{5}$ | $\frac{1}{5}$ | $\frac{1}{5}$ | $\frac{1}{5}$ |

2 one-fifths and 1 one-fifth is shaded.

It is shown in the shaded regions of the diagram that

(2 one-fifths) + (1 one-fifth) = (3 one-fifths)

That is,

$$\frac{2}{5} + \frac{1}{5} = \frac{3}{5}$$

From this observation, we can suggest the following rule.

Method of Adding Fractions Having Like Denominators

To add two or more fractions that have the same denominators, add the numerators and place the resulting sum over the common denominator. Reduce, if necessary.

☆ SAMPLE SET A

Find the following sums.

1. $\frac{3}{7} + \frac{2}{7}$. The denominators are the same. Add the numerators and place that sum over **7**.

$$\frac{3}{7} + \frac{2}{7} = \frac{3+2}{7} = \frac{5}{7}$$

2. $\frac{1}{8} + \frac{3}{8}$. The denominators are the same. Add the numerators and place the sum over **8**. Reduce.

$$\frac{1}{8} + \frac{3}{8} = \frac{1+3}{8} = \frac{4}{8} = \frac{1}{2}$$

3. $\frac{4}{9} + \frac{5}{9}$. The denominators are the same. Add the numerators and place the sum over **9**.

$$\frac{4}{9} + \frac{5}{9} = \frac{4+5}{9} = \frac{9}{9} = 1$$

4. $\dfrac{7}{8} + \dfrac{5}{8}$.   **The denominators are the same. Add the numerators and place the sum over 8.**

$$\frac{7}{8} + \frac{5}{8} = \frac{7+5}{8} = \frac{12}{8} = \frac{3}{2}$$

5. To see what happens if we *mistakenly add the denominators* as well as the numerators, let's add

$$\frac{1}{2} + \frac{1}{2}$$

Adding the numerators and *mistakenly* adding the denominators produces

$$\frac{1}{2} + \frac{1}{2} = \frac{1+1}{2+2} = \frac{2}{4} = \frac{1}{2}$$

This means that two $\dfrac{1}{2}$'s is the same as one $\dfrac{1}{2}$. Preposterous! We **do not add denominators.**

★ **PRACTICE SET A**

Find the following sums.

1. $\dfrac{1}{10} + \dfrac{3}{10}$ 2. $\dfrac{1}{4} + \dfrac{1}{4}$ 3. $\dfrac{7}{11} + \dfrac{4}{11}$ 4. $\dfrac{3}{5} + \dfrac{1}{5}$

5. Show why adding both the numerators and denominators is preposterous by adding $\dfrac{3}{4}$ and $\dfrac{3}{4}$ and examining the result.

☐ SUBTRACTION OF FRACTIONS WITH LIKE DENOMINATORS

We can picture the concept of subtraction of fractions in much the same way we pictured addition.

From this observation, we can suggest the following rule for subtracting fractions having like denominators:

Subtraction of Fractions with Like Denominators

To subtract two fractions that have like denominators, subtract the numerators and place the resulting difference over the common denominator. Reduce, if possible.

☆ **SAMPLE SET B**

Find the following differences.

1. $\frac{3}{5} - \frac{1}{5}$.

The denominators are the same. Subtract the numerators. Place the difference over 5.

$$\frac{3}{5} - \frac{1}{5} = \frac{3-1}{5} = \frac{2}{5}$$

2. $\frac{8}{6} - \frac{2}{6}$.

The denominators are the same. Subtract the numerators. Place the difference over 6.

$$\frac{8}{6} - \frac{2}{6} = \frac{8-2}{6} = \frac{6}{6} = 1$$

3. $\frac{16}{9} - \frac{2}{9}$.

The denominators are the same. Subtract numerators and place the difference over 9.

$$\frac{16}{9} - \frac{2}{9} = \frac{16-2}{9} = \frac{14}{9}$$

4. To see what happens if we *mistakenly* subtract the denominators, let's consider

$$\frac{7}{15} - \frac{4}{15} = \frac{7-4}{15-15} = \frac{3}{0}$$

We get division by zero, which is undefined. We **do not subtract denominators.**

★ **PRACTICE SET B**

Find the following differences.

1. $\frac{10}{13} - \frac{8}{13}$
2. $\frac{5}{12} - \frac{1}{12}$
3. $\frac{1}{2} - \frac{1}{2}$
4. $\frac{26}{10} - \frac{14}{10}$

5. Show why subtracting both the numerators and the denominators is in error by performing the subtraction $\frac{5}{9} - \frac{2}{9}$.

Answers to Practice Sets are on p. 207.

Section 5.1 EXERCISES

For problems 1–25, find the sums and differences. Be sure to reduce.

1. $\dfrac{3}{8} + \dfrac{2}{8}$

2. $\dfrac{1}{6} + \dfrac{2}{6}$

3. $\dfrac{9}{10} + \dfrac{1}{10}$

4. $\dfrac{3}{11} + \dfrac{4}{11}$

5. $\dfrac{9}{15} + \dfrac{4}{15}$

6. $\dfrac{3}{10} + \dfrac{2}{10}$

7. $\dfrac{5}{12} + \dfrac{7}{12}$

8. $\dfrac{11}{16} - \dfrac{2}{16}$

9. $\dfrac{3}{16} - \dfrac{3}{16}$

10. $\dfrac{15}{23} - \dfrac{2}{23}$

11. $\dfrac{1}{6} - \dfrac{1}{6}$

12. $\dfrac{1}{4} + \dfrac{1}{4} + \dfrac{1}{4}$

13. $\dfrac{3}{11} + \dfrac{1}{11} + \dfrac{5}{11}$

14. $\dfrac{16}{20} + \dfrac{1}{20} + \dfrac{2}{20}$

15. $\dfrac{12}{8} + \dfrac{2}{8} + \dfrac{1}{8}$

16. $\dfrac{1}{15} + \dfrac{8}{15} + \dfrac{6}{15}$

17. $\dfrac{3}{8} + \dfrac{2}{8} - \dfrac{1}{8}$

18. $\dfrac{11}{16} + \dfrac{9}{16} - \dfrac{5}{16}$

19. $\dfrac{4}{20} - \dfrac{1}{20} + \dfrac{9}{20}$

20. $\dfrac{7}{10} - \dfrac{3}{10} + \dfrac{11}{10}$

21. $\dfrac{16}{5} - \dfrac{1}{5} - \dfrac{2}{5}$

22. $\dfrac{21}{35} - \dfrac{17}{35} + \dfrac{31}{35}$

23. $\dfrac{5}{2} + \dfrac{16}{2} - \dfrac{1}{2}$

24. $\dfrac{1}{18} + \dfrac{3}{18} + \dfrac{1}{18} + \dfrac{4}{18} - \dfrac{5}{18}$

28. Find the total length of the screw.

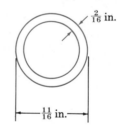

$\frac{3}{32}$ in.

$\frac{16}{32}$ in.

25. $\dfrac{6}{22} - \dfrac{2}{22} + \dfrac{4}{22} - \dfrac{1}{22} + \dfrac{11}{22}$

29. Two months ago, a woman paid off $\dfrac{3}{24}$ of a loan. One month ago, she paid off $\dfrac{5}{24}$ of the total loan. This month she will again pay off $\dfrac{5}{24}$ of the total loan. At the end of the month, how much of her total loan will she have paid off?

The following rule for addition and subtraction of two fractions is preposterous. Show why by performing the operations using the rule for problems 26 and 27.

PREPOSTEROUS RULE: To add or subtract two fractions, simply add or subtract the numerators and place this result over the sum or difference of the denominators.

26. $\dfrac{3}{10} - \dfrac{3}{10}$

30. Find the inside diameter of the pipe.

$\frac{2}{16}$ in.

$\frac{11}{16}$ in.

27. $\dfrac{8}{15} + \dfrac{8}{15}$

EXERCISES FOR REVIEW

(1.3)   **31.** Round 2,650 to the nearest hundred.

(2.5)   **32.** Use the numbers 2, 4, and 8 to illustrate the associative property of addition.

(3.3)   **33.** Find the prime factors of 495.

(4.4)   **34.** Find the value of $\dfrac{3}{4} \cdot \dfrac{16}{25} \cdot \dfrac{5}{9}$.

(4.6)   **35.** $\dfrac{8}{3}$ of what number is $1\dfrac{7}{9}$?

★ **Answers to Practice Sets (5.1)**

A. 1. $\dfrac{2}{5}$ 2. $\dfrac{1}{2}$ 3. 1 4. $\dfrac{4}{5}$ 5. $\dfrac{3}{4} + \dfrac{3}{4} = \dfrac{3+3}{4+4} = \dfrac{6}{8} = \dfrac{3}{4}$, so two $\dfrac{3}{4}$'s = one $\dfrac{3}{4}$ which is preposterous.

B. 1. $\dfrac{2}{13}$ 2. $\dfrac{1}{3}$ 3. 0 4. $\dfrac{6}{5}$ 5. $\dfrac{5}{9} - \dfrac{2}{9} = \dfrac{5-2}{9-9} = \dfrac{3}{0}$, which is undefined.

5.2 Addition and Subtraction of Fractions with Unlike Denominators

Section Overview

- ☐ **A BASIC RULE**
- ☐ **ADDITION AND SUBTRACTION OF FRACTIONS**

☐ A BASIC RULE

There is a basic rule that must be followed when adding or subtracting fractions.

A Basic Rule

> Fractions can only be added or subtracted conveniently if they have like denominators.

To see why this rule makes sense, let's consider the problem of adding a quarter and a dime.

1 quarter + 1 dime = 35 cents

Now,

$$\left. \begin{array}{l} 1 \text{ quarter} = \dfrac{25}{100} \\[2mm] 1 \text{ dime} = \dfrac{10}{100} \end{array} \right\} \text{ same denominations}$$

$$35¢ = \dfrac{35}{100}$$

$$\dfrac{25}{100} + \dfrac{10}{100} = \dfrac{25 + 10}{100} = \dfrac{35}{100}$$

In order to combine a quarter and a dime to produce 35¢, we convert them to quantities of the same denomination.

> same denomination ⟶ same denominator

☐ ADDITION AND SUBTRACTION OF FRACTIONS

Least Common Multiple (LCM) and Least Common Denominator (LCD)

In Section 3.5, we examined the least common multiple (LCM) of a collection of numbers. If these numbers are used as denominators of fractions, we call the least common multiple, the least common denominator (LCD).

Method of Adding or Subtracting Fractions with Unlike Denominators

> To add or subtract fractions having unlike denominators, convert each fraction to an equivalent fraction having as a denominator the least common denominator (LCD) of the original denominators.

Find the following sums and differences.

1. $\frac{1}{6} + \frac{3}{4}$.   **The denominators are not the same.**
Find the LCD of 6 and 4.

$\left. \begin{array}{l} 6 = 2 \cdot 3 \\ 4 = 2^2 \end{array} \right\}$ **The LCD $= 2^2 \cdot 3 = 4 \cdot 3 = 12$.**

Write each of the original fractions as a new, equivalent fraction having the common denominator 12.

$$\frac{1}{6} + \frac{3}{4} = \frac{}{12} + \frac{}{12}$$

To find a new numerator, we divide the original denominator into the LCD. Since the original denominator is being multiplied by this quotient, we must multiply the original numerator by this quotient.

$12 \div 6 = 2$ **Multiply 1 by 2: $1 \cdot 2 = 2$.**

original numerator

new numerator ——————

$12 \div 4 = 3$ **Multiply 3 by 3: $3 \cdot 3 = 9$.**

original numerator

new numerator ——————

$$\frac{1}{6} + \frac{3}{4} = \frac{1 \cdot 2}{12} + \frac{3 \cdot 3}{12}$$

$$= \frac{2}{12} + \frac{9}{12}$$ **Now the denominators are the same.**

$$= \frac{2 + 9}{12}$$ **Add the numerators and place the sum over the common denominator.**

$$= \frac{11}{12}$$

2. $\frac{1}{2} + \frac{2}{3}$.   **The denominators are not the same.**
Find the LCD of 2 and 3.

$LCD = 2 \cdot 3 = 6$

Write each of the original fractions as a new, equivalent fraction having the common denominator 6.

$$\frac{1}{2} + \frac{2}{3} = \frac{}{6} + \frac{}{6}$$

To find a new numerator, we divide the original denominator into the LCD. Since the original denominator is being multiplied by this quotient, we must multiply the original numerator by this quotient.

$6 \div 2 = 3$ **Multiply the numerator 1 by 3.**

$6 \div 3 = 2$ **Multiply the numerator 2 by 2.**

$$\frac{1}{2} + \frac{2}{3} = \frac{1 \cdot 3}{6} + \frac{2 \cdot 2}{6}$$

$$= \frac{3}{6} + \frac{4}{6}$$

$$= \frac{3 + 4}{6}$$

$$= \frac{7}{6} \text{ or } 1\frac{1}{6}$$

3. $\frac{5}{9} - \frac{5}{12}$.

The denominators are not the same.
Find the LCD of 9 and 12.

$$\left. \begin{array}{l} 9 = 3 \cdot 3 = 3^2 \\ 12 = 2 \cdot 6 = 2 \cdot 2 \cdot 3 = 2^2 \cdot 3 \end{array} \right\}$$

$LCD = 2^2 \cdot 3^2 = 4 \cdot 9 = 36.$

$$\frac{5}{9} - \frac{5}{12} = \frac{}{36} - \frac{}{36}$$

$36 \div 9 = 4$

Multiply the numerator 5 by 4.

$36 \div 12 = 3$

Multiply the numerator 5 by 3.

$$\frac{5}{9} - \frac{5}{12} = \frac{5 \cdot 4}{36} - \frac{5 \cdot 3}{36}$$

$$= \frac{20}{36} - \frac{15}{36}$$

$$= \frac{20 - 15}{36}$$

$$= \frac{5}{36}$$

4. $\frac{5}{6} - \frac{1}{8} + \frac{7}{16}$

The denominators are not the same.
Find the LCD of 6, 8, and 16.

$$\left. \begin{array}{l} 6 = 2 \cdot 3 \\ 8 = 2 \cdot 4 = 2 \cdot 2 \cdot 2 = 2^3 \\ 16 = 2 \cdot 8 = 2 \cdot 2 \cdot 4 = 2 \cdot 2 \cdot 2 \cdot 2 = 2^4 \end{array} \right\}$$

The LCD is $2^4 \cdot 3 = 48$

$$\frac{5}{6} - \frac{1}{8} + \frac{7}{16} = \frac{}{48} - \frac{}{48} + \frac{}{48}$$

$48 \div 6 = 8$

Multiply the numerator 5 by 8.

$48 \div 8 = 6$

Multiply the numerator 1 by 6.

$48 \div 16 = 3$

Multiply the numerator 7 by 3.

$$\frac{5}{6} - \frac{1}{8} + \frac{7}{16} = \frac{5 \cdot 8}{48} - \frac{1 \cdot 6}{48} + \frac{7 \cdot 3}{48}$$

$$= \frac{40}{48} - \frac{6}{48} + \frac{21}{48}$$

$$= \frac{40 - 6 + 21}{48}$$

$$= \frac{55}{48} \quad \text{or} \quad 1\frac{7}{48}$$

★ **PRACTICE SET A**

Find the following sums and differences.

1. $\dfrac{3}{4} + \dfrac{1}{12}$ **2.** $\dfrac{1}{2} - \dfrac{3}{7}$ **3.** $\dfrac{7}{10} - \dfrac{5}{8}$ **4.** $\dfrac{15}{16} + \dfrac{1}{2} - \dfrac{3}{4}$ **5.** $\dfrac{1}{32} - \dfrac{1}{48}$

Answers to the Practice Set are on p. 215.

Section 5.2 EXERCISES

1. A most basic rule of arithmetic states that two fractions may be added or subtracted conveniently only if they have _____.

For problems 2–43, find the sums and differences.

2. $\dfrac{1}{2} + \dfrac{1}{6}$

3. $\dfrac{1}{8} + \dfrac{1}{2}$

4. $\dfrac{3}{4} + \dfrac{1}{3}$

5. $\dfrac{5}{8} + \dfrac{2}{3}$

6. $\dfrac{1}{12} + \dfrac{1}{3}$

7. $\dfrac{6}{7} - \dfrac{1}{4}$

8. $\dfrac{9}{10} - \dfrac{2}{5}$

9. $\dfrac{7}{9} - \dfrac{1}{4}$

10. $\dfrac{8}{15} - \dfrac{3}{10}$

11. $\dfrac{8}{13} - \dfrac{5}{39}$

12. $\dfrac{11}{12} - \dfrac{2}{5}$

13. $\dfrac{1}{15} + \dfrac{5}{12}$

14. $\dfrac{13}{88} - \dfrac{1}{4}$

15. $\dfrac{1}{9} - \dfrac{1}{81}$

16. $\dfrac{19}{40} + \dfrac{5}{12}$

17. $\dfrac{25}{26} - \dfrac{7}{10}$

18. $\dfrac{9}{28} - \dfrac{4}{45}$

19. $\dfrac{22}{45} - \dfrac{16}{35}$

20. $\dfrac{56}{63} + \dfrac{22}{33}$

21. $\dfrac{1}{16} + \dfrac{3}{4} - \dfrac{3}{8}$

22. $\dfrac{5}{12} - \dfrac{1}{120} + \dfrac{19}{20}$

23. $\dfrac{8}{3} - \dfrac{1}{4} + \dfrac{7}{36}$

24. $\dfrac{11}{9} - \dfrac{1}{7} + \dfrac{16}{63}$

29. $\dfrac{27}{40} + \dfrac{47}{48} - \dfrac{119}{126}$

25. $\dfrac{12}{5} - \dfrac{2}{3} + \dfrac{17}{10}$

30. $\dfrac{41}{44} - \dfrac{5}{99} - \dfrac{11}{175}$

26. $\dfrac{4}{9} + \dfrac{13}{21} - \dfrac{9}{14}$

31. $\dfrac{5}{12} + \dfrac{1}{18} + \dfrac{1}{24}$

27. $\dfrac{3}{4} - \dfrac{3}{22} + \dfrac{5}{24}$

32. $\dfrac{5}{9} + \dfrac{1}{6} + \dfrac{7}{15}$

28. $\dfrac{25}{48} - \dfrac{7}{88} + \dfrac{5}{24}$

33. $\dfrac{21}{25} + \dfrac{1}{6} + \dfrac{7}{15}$

34. $\dfrac{5}{18} - \dfrac{1}{36} + \dfrac{7}{9}$

39. $\dfrac{7}{15} + \dfrac{3}{10} - \dfrac{34}{60}$

35. $\dfrac{11}{14} - \dfrac{1}{36} - \dfrac{1}{32}$

40. $\dfrac{14}{15} - \dfrac{3}{10} - \dfrac{6}{25} + \dfrac{7}{20}$

36. $\dfrac{21}{33} + \dfrac{12}{22} + \dfrac{15}{55}$

41. $\dfrac{11}{6} - \dfrac{5}{12} + \dfrac{17}{30} + \dfrac{25}{18}$

37. $\dfrac{5}{51} + \dfrac{2}{34} + \dfrac{11}{68}$

42. $\dfrac{1}{9} + \dfrac{22}{21} - \dfrac{5}{18} - \dfrac{1}{45}$

38. $\dfrac{8}{7} - \dfrac{16}{14} + \dfrac{19}{21}$

43. $\dfrac{7}{26} + \dfrac{28}{65} - \dfrac{51}{104} + 0$

44. A morning trip from San Francisco to Los Angeles took $\frac{13}{12}$ hours. The return trip took $\frac{57}{60}$ hours. How much longer did the morning trip take?

45. At the beginning of the week, Starlight Publishing Company's stock was selling for $\frac{115}{8}$ dollars per share. At the end of the week, analysts had noted that the stock had gone up $\frac{11}{4}$ dollars per share. What was the price of the stock, per share, at the end of the week?

46. A recipe for fruit punch calls for $\frac{23}{3}$ cups of pineapple juice, $\frac{1}{4}$ cup of lemon juice, $\frac{15}{2}$ cups of orange juice, 2 cups of sugar, 6 cups of water, and 8 cups of carbonated non-cola soft drink. How many cups of ingredients will be in the final mixture?

47. The side of a particular type of box measures $8\frac{3}{4}$ inches in length. Is it possible to place three such boxes next to each other on a shelf that is $26\frac{1}{5}$ inches in length? Why or why not?

48. Four resistors, $\frac{3}{8}$ ohm, $\frac{1}{4}$ ohm, $\frac{3}{5}$ ohm, and $\frac{7}{8}$ ohm, are connected in series in an electrical circuit. What is the total resistance in the circuit due to these resistors? ("In series" implies addition.)

49. A copper pipe has an inside diameter of $2\frac{3}{16}$ inches and an outside diameter of $2\frac{5}{34}$ inches. How thick is the pipe?

50. The probability of an event was originally thought to be $\frac{15}{32}$. Additional information decreased the probability by $\frac{3}{14}$. What is the updated probability?

EXERCISES FOR REVIEW

(1.5) **51.** Find the difference between 867 and 418.

(2.4) **52.** Is 81,147 divisible by 3?

(3.5) **53.** Find the LCM of 11, 15, and 20.

(4.4) **54.** Find $\frac{3}{4}$ of $4\frac{2}{9}$.

(5.1) **55.** Find the value of $\frac{8}{15} - \frac{3}{15} + \frac{2}{15}$.

★ **Answers to Practice Set (5.2)**

A. 1. $\dfrac{5}{6}$ 2. $\dfrac{1}{14}$ 3. $\dfrac{3}{40}$ 4. $\dfrac{11}{16}$ 5. $\dfrac{1}{96}$

5.3 Addition and Subtraction of Mixed Numbers

Section Overview

☐ **THE METHOD OF CONVERTING TO IMPROPER FRACTIONS**

☐ **THE METHOD OF CONVERTING TO IMPROPER FRACTIONS**

> To add or subtract mixed numbers, convert each mixed number to an improper fraction, then add or subtract the resulting improper fractions.

☆ **SAMPLE SET A**

Find the following sums and differences.

1. $8\dfrac{3}{5} + 5\dfrac{1}{4}$.   Convert each mixed number to an improper fraction.

$$8\frac{3}{5} = \frac{5 \cdot 8 + 3}{5} = \frac{40 + 3}{5} = \frac{43}{5}$$

$$5\frac{1}{4} = \frac{4 \cdot 5 + 1}{4} = \frac{20 + 1}{4} = \frac{21}{4}$$
     Now add the improper fractions $\dfrac{43}{5}$ and $\dfrac{21}{4}$.

$$\frac{43}{5} + \frac{21}{4}$$
     The LCD = 20.

$$\frac{43}{5} + \frac{21}{4} = \frac{43 \cdot 4}{20} + \frac{21 \cdot 5}{20}$$

$$= \frac{172}{20} + \frac{105}{20}$$

$$= \frac{172 + 105}{20}$$

$$= \frac{277}{20}$$
     Convert this improper fraction to a mixed number.

$$= 13\frac{17}{20}$$

Thus, $8\dfrac{3}{5} + 5\dfrac{1}{4} = 13\dfrac{17}{20}$.

Continued

2. $3\dfrac{1}{8} - \dfrac{5}{6}$.   Convert the mixed number to an improper fraction.

$$3\dfrac{1}{8} = \dfrac{3 \cdot 8 + 1}{8} = \dfrac{24 + 1}{8} = \dfrac{25}{8}$$

$$\dfrac{25}{8} - \dfrac{5}{6}$$ The LCD = 24.

$$\dfrac{25}{8} - \dfrac{5}{6} = \dfrac{25 \cdot 3}{24} - \dfrac{5 \cdot 4}{24}$$

$$= \dfrac{75}{24} - \dfrac{20}{24}$$

$$= \dfrac{75 - 20}{24}$$

$$= \dfrac{55}{24}$$ Convert this improper fraction to a mixed number.

$$= 2\dfrac{7}{24}$$

Thus, $3\dfrac{1}{8} - \dfrac{5}{6} = 2\dfrac{7}{24}$.

★ PRACTICE SET A

Find the following sums and differences.

1. $1\dfrac{5}{9} + 3\dfrac{2}{9}$ **2.** $10\dfrac{3}{4} - 2\dfrac{1}{2}$ **3.** $2\dfrac{7}{8} + 5\dfrac{1}{4}$ **4.** $8\dfrac{3}{5} - \dfrac{3}{10}$ **5.** $16 + 2\dfrac{9}{16}$

Answers to the Practice Set are on p. 220.

Section 5.3 EXERCISES

For problems 1–36, perform each indicated operation.

1. $3\dfrac{1}{8} + 4\dfrac{3}{8}$ **3.** $10\dfrac{5}{12} + 2\dfrac{1}{12}$

2. $5\dfrac{1}{3} + 6\dfrac{1}{3}$ **4.** $15\dfrac{1}{5} - 11\dfrac{3}{5}$

5. $9\dfrac{3}{11} + 12\dfrac{3}{11}$

12. $5\dfrac{1}{3} + 2\dfrac{1}{4}$

6. $1\dfrac{1}{6} + 3\dfrac{2}{6} + 8\dfrac{1}{6}$

13. $6\dfrac{2}{7} - 1\dfrac{1}{3}$

7. $5\dfrac{3}{8} + 1\dfrac{1}{8} - 2\dfrac{5}{8}$

14. $8\dfrac{2}{5} + 4\dfrac{1}{10}$

8. $\dfrac{3}{5} + 5\dfrac{1}{5}$

15. $1\dfrac{1}{3} + 12\dfrac{3}{8}$

9. $2\dfrac{2}{9} - \dfrac{5}{9}$

16. $3\dfrac{1}{4} + 1\dfrac{1}{3} - 2\dfrac{1}{2}$

10. $6 + 11\dfrac{2}{3}$

17. $4\dfrac{3}{4} - 3\dfrac{5}{6} + 1\dfrac{2}{3}$

11. $17 - 8\dfrac{3}{14}$

18. $3\dfrac{1}{12} + 4\dfrac{1}{3} + 1\dfrac{1}{4}$

19. $5\dfrac{1}{15} + 8\dfrac{3}{10} - 5\dfrac{4}{5}$

24. $\dfrac{5}{2} + 2\dfrac{1}{6} + 11\dfrac{1}{3} - \dfrac{11}{6}$

20. $7\dfrac{1}{3} + 8\dfrac{5}{6} - 2\dfrac{1}{4}$

25. $1\dfrac{1}{8} + \dfrac{9}{4} - \dfrac{1}{16} - \dfrac{1}{32} + \dfrac{19}{8}$

21. $19\dfrac{20}{21} + 42\dfrac{6}{7} - \dfrac{5}{14} + 12\dfrac{1}{7}$

26. $22\dfrac{3}{8} - 16\dfrac{1}{7}$

22. $\dfrac{1}{16} + 4\dfrac{3}{4} + 10\dfrac{3}{8} - 9$

27. $15\dfrac{4}{9} + 4\dfrac{9}{16}$

23. $11 - \dfrac{2}{9} + 10\dfrac{1}{3} - \dfrac{2}{3} - 5\dfrac{1}{6} + 6\dfrac{1}{18}$

28. $4\dfrac{17}{88} + 5\dfrac{9}{110}$

29. $6\dfrac{11}{12} + \dfrac{2}{3}$

34. $11\dfrac{11}{24} - 7\dfrac{13}{18}$

30. $8\dfrac{9}{16} - \dfrac{7}{9}$

35. $5\dfrac{27}{84} - 3\dfrac{5}{42} + 1\dfrac{1}{21}$

31. $5\dfrac{2}{11} - \dfrac{1}{12}$

36. $16\dfrac{1}{48} - 16\dfrac{1}{96} + \dfrac{1}{144}$

32. $18\dfrac{15}{16} - \dfrac{33}{34}$

37. A man pours $2\dfrac{5}{8}$ gallons of paint from a bucket into a tray. After he finishes pouring, there are $1\dfrac{1}{4}$ gallons of paint left in his bucket. How much paint did the man pour into the tray? (Hint: Think about the wording.)

33. $1\dfrac{89}{112} - \dfrac{21}{56}$

38. A particular computer stock opened at $37\frac{3}{8}$ and closed at $38\frac{1}{4}$. What was the net gain for this stock?

40. If a person who weighs $145\frac{3}{4}$ pounds goes on the diet program described in problem 39, how much would he weigh at the end of 3 months?

39. A particular diet program claims that $4\frac{3}{16}$ pounds can be lost the first month, $3\frac{1}{4}$ pounds can be lost the second month, and $1\frac{1}{2}$ pounds can be lost the third month. How many pounds does this diet program claim a person can lose over a 3-month period? .

41. If the diet program described in problem 39 makes the additional claim that from the fourth month on, a person will lose $1\frac{1}{8}$ pounds a month, how much will a person who begins the program weighing $208\frac{3}{4}$ pounds weight after 8 months?

EXERCISES FOR REVIEW

(3.1)   **42.** Use exponents to write $4 \cdot 4 \cdot 4$.

(3.4)   **43.** Find the greatest common factor of 14 and 20.

(4.2)   **44.** Convert $\frac{16}{5}$ to a mixed number.

(5.1)   **45.** Find the sum. $\frac{4}{9} + \frac{1}{9} + \frac{2}{9}$.

(5.2)   **46.** Find the difference. $\frac{15}{26} - \frac{3}{10}$.

★ **Answers to Practice Set (5.3)**

A. **1.** $4\frac{7}{9}$ **2.** $8\frac{1}{4}$ **3.** $8\frac{1}{8}$ **4.** $8\frac{3}{10}$ **5.** $18\frac{9}{16}$

5.4 Comparing Fractions

Section Overview

☐ **ORDER AND THE INEQUALITY SYMBOLS**
☐ **COMPARING FRACTIONS**

☐ ORDER AND THE INEQUALITY SYMBOLS

Our number system is called an **ordered number system** because the numbers in the system can be placed in order from smaller to larger. This is easily seen on the number line.

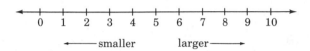

On the number line, a number that appears to the right of another number is larger than that other number. For example, 5 is greater than 2 because 5 is located to the right of 2 on the number line. We may also say that 2 is less than 5.

To make the inequality phrases "greater than" and "less than" more brief, mathematicians represent them with the symbols $>$ and $<$, respectively.

Symbols for Greater Than $>$ Less Than $<$

> $>$ represents the phrase "greater than."
> $<$ represents the phrase "less than."

$5 > 2$ represents "5 is greater than 2."
$2 < 5$ represents "2 is less than 5."

☐ COMPARING FRACTIONS

Recall that the fraction $\frac{4}{5}$ indicates that we have 4 of 5 parts of some whole quantity, and the fraction $\frac{3}{5}$ indicates that we have 3 of 5 parts. Since 4 of 5 parts is more than 3 of 5 parts, $\frac{4}{5}$ is greater than $\frac{3}{5}$; that is,

$$\frac{4}{5} > \frac{3}{5}$$

We have just observed that when two fractions have the same denominator, we can determine which is larger by comparing the numerators.

Comparing Fractions

> If two fractions have the same denominators, the fraction with the larger numerator is the larger fraction.

Thus, to compare the sizes of two or more fractions, we need only convert each of them to equivalent fractions that have a common denominator. We then compare the numerators. It is convenient if the common denominator is the LCD. The fraction with the larger numerator is the larger fraction.

1. Compare $\dfrac{8}{9}$ and $\dfrac{14}{15}$.

Convert each fraction to an equivalent fraction with the LCD as the denominator.

Find the LCD.

$$\left.\begin{array}{l} 9 = 3^2 \\ 15 = 3 \cdot 5 \end{array}\right\} \quad \textbf{The LCD} = 3^2 \cdot 5 = 9 \cdot 5 = 45.$$

$$\frac{8}{9} = \frac{8 \cdot 5}{45} = \frac{40}{45}$$

$$\frac{14}{15} = \frac{14 \cdot 3}{45} = \frac{42}{45}$$

Since $40 < 42$,

$$\frac{40}{45} < \frac{42}{45}$$

Thus, $\dfrac{8}{9} < \dfrac{14}{15}$.

2. Write $\dfrac{5}{6}, \dfrac{7}{10}$, and $\dfrac{13}{15}$ in order from smallest to largest.

Convert each fraction to an equivalent fraction with the LCD as the denominator.

Find the LCD.

$$\left.\begin{array}{l} 6 = 2 \cdot 3 \\ 10 = 2 \cdot 5 \\ 15 = 3 \cdot 5 \end{array}\right\} \quad \textbf{The LCD} = 2 \cdot 3 \cdot 5 = 30.$$

$$\frac{5}{6} = \frac{5 \cdot 5}{30} = \frac{25}{30}$$

$$\frac{7}{10} = \frac{7 \cdot 3}{30} = \frac{21}{30}$$

$$\frac{13}{15} = \frac{13 \cdot 2}{30} = \frac{26}{30}$$

Since $21 < 25 < 26$,

$$\frac{21}{30} < \frac{25}{30} < \frac{26}{30}$$

$$\frac{7}{10} < \frac{5}{6} < \frac{13}{15}$$

Writing these numbers in order from smallest to largest, we get $\dfrac{7}{10}, \dfrac{5}{6}, \dfrac{13}{15}$.

3. Compare $8\dfrac{6}{7}$ and $6\dfrac{3}{4}$.

To compare mixed numbers that have different whole number parts, we need only compare whole number parts. Since $6 < 8$,

$$6\frac{3}{4} < 8\frac{6}{7}$$

4. Compare $4\frac{5}{8}$ and $4\frac{7}{12}$.

To compare mixed numbers that have the same whole number parts, we need only compare fractional parts.

$$\left.\begin{array}{l} 8 = 2^3 \\ 12 = 2^2 \cdot 3 \end{array}\right\} \quad \textbf{The LCD} = \textbf{2}^3 \cdot \textbf{3} = \textbf{8} \cdot \textbf{3} = \textbf{24}.$$

$$\frac{5}{8} = \frac{5 \cdot 3}{24} = \frac{15}{24}$$

$$\frac{7}{12} = \frac{7 \cdot 2}{24} = \frac{14}{24}$$

Since $14 < 15$,

$$\frac{14}{24} < \frac{15}{24}$$

$$\frac{7}{12} < \frac{5}{8}$$

Hence, $4\frac{7}{12} < 4\frac{5}{8}$.

★ **PRACTICE SET A**

1. Compare $\frac{3}{4}$ and $\frac{4}{5}$.

2. Compare $\frac{9}{10}$ and $\frac{13}{15}$.

3. Write $\frac{13}{16}, \frac{17}{20},$ and $\frac{33}{40}$ in order from smallest to largest.

4. Compare $11\frac{1}{6}$ and $9\frac{2}{5}$.

5. Compare $1\frac{9}{14}$ and $1\frac{11}{16}$.

Answers to the Practice Set are on p. 225.

Section 5.4 EXERCISES

For problems 1–20, arrange each collection of numbers in order from smallest to largest.

1. $\dfrac{3}{5}, \dfrac{5}{8}$

2. $\dfrac{1}{6}, \dfrac{2}{7}$

3. $\dfrac{3}{4}, \dfrac{5}{6}$

4. $\dfrac{7}{9}, \dfrac{11}{12}$

5. $\dfrac{3}{8}, \dfrac{2}{5}$

6. $\dfrac{1}{2}, \dfrac{5}{8}, \dfrac{7}{16}$

7. $\dfrac{1}{2}, \dfrac{3}{5}, \dfrac{4}{7}$

8. $\dfrac{3}{4}, \dfrac{2}{3}, \dfrac{5}{6}$

9. $\dfrac{3}{4}, \dfrac{7}{9}, \dfrac{5}{4}$

10. $\dfrac{7}{8}, \dfrac{15}{16}, \dfrac{11}{12}$

11. $\dfrac{3}{14}, \dfrac{2}{7}, \dfrac{3}{4}$

12. $\dfrac{17}{32}, \dfrac{25}{48}, \dfrac{13}{16}$

13. $5\dfrac{3}{5}, 5\dfrac{4}{7}$

14. $11\dfrac{3}{16}, 11\dfrac{1}{12}$

15. $9\frac{2}{3}, 9\frac{4}{5}$

18. $20\frac{15}{16}, 20\frac{23}{24}$

16. $7\frac{2}{3}, 8\frac{5}{6}$

19. $2\frac{2}{9}, 2\frac{3}{7}$

17. $1\frac{9}{16}, 2\frac{1}{20}$

20. $5\frac{8}{13}, 5\frac{9}{20}$

EXERCISES FOR REVIEW

(1.3) **21.** Round 267,006,428 to the nearest ten million.

(2.4) **22.** Is the number 82,644 divisible by 2? by 3? by 4?

(4.2) **23.** Convert $3\frac{2}{7}$ to an improper fraction.

(5.2) **24.** Find the value of $\frac{5}{6} + \frac{3}{10} - \frac{2}{5}$.

(5.3) **25.** Find the value of $8\frac{3}{8} + 5\frac{1}{4}$.

★ **Answers to Practice Set (5.4)**

A. **1.** $\frac{3}{4} < \frac{4}{5}$ **2.** $\frac{13}{15} < \frac{9}{10}$ **3.** $\frac{13}{16}, \frac{33}{40}, \frac{17}{20}$ **4.** $9\frac{2}{5} < 11\frac{1}{6}$ **5.** $1\frac{9}{14} < 1\frac{11}{16}$

5.5 Complex Fractions

Section Overview	☐ SIMPLE FRACTIONS AND COMPLEX FRACTIONS ☐ CONVERTING COMPLEX FRACTIONS TO SIMPLE FRACTIONS

☐ SIMPLE FRACTIONS AND COMPLEX FRACTIONS

Simple Fraction

A **simple fraction** is any fraction in which the numerator is any whole number and the denominator is any nonzero whole number. Some examples are the following:

$$\frac{1}{2}, \quad \frac{4}{3}, \quad \frac{763}{1,000}$$

Complex Fraction

A **complex fraction** is any fraction in which the numerator and/or the denominator is a fraction; it is a fraction of fractions. Some examples of complex fractions are the following:

$$\frac{\frac{3}{4}}{\frac{5}{6}}, \quad \frac{\frac{1}{3}}{\frac{2}{2}}, \quad \frac{6}{\frac{9}{10}}, \quad \frac{4 + \frac{3}{8}}{7 - \frac{5}{6}}$$

☐ CONVERTING COMPLEX FRACTIONS TO SIMPLE FRACTIONS

The goal here is to convert a complex fraction to a simple fraction. We can do so by employing the methods of adding, subtracting, multiplying, and dividing fractions. Recall from Section 4.1 that a fraction bar serves as a grouping symbol separating the fractional quantity into two individual groups. We proceed in simplifying a complex fraction to a simple fraction by simplifying the numerator and the denominator of the complex fraction separately. We will simplify the numerator and denominator *completely* before removing the fraction bar by dividing. This technique is illustrated in problems 3, 4, 5, and 6 of Sample Set A.

☆ SAMPLE SET A

Convert each of the following complex fractions to a simple fraction.

1. $\dfrac{\frac{3}{8}}{\frac{15}{16}}$ Convert this complex fraction to a simple fraction by performing the indicated division.

$$\frac{\frac{3}{8}}{\frac{15}{16}} = \frac{3}{8} \div \frac{15}{16} \qquad \text{The divisor is } \frac{15}{16}. \text{ Invert } \frac{15}{16} \text{ and multiply.}$$

$$= \frac{\overset{1}{\cancel{3}}}{\underset{1}{\cancel{8}}} \cdot \frac{\overset{2}{\cancel{16}}}{\underset{5}{\cancel{15}}} = \frac{1 \cdot 2}{1 \cdot 5} = \frac{2}{5}$$

2. $\dfrac{\frac{4}{9}}{6}$ Write 6 as $\frac{6}{1}$ and divide.

$$\frac{\frac{4}{9}}{\frac{6}{1}} = \frac{4}{9} \div \frac{6}{1}$$

$$= \frac{\overset{2}{\cancel{4}}}{9} \cdot \frac{1}{\underset{3}{\cancel{6}}} = \frac{2 \cdot 1}{9 \cdot 3} = \frac{2}{27}$$

3. $\dfrac{5 + \frac{3}{4}}{46}$ Simplify the numerator.

$$\frac{\dfrac{4 \cdot 5 + 3}{4}}{46} = \frac{\dfrac{20 + 3}{4}}{46} = \frac{\dfrac{23}{4}}{46} \qquad \textbf{Write 46 as } \frac{46}{1}.$$

$$\frac{\dfrac{23}{4}}{\dfrac{46}{1}} = \frac{23}{4} \div \frac{46}{1}$$

$$= \frac{\overset{1}{\cancel{23}}}{4} \cdot \frac{1}{\underset{2}{\cancel{46}}} = \frac{1 \cdot 1}{4 \cdot 2} = \frac{1}{8}$$

4. $\dfrac{\dfrac{1}{4} + \dfrac{3}{8}}{\dfrac{1}{2} + \dfrac{13}{24}} = \dfrac{\dfrac{2}{8} + \dfrac{3}{8}}{\dfrac{12}{24} + \dfrac{13}{24}} = \dfrac{\dfrac{2 + 3}{8}}{\dfrac{12 + 13}{24}} = \dfrac{\dfrac{5}{8}}{\dfrac{25}{24}} = \dfrac{5}{8} \div \dfrac{25}{24}$

$$\frac{5}{8} \div \frac{25}{24} = \frac{\overset{1}{\cancel{5}}}{\underset{1}{\cancel{8}}} \cdot \frac{\overset{3}{\cancel{24}}}{\underset{5}{\cancel{25}}} = \frac{1 \cdot 3}{1 \cdot 5} = \frac{3}{5}$$

5. $\dfrac{4 + \dfrac{5}{6}}{7 - \dfrac{1}{3}} = \dfrac{\dfrac{4 \cdot 6 + 5}{6}}{\dfrac{7 \cdot 3 - 1}{3}} = \dfrac{\dfrac{29}{6}}{\dfrac{20}{3}} = \dfrac{29}{6} \div \dfrac{20}{3}$

$$= \frac{29}{\underset{2}{\cancel{6}}} \cdot \frac{\overset{1}{\cancel{3}}}{20} = \frac{29}{40}$$

6. $\dfrac{11 + \dfrac{3}{10}}{4\dfrac{4}{5}} = \dfrac{\dfrac{11 \cdot 10 + 3}{10}}{\dfrac{4 \cdot 5 + 4}{5}} = \dfrac{\dfrac{110 + 3}{10}}{\dfrac{20 + 4}{5}} = \dfrac{\dfrac{113}{10}}{\dfrac{24}{5}} = \dfrac{113}{10} \div \dfrac{24}{5}$

$$\frac{113}{10} \div \frac{24}{5} = \frac{113}{\underset{2}{\cancel{10}}} \cdot \frac{\overset{1}{\cancel{5}}}{24} = \frac{113 \cdot 1}{2 \cdot 24} = \frac{113}{48} = 2\frac{17}{48}$$

★ PRACTICE SET A

Convert each of the following complex fractions to a simple fraction.

1. $\dfrac{\dfrac{4}{9}}{\dfrac{8}{15}}$ **2.** $\dfrac{\dfrac{7}{10}}{28}$ **3.** $\dfrac{5 + \dfrac{2}{5}}{3 + \dfrac{3}{5}}$ **4.** $\dfrac{\dfrac{1}{8} + \dfrac{7}{8}}{6 - \dfrac{3}{10}}$ **5.** $\dfrac{\dfrac{1}{6} + \dfrac{5}{8}}{\dfrac{5}{9} - \dfrac{1}{4}}$ **6.** $\dfrac{16 - 10\dfrac{2}{3}}{11\dfrac{5}{6} - 7\dfrac{7}{6}}$

Answers to the Practice Set are on p. 230.

Section 5.5 EXERCISES

For problems 1–23, simplify each fraction.

1. $\dfrac{\frac{3}{5}}{\frac{9}{15}}$

2. $\dfrac{\frac{1}{3}}{\frac{1}{9}}$

3. $\dfrac{\frac{1}{4}}{\frac{5}{12}}$

4. $\dfrac{\frac{8}{9}}{\frac{4}{15}}$

5. $\dfrac{6+\frac{1}{4}}{11+\frac{1}{4}}$

6. $\dfrac{2+\frac{1}{2}}{7+\frac{1}{2}}$

7. $\dfrac{5+\frac{1}{3}}{2+\frac{2}{15}}$

8. $\dfrac{9+\frac{1}{2}}{1+\frac{8}{11}}$

9. $\dfrac{4+\frac{10}{13}}{\frac{12}{39}}$

10. $\dfrac{\frac{1}{3}+\frac{2}{7}}{\frac{26}{21}}$

11. $\dfrac{\frac{5}{6}-\frac{1}{4}}{\frac{1}{12}}$

12. $\dfrac{\dfrac{3}{10} + \dfrac{4}{12}}{\dfrac{19}{90}}$

17. $\dfrac{\dfrac{9}{70} + \dfrac{5}{42}}{\dfrac{13}{30} - \dfrac{1}{21}}$

13. $\dfrac{\dfrac{9}{16} + \dfrac{7}{3}}{\dfrac{139}{48}}$

18. $\dfrac{\dfrac{1}{16} + \dfrac{1}{14}}{\dfrac{2}{3} - \dfrac{13}{60}}$

14. $\dfrac{\dfrac{1}{288}}{\dfrac{8}{9} - \dfrac{3}{16}}$

19. $\dfrac{\dfrac{3}{20} + \dfrac{11}{12}}{\dfrac{19}{7} - 1\dfrac{11}{35}}$

15. $\dfrac{\dfrac{27}{429}}{\dfrac{5}{11} - \dfrac{1}{13}}$

20. $\dfrac{2\dfrac{2}{3} - 1\dfrac{1}{2}}{\dfrac{1}{4} + 1\dfrac{1}{16}}$

16. $\dfrac{\dfrac{1}{3} + \dfrac{2}{5}}{\dfrac{3}{5} + \dfrac{17}{45}}$

21. $\dfrac{3\dfrac{1}{5} + 3\dfrac{1}{3}}{\dfrac{6}{5} - \dfrac{15}{63}}$

22. $\dfrac{\dfrac{1\frac{1}{2}+15}{5\frac{1}{4}-3\frac{5}{12}}}{\dfrac{8\frac{1}{3}-4\frac{1}{2}}{11\frac{2}{3}-5\frac{11}{12}}}$

23. $\dfrac{\dfrac{5\frac{3}{4}+3\frac{1}{5}}{2\frac{1}{5}+15\frac{7}{10}}}{\dfrac{9\frac{1}{2}-4\frac{1}{6}}{\frac{1}{8}+2\frac{1}{120}}}$

EXERCISES FOR REVIEW

(3.3)   **24.** Find the prime factorization of 882.

(4.2)   **25.** Convert $\dfrac{62}{7}$ to a mixed number.

(4.3)   **26.** Reduce $\dfrac{114}{342}$ to lowest terms.

(5.3)   **27.** Find the value of $6\frac{3}{8}-4\frac{5}{6}$.

(5.4)   **28.** Arrange from smallest to largest: $\dfrac{1}{2},\dfrac{3}{5},\dfrac{4}{7}$.

★ **Answers to Practice Set (5.5)**

A. **1.** $\dfrac{5}{6}$ **2.** $\dfrac{1}{40}$ **3.** $\dfrac{3}{2}$ **4.** $\dfrac{10}{57}$ **5.** $2\frac{13}{22}$ **6.** $1\frac{5}{11}$

5.6 Combinations of Operations with Fractions

Section Overview

☐ **THE ORDER OF OPERATIONS**

☐ THE ORDER OF OPERATIONS

To determine the value of a quantity such as

$$\frac{1}{4}+\frac{5}{8}\cdot\frac{2}{15}$$

where we have a combination of operations (more than one operation occurs), we must use the accepted order of operations.

The Order of Operations

The order of operations:

1. In the order (2), (3), (4) described below, perform all operations inside grouping symbols: (   ), [   ], {   }, —. Work from the innermost set to the outermost set.
2. Perform exponential and root operations.
3. Perform all multiplications and divisions moving left to right.
4. Perform all additions and subtractions moving left to right.

☆ SAMPLE SET A

Determine the value of each of the following quantities.

1. $\dfrac{1}{4} + \dfrac{5}{8} \cdot \dfrac{2}{15}$

   (a) Multiply first.

$$\frac{1}{4} + \frac{\overset{1}{\cancel{5}}}{\underset{4}{\cancel{8}}} \cdot \frac{\overset{1}{\cancel{2}}}{\underset{3}{\cancel{15}}} = \frac{1}{4} + \frac{1 \cdot 1}{4 \cdot 3} = \frac{1}{4} + \frac{1}{12}$$

   (b) Now perform this addition. Find the LCD.

$$\left.\begin{array}{l} 4 = 2^2 \\ 12 = 2^2 \cdot 3 \end{array}\right\} \qquad \textbf{The LCD} = 2^2 \cdot 3 = 12.$$

$$\frac{1}{4} + \frac{1}{12} = \frac{1 \cdot 3}{12} + \frac{1}{12} = \frac{3}{12} + \frac{1}{12}$$

$$= \frac{3+1}{12} = \frac{4}{12} = \frac{1}{3}$$

   Thus, $\dfrac{1}{4} + \dfrac{5}{8} \cdot \dfrac{2}{15} = \dfrac{1}{3}$

2. $\dfrac{3}{5} + \dfrac{9}{44}\left(\dfrac{5}{9} - \dfrac{1}{4}\right)$

   (a) Operate within the parentheses first, $\left(\dfrac{5}{9} - \dfrac{1}{4}\right)$.

$$\left.\begin{array}{l} 9 = 3^2 \\ 4 = 2^2 \end{array}\right\} \qquad \textbf{The LCD} = 2^2 \cdot 3^2 = 4 \cdot 9 = 36.$$

$$\frac{5 \cdot 4}{36} - \frac{1 \cdot 9}{36} = \frac{20}{36} - \frac{9}{36} = \frac{20 - 9}{36} = \frac{11}{36}$$

   Now we have

$$\frac{3}{5} + \frac{9}{44}\left(\frac{11}{36}\right)$$

   (b) Perform the multiplication.

$$\frac{3}{5} + \frac{\overset{1}{\cancel{9}}}{\underset{4}{\cancel{44}}} \cdot \frac{\overset{1}{\cancel{11}}}{\underset{4}{\cancel{36}}} = \frac{3}{5} + \frac{1 \cdot 1}{4 \cdot 4} = \frac{3}{5} + \frac{1}{16}$$

Continued

(c) Now perform the addition. The LCD = 80.

$$\frac{3}{5} + \frac{1}{16} = \frac{3 \cdot 16}{80} + \frac{1 \cdot 5}{80} = \frac{48}{80} + \frac{5}{80} = \frac{48 + 5}{80} = \frac{53}{80}$$

Thus, $\frac{3}{5} + \frac{9}{44}\left(\frac{5}{9} - \frac{1}{4}\right) = \frac{53}{80}$

3. $8 - \frac{15}{426}\left(2 - 1\frac{4}{15}\right)\left(3\frac{1}{5} + 2\frac{1}{8}\right)$

(a) Work within each set of parentheses individually.

$$2 - 1\frac{4}{15} = 2 - \frac{1 \cdot 15 + 4}{15} = 2 - \frac{19}{15}$$

$$= \frac{30}{15} - \frac{19}{15} = \frac{30 - 19}{15} = \frac{11}{15}$$

$$3\frac{1}{5} + 2\frac{1}{8} = \frac{3 \cdot 5 + 1}{5} + \frac{2 \cdot 8 + 1}{8}$$

$$= \frac{16}{5} + \frac{17}{8} \qquad \text{LCD} = 40.$$

$$= \frac{16 \cdot 8}{40} + \frac{17 \cdot 5}{40}$$

$$= \frac{128}{40} + \frac{85}{40}$$

$$= \frac{128 + 85}{40}$$

$$= \frac{213}{40}$$

Now we have

$$8 - \frac{15}{426}\left(\frac{11}{15}\right)\left(\frac{213}{40}\right)$$

(b) Now multiply.

$$8 - \frac{\overset{1}{\cancel{15}}}{\underset{2}{\cancel{426}}} \cdot \frac{11}{\underset{1}{\cancel{15}}} \cdot \frac{\overset{1}{\cancel{213}}}{40} = 8 - \frac{1 \cdot 11 \cdot 1}{2 \cdot 1 \cdot 40} = 8 - \frac{11}{80}$$

(c) Now subtract.

$$8 - \frac{11}{80} = \frac{80 \cdot 8}{80} - \frac{11}{80} = \frac{640}{80} - \frac{11}{80} = \frac{640 - 11}{80} = \frac{629}{80} \text{ or } 7\frac{69}{80}$$

Thus, $8 - \frac{15}{426}\left(2 - 1\frac{4}{15}\right)\left(3\frac{1}{5} + 2\frac{1}{8}\right) = 7\frac{69}{80}$

4. $\left(\frac{3}{4}\right)^2 \cdot \frac{8}{9} - \frac{5}{12}$

(a) Square $\frac{3}{4}$.

$$\left(\frac{3}{4}\right)^2 = \frac{3}{4} \cdot \frac{3}{4} = \frac{3 \cdot 3}{4 \cdot 4} = \frac{9}{16}$$

Now we have

$$\frac{9}{16} \cdot \frac{8}{9} - \frac{5}{12}$$

(b) Perform the multiplication.

$$\frac{\overset{1}{\cancel{9}}}{\underset{2}{\cancel{16}}} \cdot \frac{\overset{1}{\cancel{8}}}{\underset{1}{\cancel{9}}} - \frac{5}{12} = \frac{1 \cdot 1}{2 \cdot 1} - \frac{5}{12} = \frac{1}{2} - \frac{5}{12}$$

(c) Now perform the subtraction.

$$\frac{1}{2} - \frac{5}{12} = \frac{6}{12} - \frac{5}{12} = \frac{6-5}{12} = \frac{1}{12}$$

Thus, $\left(\frac{4}{3}\right)^2 \cdot \frac{8}{9} - \frac{5}{12} = \frac{1}{12}$

5. $2\frac{7}{8} + \sqrt{\frac{25}{36}} \div \left(2\frac{1}{2} - 1\frac{1}{3}\right)$

(a) Begin by operating inside the parentheses.

$$2\frac{1}{2} - 1\frac{1}{3} = \frac{2 \cdot 2 + 1}{2} - \frac{1 \cdot 3 + 1}{3} = \frac{5}{2} - \frac{4}{3}$$

$$= \frac{15}{6} - \frac{8}{6} = \frac{15-8}{6} = \frac{7}{6}$$

(b) Now simplify the square root.

$$\sqrt{\frac{25}{36}} - \frac{5}{6} \quad \left(\text{since } \left(\frac{5}{6}\right)^2 = \frac{25}{36}\right)$$

Now we have

$$2\frac{7}{8} + \frac{5}{6} \div \frac{7}{6}$$

(c) Perform the division.

$$2\frac{7}{8} + \frac{5}{\underset{1}{\cancel{6}}} \cdot \frac{\overset{1}{\cancel{6}}}{7} = 2\frac{7}{8} + \frac{5 \cdot 1}{1 \cdot 7} = 2\frac{7}{8} + \frac{5}{7}$$

(d) Now perform the addition.

$$2\frac{7}{8} + \frac{5}{7} = \frac{2 \cdot 8 + 7}{8} + \frac{5}{7} = \frac{23}{8} + \frac{5}{7} \qquad \textbf{LCD = 56.}$$

$$= \frac{23 \cdot 7}{56} + \frac{5 \cdot 8}{56} = \frac{161}{56} + \frac{40}{56}$$

$$= \frac{161 + 40}{56} = \frac{201}{56} \text{ or } 3\frac{33}{56}$$

Thus, $2\frac{7}{8} + \sqrt{\frac{25}{36}} \div \left(2\frac{1}{2} - 1\frac{1}{3}\right) = 3\frac{33}{56}$

★ PRACTICE SET A

Find the value of each of the following quantities.

1. $\dfrac{5}{16} \cdot \dfrac{1}{10} - \dfrac{1}{32}$

2. $\dfrac{6}{7} \cdot \dfrac{21}{40} \div \dfrac{9}{10} + 5\dfrac{1}{3}$

3. $8\dfrac{7}{10} - 2\left(4\dfrac{1}{2} - 3\dfrac{2}{3}\right)$

4. $\dfrac{17}{18} - \dfrac{58}{30}\left(\dfrac{1}{4} - \dfrac{3}{32}\right)\left(1 - \dfrac{13}{29}\right)$

5. $\left(\dfrac{1}{10} + 1\dfrac{1}{2}\right) \div \left(1\dfrac{4}{5} - 1\dfrac{6}{25}\right)$

6. $\dfrac{\dfrac{2}{3} - \dfrac{3}{8} \cdot \dfrac{4}{9}}{\dfrac{7}{16} \cdot 1\dfrac{1}{3} + 1\dfrac{1}{4}}$

7. $\left(\dfrac{3}{8}\right)^2 + \dfrac{3}{4} \cdot \dfrac{1}{8}$

8. $\dfrac{2}{3} \cdot 2\dfrac{1}{4} - \sqrt{\dfrac{4}{25}}$

Answers to the Practice Set are on p. 237.

Section 5.6 EXERCISES

For problems 1–25, find each value.

1. $\dfrac{4}{3} - \dfrac{1}{6} \cdot \dfrac{1}{2}$

3. $2\dfrac{2}{7} + \dfrac{5}{8} \div \dfrac{5}{16}$

2. $\dfrac{7}{9} - \dfrac{4}{5} \cdot \dfrac{5}{36}$

4. $\dfrac{3}{16} \div \dfrac{9}{14} \cdot \dfrac{12}{21} + \dfrac{5}{6}$

5. $\dfrac{4}{25} \div \dfrac{8}{15} - \dfrac{7}{20} \div 2\dfrac{1}{10}$

11. $\left(\dfrac{1}{2}\right)^2 + \dfrac{1}{8}$

6. $\dfrac{2}{5} \cdot \left(\dfrac{1}{19} + \dfrac{3}{38}\right)$

12. $\left(\dfrac{3}{5}\right)^2 - \dfrac{3}{10}$

7. $\dfrac{3}{7} \cdot \left(\dfrac{3}{10} - \dfrac{1}{15}\right)$

8. $\dfrac{10}{11} \cdot \left(\dfrac{8}{9} - \dfrac{2}{5}\right) + \dfrac{3}{25} \cdot \left(\dfrac{5}{3} + \dfrac{1}{4}\right)$

13. $\sqrt{\dfrac{36}{81}} + \dfrac{1}{3} \cdot \dfrac{2}{9}$

9. $\dfrac{2}{7} \cdot \left(\dfrac{6}{7} - \dfrac{3}{28}\right) + 5\dfrac{1}{3} \cdot \left(1\dfrac{1}{4} - \dfrac{1}{8}\right)$

14. $\sqrt{\dfrac{49}{64}} - \sqrt{\dfrac{9}{4}}$

10. $\dfrac{\left(\dfrac{6}{11} - \dfrac{1}{3}\right) \cdot \left(\dfrac{1}{21} + 2\dfrac{13}{42}\right)}{1\dfrac{1}{5} + \dfrac{7}{40}}$

15. $\dfrac{2}{3} \cdot \sqrt{\dfrac{9}{4}} - \dfrac{15}{4} \cdot \sqrt{\dfrac{16}{225}}$

16. $\left(\dfrac{3}{4}\right)^2 + \sqrt{\dfrac{25}{16}}$

17. $\left(\dfrac{1}{3}\right)^2 \cdot \sqrt{\dfrac{81}{25}} + \dfrac{1}{40} \div \dfrac{1}{8}$

18. $\left(\sqrt{\dfrac{4}{49}}\right)^2 + \dfrac{3}{7} \div 1\dfrac{3}{4}$

19. $\left(\sqrt{\dfrac{100}{121}}\right)^2 + \dfrac{21}{(11)^2}$

20. $\sqrt{\dfrac{3}{8} + \dfrac{1}{64}} - \dfrac{1}{2} \div 1\dfrac{1}{3}$

21. $\sqrt{\dfrac{1}{4}} \cdot \left(\dfrac{5}{6}\right)^2 + \dfrac{9}{14} \cdot 2\dfrac{1}{3} - \sqrt{\dfrac{1}{81}}$

22. $\sqrt{\dfrac{1}{9}} \cdot \sqrt{\dfrac{6\dfrac{3}{8} + 2\dfrac{5}{8}}{16}} + 7\dfrac{7}{10}$

23. $\dfrac{3\dfrac{3}{4} + \dfrac{4}{5} \cdot \left(\dfrac{1}{2}\right)^3}{\dfrac{67}{240} + \left(\dfrac{1}{3}\right)^4 \cdot \left(\dfrac{9}{10}\right)}$

24. $\sqrt{\sqrt{\dfrac{16}{81}} + \dfrac{1}{4} \cdot 6}$

25. $\sqrt{\sqrt{\dfrac{81}{256}} - \dfrac{3}{32} \cdot 1\dfrac{1}{8}}$

EXERCISES FOR REVIEW

(1.1) **26.** True or false: Our number system, the Hindu-Arabic number system, is a positional number system with base ten.

(2.5) **27.** The fact that

1 times any whole number = that particular whole number

illustrates which property of multiplication?

(4.2) **28.** Convert $8\frac{6}{7}$ to an improper fraction.

(5.2) **29.** Find the sum. $\frac{3}{8} + \frac{4}{5} + \frac{5}{6}$.

(5.5) **30.** Simplify $\dfrac{6 + \dfrac{1}{8}}{6 - \dfrac{1}{8}}$.

★ **Answers to Practice Set (5.6)**

A. **1.** 0   **2.** $\frac{35}{6}$ or $5\frac{5}{6}$ **3.** $\frac{211}{30}$ or $7\frac{1}{30}$ **4.** $\frac{7}{9}$ **5.** $2\frac{6}{7}$ **6.** $\frac{3}{11}$ **7.** $\frac{15}{64}$ **8.** $\frac{11}{10}$

Addition and Subtraction of Fractions with Like Denominators (5.1)

To *add or subtract two fractions that have the same denominators,* add or subtract the numerators and place the resulting sum or difference over the common denominator. Reduce, if necessary. Do not add or subtract the denominators.

$$\frac{1}{8} + \frac{5}{8} = \frac{1+5}{8} = \frac{6}{8} = \frac{3}{4}$$

Basic Rule for Adding and Subtracting Fractions (5.2)

Fractions can be added or subtracted conveniently only if they have like denominators.

Addition and Subtraction of Fractions with Unlike Denominators (5.2)

To *add or subtract fractions having unlike denominators,* convert each fraction to an equivalent fraction having as denominator the LCD of the original denominators.

Addition and Subtraction of Mixed Numbers (5.3)

1. To *add or subtract mixed numbers,* convert each mixed number to an improper fraction, then add or subtract the fractions.

Ordered Number System (5.4)

Our number system is *ordered* because the numbers in the system can be placed in order from smaller to larger.

Inequality Symbols (5.4)

$>$ represents the phrase "greater than."
$<$ represents the phrase "less than."

Comparing Fractions (5.4)

If two fractions have the same denominators, the fraction with the larger numerator is the larger fraction.

$$\frac{5}{8} > \frac{3}{8}$$

Simple Fractions (5.5)

A *simple fraction* is any fraction in which the numerator is any whole number and the denominator is any nonzero whole number.

Complex Fractions (5.5)

A *complex fraction* is any fraction in which the numerator and/or the denominator is a fraction.

Complex fractions can be converted to simple fractions by employing the methods of adding, subtracting, multiplying, and dividing fractions.

EXERCISE SUPPLEMENT

For problems 1–53, perform each indicated operation and write the result in simplest form.

1. $\dfrac{3}{4} + \dfrac{5}{8}$

2. $\dfrac{9}{16} + \dfrac{1}{4}$

3. $\dfrac{1}{8} + \dfrac{3}{8}$

4. $\dfrac{5}{7} + \dfrac{1}{14} + \dfrac{5}{21}$

5. $\dfrac{5}{6} + \dfrac{1}{3} + \dfrac{5}{21}$

6. $\dfrac{2}{5} + \dfrac{1}{8}$

7. $\dfrac{1}{4} + \dfrac{1}{8} + \dfrac{1}{4}$

8. $\dfrac{1}{16} + \dfrac{1}{10}$

9. $\dfrac{2}{7} + \dfrac{1}{3}$

10. $2\dfrac{1}{3} + \dfrac{1}{6}$

11. $3\dfrac{11}{16} + \dfrac{3}{4}$

12. $5\dfrac{1}{12} + 3\dfrac{1}{8}$

13. $16\dfrac{2}{5} - 8\dfrac{1}{4}$

14. $1\dfrac{1}{7} + 2\dfrac{4}{7}$

15. $1\dfrac{3}{8} + 0$

16. $3\dfrac{1}{10} + 4$

17. $18\dfrac{2}{3} + 6$

18. $1\dfrac{4}{3} + 5\dfrac{5}{4}$

19. $\dfrac{21}{4} + \dfrac{2}{3}$

20. $\dfrac{15}{16} - \dfrac{1}{8}$

21. $\dfrac{9}{11} - \dfrac{5}{22}$

22. $6\dfrac{2}{15} - 1\dfrac{3}{10}$

23. $5\dfrac{2}{3} + 8\dfrac{1}{5} - 2\dfrac{1}{4}$

24. $8\dfrac{3}{10} - 4\dfrac{5}{6} - 3\dfrac{1}{15}$

25. $\dfrac{11}{12} + \dfrac{1}{9} - \dfrac{1}{16}$

26. $7\dfrac{2}{9} - 5\dfrac{5}{6} - 1\dfrac{1}{3}$

27. $16\dfrac{2}{5} - 8\dfrac{1}{6} - 3\dfrac{2}{15}$

28. $\dfrac{3}{100} + \dfrac{4}{10} - \dfrac{1}{1000}$

29. $4\dfrac{1}{8} + 0 - \dfrac{32}{8}$

30. $8 - 2\dfrac{1}{3}$

31. $4 - 3\dfrac{5}{16}$

32. $6\dfrac{3}{7} + 4$

33. $11\dfrac{2}{11} - 3$

34. $21\dfrac{5}{8} - \dfrac{5}{8}$

35. $\dfrac{3}{4} + \dfrac{5}{16} \cdot \dfrac{4}{5}$

36. $\dfrac{11}{12} + \dfrac{15}{16} \div 2\dfrac{1}{2}$

37. $1\dfrac{3}{10} + 2\dfrac{2}{3} \div \dfrac{4}{9}$

38. $8\dfrac{3}{5} - 1\dfrac{1}{14} \cdot \dfrac{3}{7}$

39. $2\dfrac{3}{8} \div 3\dfrac{9}{16} - \dfrac{1}{9}$

40. $15\dfrac{2}{5} \div 50 - \dfrac{1}{10}$

Sections 5.5 and 5.6

41. $\dfrac{\dfrac{9}{16}}{\dfrac{21}{32}}$

42. $\dfrac{\dfrac{10}{21}}{\dfrac{11}{14}}$

43. $\dfrac{1\dfrac{7}{9}}{1\dfrac{5}{27}}$

44. $\dfrac{\dfrac{15}{17}}{\dfrac{50}{51}}$

45. $\dfrac{1\dfrac{9}{16}}{2\dfrac{11}{12}}$

46. $\dfrac{8\dfrac{4}{15}}{3}$

47. $\dfrac{9\dfrac{1}{18}}{6}$

48. $\dfrac{3\dfrac{1}{4} + 2\dfrac{1}{8}}{5\dfrac{1}{6}}$

49. $\dfrac{3 + 2\dfrac{1}{2}}{\dfrac{1}{4} + \dfrac{5}{6}}$

50. $\dfrac{4 + 1\frac{7}{10}}{9 - 2\frac{1}{5}}$

51. $\dfrac{1\frac{2}{5}}{9 - \frac{2}{2}}$

52. $\dfrac{1\frac{2}{3} \cdot \left(\frac{1}{4} + \frac{1}{5}\right)}{1\frac{1}{2}}$

53. $\dfrac{\frac{10}{23} \cdot \left(\frac{5}{6} + 2\right)}{\frac{8}{9}}$

Section 5.4

For problems 54 – 65, place each collection of fractions in order from smallest to largest.

54. $\dfrac{1}{8}, \dfrac{3}{16}$

55. $\dfrac{3}{32}, \dfrac{1}{8}$

56. $\dfrac{5}{16}, \dfrac{3}{24}$

57. $\dfrac{3}{10}, \dfrac{5}{6}$

58. $\dfrac{2}{9}, \dfrac{1}{3}, \dfrac{1}{6}$

59. $\dfrac{3}{8}, \dfrac{8}{3}, \dfrac{19}{6}$

60. $\dfrac{3}{5}, \dfrac{2}{10}, \dfrac{7}{20}$

61. $\dfrac{4}{7}, \dfrac{5}{9}$

62. $\dfrac{4}{5}, \dfrac{5}{7}$

63. $\dfrac{5}{12}, \dfrac{4}{9}, \dfrac{7}{15}$

64. $\dfrac{7}{36}, \dfrac{1}{24}, \dfrac{5}{12}$

65. $\dfrac{5}{8}, \dfrac{13}{16}, \dfrac{3}{4}$

For problems 1–12, perform each indicated operation and write the result in simplest form.

1. _____ $-\dfrac{1}{4}$

1. **(5.2)** $\dfrac{3}{16} + \dfrac{1}{8} = \dfrac{4}{16} = \dfrac{1}{4}$

2. _____ $7\dfrac{1}{2}$

2. **(5.3)** $2\dfrac{2}{3} + 5\dfrac{1}{6} = 7\dfrac{3}{6} = 7\dfrac{1}{2}$

3. _____

3. **(5.6)** $\dfrac{7}{15} \cdot \dfrac{20}{21} + \dfrac{5}{9}$

4. _____ $8/11$

4. **(5.1)** $\dfrac{3}{11} + \dfrac{5}{11}$

5. _____

5. **(5.6)** $6\dfrac{2}{9} \cdot 1\dfrac{17}{28} - \left(3\dfrac{4}{17} - \dfrac{21}{17}\right)$

6. _____

6. **(5.3)** $5\dfrac{1}{8} - 2\dfrac{4}{5}$

7. _____

7. **(5.5)** $\dfrac{\dfrac{7}{12}}{\dfrac{8}{21}}$

243

8. _____

8. **(5.5)** $\dfrac{\dfrac{1}{8}+\dfrac{3}{4}}{1\dfrac{7}{8}}$

9. _____

9. **(5.3)** $4\dfrac{5}{16}+1\dfrac{1}{3}-2\dfrac{5}{24}$

10. _____

10. **(5.6)** $\dfrac{5}{18}\cdot\left(\dfrac{15}{16}-\dfrac{3}{8}\right)$

11. _____

11. **(5.3)** $4+2\dfrac{1}{3}$

12. _____

12. **(5.3)** $8\dfrac{3}{7}-5$

For problems 13–15, specify the fractions that are equivalent.

13. _____

13. **(5.4)** $\dfrac{4}{5},\dfrac{12}{15}$

14. _____

14. **(5.4)** $\dfrac{5}{8},\dfrac{24}{40}$

244

15. _____   **15. (5.4)** $\dfrac{5}{12}, \dfrac{80}{192}$

For problems 16–20, place each collection of fractions in order from smallest to largest.

16. _____   **16. (5.4)** $\dfrac{8}{9}, \dfrac{6}{7}$

17. _____   **17. (5.4)** $\dfrac{5}{8}, \dfrac{7}{9}$

18. _____   **18. (5.4)** $11\dfrac{5}{16}, 11\dfrac{5}{12}$

19. _____   **19. (5.4)** $\dfrac{2}{15}, \dfrac{3}{10}, \dfrac{1}{6}$

20. _____   **20. (5.4)** $\dfrac{19}{32}, \dfrac{9}{16}, \dfrac{5}{8}$

6 Decimals

After completing this chapter, you should

Section 6.1 Reading and Writing Decimals
- understand the meaning of digits occurring to the right of the ones position
- be familiar with the meaning of decimal fractions
- be able to read and write a decimal fraction

Section 6.2 Converting a Decimal to a Fraction
- be able to convert an ordinary decimal and a complex decimal to a fraction

Section 6.3 Rounding Decimals
- be able to round a decimal number to a specified position

Section 6.4 Addition and Subtraction of Decimals
- understand the method used for adding and subtracting decimals
- be able to add and subtract decimals
- be able to use the calculator to add and subtract decimals

Section 6.5 Multiplication of Decimals
- understand the method used for multiplying decimals
- be able to multiply decimals
- be able to simplify a multiplication of a decimal by a power of 10
- understand how to use the word "of" in multiplication

Section 6.6 Division of Decimals
- understand the method used for dividing decimals
- be able to divide a decimal number by a nonzero whole number and by another, nonzero, decimal number
- be able to simplify a division of a decimal by a power of 10

Section 6.7 Nonterminating Divisions
- understand the meaning of a nonterminating division
- be able to recognize a nonterminating number by its notation

Section 6.8 Converting a Fraction to a Decimal
- be able to convert a fraction to a decimal

Section 6.9 Combinations of Operations with Decimals and Fractions
- be able to combine operations with decimals

6.1 Reading and Writing Decimals

Section
Overview

- ❑ **DIGITS TO THE RIGHT OF THE ONES POSITION**
- ❑ **DECIMAL FRACTIONS**
- ❑ **READING DECIMAL FRACTIONS**
- ❑ **WRITING DECIMAL FRACTIONS**

❑ DIGITS TO THE RIGHT OF THE ONES POSITION

We began our study of arithmetic (Section 1.1) by noting that our number system is called a positional number system with base ten. We also noted that each position has a particular value. We observed that each position has ten times the value of the position to its right.

$$10 \times 100,000 \quad 10 \times 10,000 \quad 10 \times 1,000 \quad 10 \times 100 \quad 10 \times 10 \quad 10 \times 1 \quad 1$$

This means that each position has $\dfrac{1}{10}$ the value of the position to its left.

$$1,000,000 \quad \frac{1}{10} \times 1,000,000 \quad \frac{1}{10} \times 100,000 \quad \frac{1}{10} \times 10,000 \quad \frac{1}{10} \times 1,000 \quad \frac{1}{10} \times 100 \quad \frac{1}{10} \times 10$$

Thus, a digit written to the right of the units position must have a value of $\dfrac{1}{10}$ of 1. Recalling that the word "of" translates to multiplication ($\cdot$), we can see that the value of the *first position* to the right of the units digit is $\dfrac{1}{10}$ of 1, or

$$\frac{1}{10} \cdot 1 = \frac{1}{10}$$

The value of the *second position* to the right of the units digit is $\dfrac{1}{10}$ of $\dfrac{1}{10}$, or

$$\frac{1}{10} \cdot \frac{1}{10} = \frac{1}{10^2} = \frac{1}{100}$$

The value of the third position to the right of the units digit is $\dfrac{1}{10}$ of $\dfrac{1}{100}$, or

$$\frac{1}{10} \cdot \frac{1}{100} = \frac{1}{10^3} = \frac{1}{1000}$$

This pattern continues.

We can now see that if we were to write digits in positions to the right of the units position, those positions have values that are fractions. Not only do the positions have fractional values, but the fractional values are all powers of 10 $(10, 10^2, 10^3, \ldots)$.

❑ DECIMAL FRACTIONS

Decimal Point

Decimal

If we are to write numbers with digits appearing to the right of the units digit, we must have a way of denoting where the whole number part ends and the fractional part begins. Mathematicians denote the separation point of the units digit and the tenths digit by writing a **decimal point.** The word *decimal* comes from the Latin prefix "deci" which means ten, and we use it because we use a base ten number system. Numbers written in this form are called **decimal fractions,** or more simply, **decimals.**

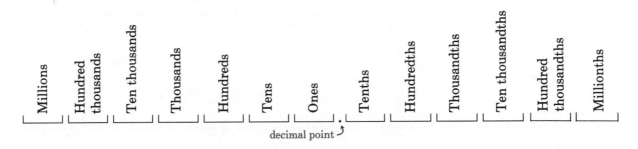

Notice that decimal numbers have the suffix "th."

Decimal Fraction

> A **decimal fraction** is a fraction in which the denominator is a power of 10.

The following numbers are examples of decimals.

1. 42.6

   The 6 is in the tenths position.

   $$42.6 = 42\frac{6}{10}$$

2. 9.8014

   The 8 is in the tenths position.
   The 0 is in the hundredths position.
   The 1 is in the thousandths position.
   The 4 is in the ten thousandths position.

   $$9.8014 = 9\frac{8014}{10,000}$$

3. 0.93

   The 9 is in the tenths position.
   The 3 is in the hundredths position.

   $$0.93 = \frac{93}{100}$$

   Note: Quite often a zero is inserted in front of a decimal point (in the units position) of a decimal fraction that has a value less than one. This zero helps keep us from overlooking the decimal point.

4. 0.7

   The 7 is in the tenths position.

   $$0.7 = \frac{7}{10}$$

Note: We can insert zeros to the right of the right-most digit in a decimal fraction without changing the value of the number.

$$\frac{7}{10} = 0.7 = 0.70 = \frac{70}{100} = \frac{7}{10}$$

☐ READING DECIMAL FRACTIONS

Reading a Decimal Fraction

To read a decimal fraction,

1. Read the whole number part as usual. (If the whole number is less than 1, omit steps 1 and 2.)
2. Read the decimal point as the word "and."
3. Read the number to the right of the decimal point as if it were a whole number.
4. Say the name of the position of the last digit.

☆ SAMPLE SET A

Read the following numbers.

1. 6.8

6. ⌐8⌐ ← tenths position
six and eight tenths

Note: Some people read this as "six point eight." This phrasing gets the message across, but technically, "six *and* eight tenths" is the correct phrasing.

2. 14.116

14.11 ⌐6⌐ ← thousandths position
fourteen and one hundred sixteen thousandths

3. 0.0019

0.001 ⌐9⌐ ← ten thousandths position
nineteen ten thousandths

4. 81

Eighty-one

In this problem, the indication is that any whole number is a decimal fraction. Whole numbers are often called *decimal numbers*.

81 = 81.0

★ PRACTICE SET A

Read the following decimal fractions.

1. 12.9   **2.** 4.86   **3.** 7.00002   **4.** 0.030405

❏ WRITING DECIMAL FRACTIONS

Writing a Decimal Fraction

To write a decimal fraction,

1. Write the whole number part.
2. Write a decimal point for the word "and."
3. Write the decimal part of the number so that the right-most digit appears in the position indicated in the word name. If necessary, insert zeros to the right of the decimal point in order that the right-most digit appears in the correct position.

☆ SAMPLE SET B

Write each number.

1. Thirty-one and twelve hundredths.

The decimal position indicated is the hundredths position.

31.12

2. Two and three hundred-thousandths.

The decimal position indicated is the hundred thousandths. We'll need to insert enough zeros to the immediate right of the decimal point in order to locate the 3 in the correct position.

2.00003

3. Six thousand twenty-seven and one hundred four millionths.

The decimal position indicated is the millionths position. We'll need to insert enough zeros to the immediate right of the decimal point in order to locate the 4 in the correct position.

6,027.000104

4. Seventeen hundredths.

The decimal position indicated is the hundredths position.

0.17

★ PRACTICE SET B

Write each decimal fraction.

1. Three hundred six and forty-nine hundredths.

2. Nine and four thousandths.

3. Sixty-one millionths.

Answers to Practice Sets are on p. 253.

Section 6.1 EXERCISES

For problems 1–3, give the decimal name of the position of the given number in each decimal fraction.

1. 3.941

   9 is in the _____ position.

   4 is in the _____ position.

   1 is in the _____ position.

2. 17.1085

   1 is in the _____ position.

   0 is in the _____ position.

   8 is in the _____ position.

   5 is in the _____ position.

3. 652.3561927

   9 is in the _____ position.

   7 is in the _____ position.

For problems 4–10, read each decimal fraction by writing it.

4. 9.2

5. 8.1

6. 10.15

7. 55.06

8. 0.78

9. 1.904

10. 10.00011

For problems 11–20, write each decimal fraction.

11. Three and twenty one-hundredths.

12. Fourteen and sixty seven-hundredths.

13. One and eight tenths.

14. Sixty-one and five tenths.

15. Five hundred eleven and four thousandths.

16. Thirty-three and twelve ten-thousandths.

17. Nine hundred forty-seven thousandths.

18. Two millionths.

19. Seventy-one hundred-thousandths.

20. One and ten ten-millionths.

25. $4 \div 25$

26. $1 \div 50$

📟 **Calculator Problems**

For problems 21–30, perform each division using a calculator. Then write the resulting decimal using words.

27. $3 \div 16$

21. $3 \div 4$

22. $1 \div 8$

28. $15 \div 8$

23. $4 \div 10$

29. $11 \div 20$

24. $2 \div 5$

30. $9 \div 40$

EXERCISES FOR REVIEW

(1.3) **31.** Round 2,614 to the nearest ten.

(2.4) **32.** Is 691,428,471 divisible by 3?

(4.3) **33.** Determine the missing numerator.

$$\frac{3}{14} = \frac{?}{56}$$

(4.6) **34.** Find $\frac{3}{16}$ of $\frac{32}{39}$.

(5.6) **35.** Find the value of $\sqrt{\frac{25}{81}} + \left(\frac{2}{3}\right)^2 + \frac{1}{9}$.

★ **Answers to Practice Sets (6.1)**

A. 1. twelve and nine tenths   2. four and eighty-six hundredths
3. seven and two hundred thousandths   4. thirty thousand four hundred five millionths

B. 1. 306.49   2. 9.004   3. 0.000061

6.2 Converting a Decimal to a Fraction

☐ **CONVERTING AN ORDINARY DECIMAL TO A FRACTION**
☐ **CONVERTING A COMPLEX DECIMAL TO A FRACTION**

☐ CONVERTING AN ORDINARY DECIMAL TO A FRACTION

We can convert a decimal fraction to a fraction, essentially, by saying it in words, then writing what we say. We may have to reduce that fraction.

☆ SAMPLE SET A

Convert each decimal fraction to a proper fraction or a mixed number.

1. 0.6
  └── tenths position

Reading: six tenths → $\frac{6}{10}$.

Reduce: $\frac{3}{5}$.

2. 0.903
  └── thousandths position

Reading: nine hundred three thousandths → $\frac{903}{1000}$.

3. 18.61
  └── hundredths position

Reading: eighteen and sixty-one hundredths → $18\frac{61}{100}$.

4. 508.0005
  └── ten thousandths position

Reading: five hundred eight and five ten thousandths → $508\frac{5}{10,000}$.

Reduce: $508\frac{1}{2,000}$.

★ PRACTICE SET A

Convert the following decimals to fractions or mixed numbers. Be sure to reduce.

1. 16.84   **2.** 0.513   **3.** 6,646.0107   **4.** 1.1

☐ CONVERTING A COMPLEX DECIMAL TO A FRACTION

Complex Decimals

Numbers such as $0.11\frac{2}{3}$ are called **complex decimals.** We can also convert complex decimals to fractions.

☆ **SAMPLE SET B**

Convert the following complex decimals to fractions.

1. $0.11\frac{2}{3}$

The $\frac{2}{3}$ appears to occur in the thousands position, but it is referring to $\frac{2}{3}$ of a hundredth.

So, we read $0.11\frac{2}{3}$ as "eleven and two-thirds hundredths."

$$0.11\frac{2}{3} = \frac{11\frac{2}{3}}{100} = \frac{\frac{11 \cdot 3 + 2}{3}}{100}$$

$$= \frac{\frac{35}{3}}{\frac{100}{1}}$$

$$= \frac{35}{3} \div \frac{100}{1}$$

$$= \frac{\overset{7}{\cancel{35}}}{3} \cdot \frac{1}{\underset{20}{\cancel{100}}}$$

$$= \frac{7}{60}$$

2. $4.006\frac{1}{4}$

Note that $4.006\frac{1}{4} = 4 + .006\frac{1}{4}$.

$$4 + .006\frac{1}{4} = 4 + \frac{6\frac{1}{4}}{1000}$$

$$= 4 + \frac{\frac{25}{4}}{\frac{1000}{1}}$$

$$= 4 + \frac{25}{4} \cdot \frac{1}{\underset{40}{\cancel{1000}}}\overset{1}{}$$

$$= 4 + \frac{1 \cdot 1}{4 \cdot 40}$$

$$= 4 + \frac{1}{160}$$

$$= 4\frac{1}{160}$$

★ **PRACTICE SET B**

Convert each complex decimal to a fraction or mixed number. Be sure to reduce.

1. $0.8\frac{3}{4}$ **2.** $0.12\frac{2}{5}$ **3.** $6.005\frac{5}{6}$ **4.** $18.1\frac{3}{17}$

Answers to Practice Sets are on p. 257.

Section 6.2 EXERCISES

For problems 1–20, convert each decimal fraction to a proper fraction or a mixed number. Be sure to reduce.

1. 0.7   **2.** 0.1

3. 0.53   **4.** 0.71

5. 0.219   **6.** 0.811

7. 4.8   **8.** 2.6

9. 16.12   **10.** 25.88

11. 6.0005   **12.** 1.355

13. 16.125   **14.** 0.375

15. 3.04   **16.** 21.1875

17. 8.225   **18.** 1.0055

19. 9.99995   **20.** 22.110

For problems 21–30, convert each complex decimal to a fraction.

21. $0.7\frac{1}{2}$ **22.** $0.012\frac{1}{2}$

23. $2.16\frac{1}{4}$ **24.** $5.18\frac{2}{3}$

25. $14.112\frac{1}{3}$ **26.** $80.0011\frac{3}{7}$

27. $1.40\frac{5}{16}$ **28.** $0.8\frac{5}{3}$

29. $1.9\frac{7}{5}$ **30.** $1.7\frac{37}{9}$

EXERCISES FOR REVIEW

(3.5) **31.** Find the greatest common factor of 70, 182, and 154.

(3.5) **32.** Find the greatest common multiple of 14, 26, and 60.

(4.4) **33.** Find the value of $\dfrac{3}{5} \cdot \dfrac{15}{18} \div \dfrac{5}{9}$.

(5.3) **34.** Find the value of $5\dfrac{2}{3} + 8\dfrac{1}{12}$.

(6.1) **35.** In the decimal number 26.10742, the digit 7 is in what position?

★ **Answers to Practice Sets (6.2)**

A. **1.** $16\dfrac{21}{25}$ **2.** $\dfrac{513}{1,000}$ **3.** $6,646\dfrac{107}{10,000}$ **4.** $1\dfrac{1}{10}$

B. **1.** $\dfrac{7}{8}$ **2.** $\dfrac{31}{250}$ **3.** $6\dfrac{7}{1,200}$ **4.** $18\dfrac{2}{17}$

6.3 Rounding Decimals

Section Overview

❑ **ROUNDING DECIMAL NUMBERS**

❑ **ROUNDING DECIMAL NUMBERS**

We first considered the concept of rounding numbers in Section 1.3 where our concern with rounding was related to whole numbers only. With a few minor changes, we can apply the same rules of rounding to decimals.

> To round a decimal to a particular position:
> 1. Mark the position of the round-off digit (with an arrow or check).
> 2. Note whether the digit to the immediate right of the marked digit is
>   (a) *less than 5.* If so, leave the round-off digit unchanged.
>   (b) *5 or greater.* If so, add 1 to the round-off digit.
> 3. If the round-off digit is
>   (a) to the right of the decimal point, eliminate all the digits to its right.
>   (b) to the left of the decimal point, replace all the digits between it and the decimal point with zeros and eliminate the decimal point and all the decimal digits.

☆ **SAMPLE SET A**

Round each decimal to the specified position. (The numbers in parentheses indicate which step is being used.)

1. Round 32.116 to the nearest hundredth.

   (1) 32.116
          ↑
          └── hundredths position

Continued

(2b) The digit immediately to the right is 6, and $6 > 5$, so we add 1 to the round-off digit:

$$1 + 1 = 2$$

(3a) The round-off digit is to the right of the decimal point, so we eliminate all digits to its right.

32.12

The number 32.116 rounded to the nearest hundredth is 32.12.

2. Round 633.14216 to the nearest hundred.

(1) 633.14216
 └── hundreds position

(2a) The digit immediately to the right is 3, and $3 < 5$ so we leave the round-off digit unchanged.

(3b) The round-off digit is to the left of 0, so we replace all the digits between it and the decimal point with zeros and eliminate the decimal point and all the decimal digits.

600

The number 633.14216 rounded to the nearest hundred is 600.

3. 1,729.63 rounded to the nearest ten is 1,730.

4. 1.0144 rounded to the nearest tenth is 1.0.

5. 60.98 rounded to the nearest one is 61.

Sometimes we hear a phrase such as "round to three decimal places." This phrase means that the round-off digit is the third decimal digit (the digit in the thousandths position).

6. 67.129 rounded to the second decimal place is 67.13.

7. 67.129558 rounded to 3 decimal places is 67.130.

★ PRACTICE SET A

Round each decimal to the specified position.

1. 4.816 to the nearest hundredth.

2. 0.35928 to the nearest ten thousandths.

3. 82.1 to the nearest one.

4. 753.98 to the nearest hundred.

5. Round 43.99446 to three decimal places.

6. Round 105.019997 to four decimal places.

7. Round 99.9999 to two decimal places.

Answers to Practice Sets are on p. 260.

Section 6.3 EXERCISES

For problems 1–10, complete the chart by rounding each decimal to the indicated positions.

	Tenth	Hundredth	Thousandth	Ten Thousandth
1. 20.01071	20	20.01	20.011	20.0107
2. 3.52612	3.5	3.53	3.526	3.5261
3. 531.21878	531.2	531.22	531.219	531.2188
4. 36.109053	36.1	36.11	36.109	36.1091
5. 1.999994	1.9	1.100	1.1000	1.
6. 7.4141998			7.414	
7. 0.000007				
8. 0.00008				0.0001
9. 9.19191919				
10. 0.0876543				

For problems 11–15, round 18.4168095 to the indicated place.

11. 3 decimal places. **12.** 1 decimal place.

13. 5 decimal places. **14.** 6 decimal places.

15. 2 decimal places.

Calculator Problems

For problems 16–22, perform each division using a calculator.

16. $4 \div 3$ and round to 2 decimal places

17. $1 \div 8$ and round to 1 decimal place.

18. $1 \div 27$ and round to 6 decimal places.

19. $51 \div 61$ and round to 5 decimal places.

20. $3 \div 16$ and round to 3 decimal places.

21. $16 \div 3$ and round to 3 decimal places.

22. $26 \div 7$ and round to 5 decimal places.

EXERCISES FOR REVIEW

(1.1) **23.** What is the value of 2 in the number 421,916,017?

(2.3) **24.** Perform the division: $378 \div 29$.

(3.1) **25.** Find the value of 4^4.

(4.2) **26.** Convert $\frac{11}{3}$ to a mixed number.

(6.2) **27.** Convert 3.16 to a mixed number fraction.

6.4 Addition and Subtraction of Decimals

Section Overview

- ❑ **THE LOGIC BEHIND THE METHOD**
- ❑ **THE METHOD OF ADDING AND SUBTRACTING DECIMALS**
- ❑ **CALCULATORS**

❑ THE LOGIC BEHIND THE METHOD

Consider the sum of 4.37 and 3.22. Changing each decimal to a fraction, we have

$4\frac{37}{100} + 3\frac{22}{100}$. Performing the addition, we get

$$4.37 + 3.22 = 4\frac{37}{100} + 3\frac{22}{100} = \frac{4 \cdot 100 + 37}{100} + \frac{3 \cdot 100 + 22}{100}$$

$$= \frac{437}{100} + \frac{322}{100}$$

$$= \frac{437 + 322}{100}$$

$$= \frac{759}{100}$$

$$= 7\frac{59}{100}$$

$$= \text{seven and fifty-nine hundredths}$$

$$= 7.59$$

Thus, $4.37 + 3.22 = 7.59$.

❑ THE METHOD OF ADDING AND SUBTRACTING DECIMALS

When writing the previous addition, we could have written the numbers in columns.

```
  4.37
+ 3.22
------
  7.59
```

This agrees with our previous result. From this observation, we can suggest a method for adding and subtracting decimal numbers.

Method of Adding and Subtracting Decimals

To add or subtract decimals:

1. Align the numbers vertically so that the decimal points line up under each other and the corresponding decimal positions are in the same column.
2. Add or subtract the numbers as if they were whole numbers.
3. Place a decimal point in the resulting sum or difference directly under the other decimal points.

☆ **SAMPLE SET A**

Find the following sums and differences.

1. $9.813 + 2.140$

$$
\begin{array}{r}
9.813 \\
+\ 2.140 \\
\hline
11.953
\end{array}
$$

The decimal points are aligned in the same column.

2. $841.0056 + 47.016 + 19.058$

$$
\begin{array}{r}
841.0056 \\
47.016 \\
+\ 19.058 \\
\hline
\end{array}
$$

To insure that the columns align properly, we can write a 0 in the position at the end of the numbers 47.016 and 19.058 without changing their values.

$$
\begin{array}{r}
{\scriptstyle 11\quad 1} \\
841.0056 \\
47.0160 \\
+\ 19.0580 \\
\hline
907.0796
\end{array}
$$

3. $1.314 - 0.58$

$$
\begin{array}{r}
1.314 \\
-0.58 \\
\hline
\end{array}
$$

Write a 0 in the thousandths position.

$$
\begin{array}{r}
{\scriptstyle 12} \\
{\scriptstyle 0\ \not{2}11} \\
1.\not{3}\not{1}4 \\
-0.580 \\
\hline
0.734
\end{array}
$$

4. $16.01 - 7.053$

$$
\begin{array}{r}
16.01 \\
-\ 7.053 \\
\hline
\end{array}
$$

Write a 0 in the thousandths position.

$$
\begin{array}{r}
{\scriptstyle 15\ 9\,10\ 10} \\
1\not{6}.\not{0}\not{1}\not{0} \\
-\ 7.053 \\
\hline
8.957
\end{array}
$$

5. Find the sum of 6.88106 and 3.5219 and round it to three decimal places.

$$
\begin{array}{r}
6.88106 \\
+3.5219 \\
\hline
\end{array}
$$

Write a 0 in the ten thousandths position.

$$
\begin{array}{r}
{\scriptstyle 1\ 1} \\
6.88106 \\
+\ 3.52190 \\
\hline
10.40296
\end{array}
$$

We need to round the sum to the thousandths position. Since the digit in the position immediately to the right is 9, and $9 > 5$, we get

10.403

Continued

6. Wendy has $643.12 in her checking account. She writes a check for $16.92. How much is her new account balance?

To find the new account balance, we need to find the difference between 643.12 and 16.92. We will subtract 16.92 from 643.12.

$$
\begin{array}{r}
{\scriptstyle 3\;12\,11}\\
643.12\\
-\quad 16.92\\
\hline
626.20
\end{array}
$$

After writing a check for $16.92, Wendy now has a balance of $626.20 in her checking account.

★ PRACTICE SET A

Find the following sums and differences.

1. $3.187 + 2.992$ **2.** $14.987 - 5.341$ **3.** $0.5261 + 1.0783$ **4.** $1.06 - 1.0535$

5. $16,521.07 + 9,256.15$ **6.** Find the sum of 11.6128 and 14.07353, and round it to two decimal places.

☐ CALCULATORS

The calculator can be useful for finding sums and differences of decimal numbers. However, calculators with an eight-digit display cannot be used when working with decimal numbers that contain more than eight digits, or when the sum results in more than eight digits. In practice, an eight-place decimal will seldom be encountered. There are some inexpensive calculators that can handle 13 decimal places.

☆ SAMPLE SET B

Use a calculator to find each sum or difference.

1. $42.0638 + 126.551$

		Display Reads
Type	42.0638	42.0638
Press	$+$	42.0638
Type	126.551	126.551
Press	$=$	168.6148

The sum is 168.6148.

2. Find the difference between 305.0627 and 14.29667.

		Display Reads
Type	305.0627	305.0627
Press	$-$	305.0627
Type	14.29667	14.29667
Press	$=$	290.76603

The difference is 290.76603.

3. 51.07 + 3,891.001786

Since 3,891.001786 contains more than eight digits, we will be unable to use an eight-digit display calculator to perform this addition. We can, however, find the sum by hand.

$$\begin{array}{r} 51.070000 \\ 3891.001786 \\ \hline 3942.071786 \end{array}$$

The sum is 3,942.071786.

★ **PRACTICE SET B**

Use a calculator to perform each operation.

1. 4.286 + 8.97   **2.** 452.0092 − 392.558   **3.** Find the sum of 0.095 and 0.001862

4. Find the difference between 0.5 and 0.025   **5.** Find the sum of 2,776.00019 and 2,009.00012.

Answers to Practice Sets are on p. 265.

Section 6.4 EXERCISES

For problems 1–15, perform each addition or subtraction. Use a calculator to check each result.

1. 1.84 + 7.11

2. 15.015 − 6.527

3. 11.842 + 28.004

4. 3.16 − 2.52

5. 3.55267 + 8.19664

6. 0.9162 − 0.0872

7. 65.512 − 8.3005

8. 761.0808 − 53.198

9. 4.305 + 2.119 − 3.817

10. 19.1161 + 27.8014 + 39.3161

11. $0.41276 - 0.0018 - 0.00011$

12. $2.181 + 6.05 + 1.167 + 8.101$

13. $1.0031 + 6.013106 + 0.00018 + 0.0092 + 2.11$

14. $27 + 42 + 9.16 - 0.1761 + 81.6$

15. $10.28 + 11.111 + 0.86 + 5.1$

For problems 16–25, solve as directed. A calculator may be useful.

16. Add 6.1121 and 4.916 and round to 2 decimal places.

17. Add 21.66418 and 18.00184 and round to 4 decimal places.

18. Subtract 5.2121 from 9.6341 and round to 1 decimal place.

19. Subtract 0.918 from 12.006 and round to 2 decimal places.

20. Subtract 7.01884 from the sum of 13.11848 and 2.108 and round to 4 decimal places.

21. A checking account has a balance of $42.51. A check is written for $19.28. What is the new balance?

22. A checking account has a balance of $82.97. One check is written for $6.49 and another for $39.95. What is the new balance?

23. A person buys $4.29 worth of hamburger and pays for it with a $10 bill. How much change does this person get?

24. A man buys $6.43 worth of stationery and pays for it with a $20 bill. After receiving his change, he realizes he forgot to buy a pen. If the total price of the pen is $2.12, and he buys it, how much of the $20 bill is left?

25. A woman starts recording a movie on her video cassette recorder with the tape counter set at 21.93. The movie runs 847.44 tape counter units. What is the final tape counter reading?

EXERCISES FOR REVIEW

(1.5) **26.** Find the difference between 11,206 and 10,884.

(2.1) **27.** Find the product. $820 \cdot 10,000$.

(3.2) **28.** Find the value of $\sqrt{121} - \sqrt{25} + 8^2 + 16 \div 2^2$.

(4.5) **29.** Find the value of $8\dfrac{1}{3} \cdot \dfrac{36}{75} \div 2\dfrac{2}{5}$.

(6.3) **30.** Round 1.08196 to the nearest hundredth.

★ **Answers to Practice Sets (6.4)**

A. **1.** 6.179 **2.** 9.646 **3.** 1.6044 **4.** 0.0065 **5.** 25,777.22 **6.** 25.69

B. **1.** 13.256 **2.** 59.4512 **3.** 0.096862 **4.** 0.475 **5.** Since each number contains more than eight digits, using some calculators may not be helpful. Adding these by "hand technology," we get 4,785.00031.

6.5 Multiplication of Decimals

Section Overview

☐ **THE LOGIC BEHIND THE METHOD**
☐ **THE METHOD OF MULTIPLYING DECIMALS**
☐ **CALCULATORS**
☐ **MULTIPLYING DECIMALS BY POWERS OF 10**
☐ **MULTIPLICATION IN TERMS OF "OF"**

☐ THE LOGIC BEHIND THE METHOD

Consider the product of 3.2 and 1.46. Changing each decimal to a fraction, we have

$$(3.2)(1.46) = 3\frac{2}{10} \cdot 1\frac{46}{100}$$

$$= \frac{32}{10} \cdot \frac{146}{100}$$

$$= \frac{32 \cdot 146}{10 \cdot 100}$$

$$= \frac{4672}{1000}$$

$$= 4\frac{672}{1000}$$

$$= \text{four and six hundred seventy-two thousandths}$$

$$= 4.672$$

Thus, $(3.2)(1.46) = 4.672$.
   Notice that the factor

$\left.\begin{array}{l} 3.2 \text{ has 1 decimal place,} \\ 1.46 \text{ has 2 decimal places,} \\ \quad \text{and the product} \\ 4.672 \text{ has 3 decimal places.} \end{array}\right\} 1 + 2 = 3$

Using this observation, we can suggest that the sum of the number of decimal places in the factors equals the number of decimal places in the product.

$$
\begin{array}{r}
1 \\
1.46 \quad\longleftarrow\quad \text{2 decimal places} \\
\times\quad 3.2 \quad\longleftarrow\quad +\text{1 decimal place} \\
\hline
292 \\
438 \\
\hline
4.672 \quad\longleftarrow\quad \text{3 decimal places}
\end{array}
$$

☐ THE METHOD OF MULTIPLYING DECIMALS

Method of Multiplying Decimals

To multiply decimals,

1. Multiply the numbers as if they were whole numbers.
2. Find the sum of the number of decimal places in the factors.
3. The number of decimal places in the product is the sum found in step 2.

☆ SAMPLE SET A

Find the following products.

1. $6.5 \cdot 4.3$

$$
\begin{array}{r}
6.5 \quad\longleftarrow\quad \text{1 decimal plate} \\
4.3 \quad\longleftarrow\quad \text{1 decimal place}
\end{array}\Big\}\; 1 + 1 = 2 \text{ decimal places in the product.}
$$

$$
\begin{array}{r}
\hline
195 \\
260 \\
\hline
27.95 \quad\longleftarrow\quad \text{2 decimal places}
\end{array}
$$

Thus, $6.5 \cdot 4.3 = 27.95$.

2. $23.4 \cdot 1.96$

$$
\begin{array}{r}
23.4 \quad\longleftarrow\quad \text{1 decimal place} \\
1.96 \quad\longleftarrow\quad \text{2 decimal places}
\end{array}\Big\}\; 1 + 2 = 3 \text{ decimal places in the product.}
$$

$$
\begin{array}{r}
\hline
1404 \\
2106 \\
234 \\
\hline
45.864 \quad\longleftarrow\quad \text{3 decimal places}
\end{array}
$$

Thus, $23.4 \cdot 1.96 = 45.864$.

3. Find the product of 0.251 and 0.00113 and round to three decimal places.

$$
\begin{array}{r}
0.251 \quad\longleftarrow\quad \text{3 decimal places} \\
0.00113 \quad\longleftarrow\quad \text{5 decimal places}
\end{array}\Big\}\; 3 + 5 = 8 \text{ decimal places in the product.}
$$

$$
\begin{array}{r}
\hline
753 \\
251 \\
251 \\
\hline
0.00028363
\end{array}
$$

— We need to add three zeros to get 8 decimal places.

Now, rounding to three decimal places, we get

$0.251 \cdot 0.00113 = 0.000$ — to three decimal places.

★ **PRACTICE SET A**

Find the following products.

1. $5.3 \cdot 8.6$ **2.** $2.12 \cdot 4.9$ **3.** $1.054 \cdot 0.16$ **4.** $0.00031 \cdot 0.002$

5. Find the product of 2.33 and 4.01 and round to one decimal place.

6. $10 \cdot 5.394$ **7.** $100 \cdot 5.394$ **8.** $1000 \cdot 5.394$ **9.** $10,000 \cdot 5.394$

❑ **CALCULATORS**

Calculators can be used to find products of decimal numbers. However, a calculator that has only an eight-digit display may not be able to handle numbers or products that result in more than eight digits. But there are plenty of inexpensive ($50 – $75) calculators with more than eight-digit displays.

☆ **SAMPLE SET B**

Find the following products, if possible, using a calculator.

1. $2.58 \cdot 8.61$

		Display Reads
Type	2.58	2.58
Press	$\boxed{\times}$	2.58
Type	8.61	8.61
Press	$\boxed{=}$	22.2138

The product is 22.2138.

2. $0.006 \cdot 0.0042$

		Display Reads
Type	.006	0.006
Press	$\boxed{\times}$	0.006
Type	.0042	0.0042
Press	$\boxed{=}$	0.0000252

We know that there will be seven decimal places in the product (since $3 + 4 = 7$). Since the display shows 7 decimal places, we can assume the product is correct. Thus, the product is 0.0000252.

Continued

3. 0.0026 · 0.11976

Since we expect $4 + 5 = 9$ decimal places in the product, we know that an eight-digit display calculator will not be able to provide us with the exact value. To obtain the exact value, we must use "hand technology." Suppose, however, that we agree to round off this product to three decimal places. We then need only four decimal places on the display.

		Display Reads
Type	.0026	0.0026
Press	$\times$	0.0026
Type	.11976	0.11976
Press	$=$	0.0003114

Rounding 0.0003114 to three decimal places we get 0.000. Thus, $0.0026 \cdot 0.11976 = 0.000$ to three decimal places.

★ **PRACTICE SET B**

Use a calculator to find each product. If the calculator will not provide the exact product, round the result to four decimal places.

1. 5.126 · 4.08   **2.** 0.00165 · 0.04   **3.** 0.5598 · 0.4281   **4.** 0.000002 · 0.06

❑ **MULTIPLYING DECIMALS BY POWERS OF 10**

There is an interesting feature of multiplying decimals by powers of 10. Consider the following multiplications.

Multiplication	Number of Zeros in the Power of 10	Number of Positions the Decimal Point Has Been Moved to the Right
$10 \cdot 8.315274 = 83.15274$	1	1
$100 \cdot 8.315274 = 831.5274$	2	2
$1,000 \cdot 8.315274 = 8,315.274$	3	3
$10,000 \cdot 8.315274 = 83,152.74$	4	4

Multiplying a Decimal by a Power of 10

To multiply a decimal by a power of 10, move the decimal place to the *right* of its current position as many places as there are zeros in the power of 10. Add zeros if necessary.

☆ **SAMPLE SET C**

Find the following products.

1. 100 · 34.876.

Since there are **2 zeros in 100**, move the decimal point in **34.876 two places to the right.**

$100 \cdot 34.876 = 3487.6$

$= 3,487.6$

2. $1,000 \cdot 4.8058$.

$$1,000 \cdot 4.8058 = 4805.8$$
$$= 4,805.8$$

Since there are 3 zeros in 1,000, move the decimal point in 4.8058 three places to the right.

3. $10,000 \cdot 56.82$.

$$10,000 \cdot 56.82 = 568200.$$

$$= 568,200$$

Since there are 4 zeros in 10,000, move the decimal point in 56.82 four places to the right. We will have to add two zeros in order to obtain the four places.

Since there is no fractional part, we can drop the decimal point.

4. $(1,000,000)(2.57) = 2570000.$
$$= 2,570,000$$

5. $(1,000)(0.0000029) = 0\,000.0029$
$$= 0.0029$$

★ PRACTICE SET C

Find the following products.

1. $100 \cdot 4.27$

2. $10,000 \cdot 16.52187$

3. $(10)(0.0188)$

4. $(10,000,000,000)(52.7)$

❏ MULTIPLICATION IN TERMS OF "OF"

Recalling that the word "of" translates to the arithmetic operation of multiplication, let's observe the following multiplications.

☆ SAMPLE SET D

1. Find 4.1 of 3.8.

Translating "of" to "×," we get

$$
\begin{array}{r}
4.1 \\
\times 3.8 \\
\hline
328 \\
123 \\
\hline
15.58
\end{array}
$$

Thus, 4.1 of 3.8 is 15.58.

2. Find 0.95 of the sum of 2.6 and 0.8.

We first find the sum of 2.6 and 0.8.

$$
\begin{array}{r}
2.6 \\
+0.8 \\
\hline
3.4
\end{array}
$$

Continued

Now find 0.95 of 3.4

$$
\begin{array}{r}
3.4 \\
\times\,0.95 \\
\hline
170 \\
306 \\
\hline
3.230
\end{array}
$$

Thus, 0.95 of (2.6 + 0.8) is 3.230.

★ **PRACTICE SET D**

1. Find 2.8 of 6.4.

2. Find 0.1 of 1.3.

3. Find 1.01 of 3.6.

4. Find 0.004 of 0.0009.

5. Find 0.83 of 12.

6. Find 1.1 of the sum of 8.6 and 4.2.

Answers to Practice Sets are on p. 273.

Section 6.5 EXERCISES

For problems 1–30, find each product and check each result with a calculator.

1. 3.4 · 9.2

2. 4.5 · 6.1

3. 8.0 · 5.9

4. 6.1 · 7

5. (0.1)(1.52)

6. (1.99)(0.05)

7. (12.52)(0.37)

8. (5.116)(1.21)

9. (31.82)(0.1)

10. (16.527)(9.16)

11. $0.0021 \cdot 0.013$

12. $1.0037 \cdot 1.00037$

13. $(1.6)(1.6)$

14. $(4.2)(4.2)$

15. $0.9 \cdot 0.9$

16. $1.11 \cdot 1.11$

17. $6.815 \cdot 4.3$

18. $9.0168 \cdot 1.2$

19. $(3.5162)(0.0000003)$

20. $(0.000001)(0.01)$

21. $(10)(4.96)$

22. $(10)(36.17)$

23. $10 \cdot 421.8842$

24. $10 \cdot 8.0107$

25. $100 \cdot 0.19621$

26. $100 \cdot 0.779$

27. $1000 \cdot 3.596168$

28. $1000 \cdot 42.7125571$

29. $1000 \cdot 25.01$

30. $100,000 \cdot 9.923$

For problems 31–35, perform each multiplication and round to the indicated position.

	Actual Product	Tenths	Hundredths	Thousandths
31. (4.6)(6.17)				
32. (8.09)(7.1)				
33. (11.1106)(12.08)				
34. 0.0083 · 1.090901				
35. 7 · 26.518				

For problems 36–50, perform the indicated operations.

36. Find 5.2 of 3.7.

37. Find 12.03 of 10.1.

38. Find 16 of 1.04.

39. Find 12 of 0.1.

40. Find 0.09 of 0.003.

41. Find 1.02 of 0.9801.

42. Find 0.01 of the sum of 3.6 and 12.18.

43. Find 0.2 of the sum of 0.194 and 1.07.

44. Find the difference of 6.1 of 2.7 and 2.7 of 4.03.

45. Find the difference of 0.071 of 42 and 0.003 of 9.2.

46. If a person earns $8.55 an hour, how much does he earn in twenty-five hundredths of an hour?

47. A man buys 14 items at $1.16 each. What is the total cost?

48. In problem 47, how much is the total cost if 0.065 sales tax is added?

49. A river rafting trip is supposed to last for 10 days and each day 6 miles is to be rafted. On the third day a person falls out of the raft after only $\frac{2}{5}$ of that day's mileage. If this person gets discouraged and quits, what fraction of the entire trip did he complete?

 **Calculator Problems**

For problems 51–60, use a calculator to determine each product. If the calculator will not provide the exact product, round the result to five decimal places.

51. $0.019 \cdot 0.321$ **52.** $0.261 \cdot 1.96$

53. $4.826 \cdot 4.827$ **54.** $(9.46)^2$

55. $(0.012)^2$ **56.** $0.00037 \cdot 0.0065$

57. $0.002 \cdot 0.0009$ **58.** $0.1286 \cdot 0.7699$

59. $0.01 \cdot 0.00000471$

60. $0.00198709 \cdot 0.03$

50. A woman starts the day with $42.28. She buys one item for $8.95 and another for $6.68. She then buys another item for sixty two-hundredths of the remaining amount. How much money does she have left?

EXERCISES FOR REVIEW

(2.2)   **61.** Find the value, if it exists, of $0 \div 15$.

(3.4)   **62.** Find the greatest common factor of 210, 231, and 357.

(4.3)   **63.** Reduce $\dfrac{280}{2,156}$ to lowest terms.

(6.1)   **64.** Write "fourteen and one hundred twenty-one ten-thousandths, using digits."

(6.4)   **65.** Subtract 6.882 from 8.661 and round the result to two decimal places.

★ **Answers to Practice Sets (6.5)**

A.   **1.** 45.58   **2.** 10.388   **3.** 0.16864   **4.** 0.00000062   **5.** 9.3   **6.** 53.94   **7.** 539.4   **8.** 5,394
   9. 59,340

B.   **1.** 20.91408   **2.** 0.000066   **3.** 0.2397   **4.** 0.0000

C.   **1.** 427   **2.** 165,218.7   **3.** 0.188   **4.** 527,000,000,000

D.   **1.** 17.92   **2.** 0.13   **3.** 3.636   **4.** 0.0000036   **5.** 9.96   **6.** 14.08

6.6 Division of Decimals

Section
Overview

☐ **THE LOGIC BEHIND THE METHOD**
☐ **A METHOD OF DIVIDING A DECIMAL BY A NONZERO WHOLE NUMBER**
☐ **A METHOD OF DIVIDING A DECIMAL BY A NONZERO DECIMAL**
☐ **DIVIDING DECIMALS BY POWERS OF 10**

☐ THE LOGIC BEHIND THE METHOD

As we have done with addition, subtraction, and multiplication of decimals, we will study a method of division of decimals by converting them to fractions, then we will make a general rule.

We will proceed by using this example: Divide 196.8 by 6.

$$
\begin{array}{r}
32 \\
6\overline{)196.8} \\
\underline{18} \\
16 \\
\underline{12} \\
4
\end{array}
$$

We have, up to this point, divided 196.8 by 6 and have gotten a quotient of 32 with a remainder of 4. If we follow our intuition and bring down the .8, we have the division $4.8 \div 6$.

$$
4.8 \div 6 = 4\frac{8}{10} \div 6
$$

$$
= \frac{48}{10} \div \frac{6}{1}
$$

$$
= \frac{\overset{8}{\cancel{48}}}{10} \cdot \frac{1}{\underset{1}{\cancel{6}}}
$$

$$
= \frac{8}{10}
$$

Thus, $4.8 \div 6 = .8$.

Now, our intuition and experience with division direct us to place the .8 immediately to the right of 32.

Notice that the decimal points appear in the same column.

$$
\begin{array}{r}
32.8 \\
6\overline{)196.8} \\
\underline{18} \\
16 \\
\underline{12} \\
4.8 \\
\underline{4.8} \\
0
\end{array}
$$

From these observations, we suggest the following method of division.

☐ A METHOD OF DIVIDING A DECIMAL BY A NONZERO WHOLE NUMBER

Method of Dividing a
Decimal by a Nonzero Whole
Number

To divide a decimal by a nonzero whole number:

1. Write a decimal point above the division line and directly over the decimal point of the dividend.
2. Proceed to divide as if both numbers were whole numbers.
3. If, in the quotient, the first nonzero digit occurs to the right of the decimal point, but not in the tenths position, place a zero in each position between the decimal point and the first nonzero digit of the quotient.

☆ SAMPLE SET A

Find the decimal representations of the following quotients.

1. $114.1 \div 7$

```
      16.3
  7) 114.1
      7
      ──
      44
      42
      ──
      2.1
      2.1
      ───
        0
```

Thus, $114.1 \div 7 = 16.3$.

Check: If $114.1 \div 7 = 16.3$, then $7 \cdot 16.3$ should equal 114.1.

```
    42
  16.3
     7
  ─────
 114.1    True.
```

2. $0.02068 \div 4$

```
    0.00517
  4) 0.02068
       20
       ──
        6
        4
       ──
       28
       28
       ──
        0
```

Place zeros in the tenths and hundredths positions. (See Step 3.)

Thus, $0.02068 \div 4 = 0.00517$.

★ PRACTICE SET A

Find the following quotients.

1. $184.5 \div 3$ **2.** $16.956 \div 9$ **3.** $0.2964 \div 4$ **4.** $0.000496 \div 8$

❏ A METHOD OF DIVIDING A DECIMAL BY A NONZERO DECIMAL

Now that we can divide decimals by nonzero whole numbers, we are in a position to divide decimals by a nonzero decimal. We will do so by converting a division by a decimal into a division by a whole number, a process with which we are already familiar. We'll illustrate the method using this example: Divide 4.32 by 1.8.

Let's look at this problem as $4\dfrac{32}{100} \div 1\dfrac{8}{10}$.

$$4\frac{32}{100} \div 1\frac{8}{10} = \frac{4\dfrac{32}{100}}{1\dfrac{8}{10}}$$

$$= \frac{\dfrac{432}{100}}{\dfrac{18}{10}}$$

The divisor is $\dfrac{18}{10}$. We can convert $\dfrac{18}{10}$ into a whole number if we multiply it by 10.

$$\frac{18}{10} \cdot 10 = \frac{18}{\overset{}{\underset{1}{\cancel{10}}}} \cdot \frac{\overset{1}{\cancel{10}}}{1} = 18$$

But, we know from our experience with fractions, that if we multiply the denominator of a fraction by a nonzero whole number, we must multiply the numerator by that same nonzero whole number. Thus, when converting $\dfrac{18}{10}$ to a whole number by multiplying it by 10, we must also multiply the numerator $\dfrac{432}{100}$ by 10.

$$\frac{432}{100} \cdot 10 = \frac{432}{\underset{10}{\cancel{100}}} \cdot \frac{\overset{1}{\cancel{10}}}{1} = \frac{432 \cdot 1}{10 \cdot 1} = \frac{432}{10}$$

$$= 43\frac{2}{10}$$

$$= 43.2$$

We have converted the division $4.32 \div 1.8$ into the division $43.2 \div 18$, that is,

$$1.8\overline{)4.32} \longrightarrow 18\overline{)43.2}$$

Notice what has occurred.

$$1.8\overline{)4.32} \longrightarrow 1\,8.\overline{)4\,3.2}$$

If we "move" the decimal point of the divisor one digit to the right, we must also "move" the decimal point of the dividend one place to the right. The word "move" actually indicates the process of multiplication by a power of 10.

Method of Dividing a
Decimal by a Decimal Number

To divide a decimal by a nonzero decimal,

1. Convert the divisor to a whole number by moving the decimal point to the position immediately to the right of the divisor's last digit.
2. Move the decimal point of the dividend to the right the same number of digits it was moved in the divisor.
3. Set the decimal point in the quotient by placing a decimal point directly above the newly located decimal point in the dividend.
4. Divide as usual.

☆ SAMPLE SET B

Find the following quotients.

1. $32.66 \div 7.1$

$$7.1 \overline{) 32.66}$$ The divisor has one decimal place.

$$71. \overline{) 326.6} \quad \begin{array}{r} 4.6 \\ \hline \end{array}$$ Move the decimal point of both the divisor and the dividend 1 place to the right.

$$\begin{array}{r} 284 \\ \hline 42.6 \end{array}$$ Set the decimal point.

$$\begin{array}{r} 42.6 \\ \hline 0 \end{array}$$ Divide as usual.

Thus, $32.66 \div 7.1 = 4.6$.

Check: $32.66 \div 7.1 = 4.6$ if $4.6 \times 7.1 = 32.66$

$$\begin{array}{r} 4.6 \\ \times 7.1 \\ \hline 46 \\ 322 \\ \hline 32.66 \end{array} \quad \text{True.}$$

2. $1.0773 \div 0.513$

The divisor has 3 decimal places.

$$.513 \overline{) 1.077\,3} \quad \begin{array}{r} 2.1 \\ \hline \end{array}$$ Move the decimal point of both the divisor and the dividend 3 places to the right.

$$\begin{array}{r} 1\,026 \\ \hline 51\,3 \\ 51\,3 \\ \hline 0 \end{array}$$ Set the decimal place and divide.

Thus, $1.0773 \div 0.513 = 2.1$.

Checking by multiplying 2.1 and 0.513 will convince us that we have obtained the correct result. (Try it.)

3. $12 \div 0.00032$

The divisor has 5 decimal places.

$$0.00032 \overline{) 12.00000}$$ Move the decimal point of both the divisor and the dividend 5 places to the right. We will need to add 5 zeros to 12.

Set the decimal place and divide.

Continued

$$0.00032 \overline{)12.00000}$$ **This is now the same as the division of whole numbers.**

```
        37500.
32) 1200000.
     96
     240
     224
      160
      160
      000
```

Checking assures us that $12 \div 0.00032 = 37,500$.

★ PRACTICE SET B

Find the decimal representation of each quotient.

1. $9.176 \div 3.1$ **2.** $5.0838 \div 1.11$ **3.** $16 \div 0.0004$ **4.** $8,162.41 \div 10$

5. $8,162.41 \div 100$ **6.** $8,162.41 \div 1,000$ **7.** $8,162.41 \div 10,000$

☐ CALCULATORS

Calculators can be useful for finding quotients of decimal numbers. As we have seen with the other calculator operations, we can sometimes expect only approximate results. We are alerted to approximate results when the calculator display is filled with digits. We know it is possible that the operation may produce more digits than the calculator has the ability to show. For example, the multiplication

$$\underbrace{0.12345}_{\substack{\text{5 decimal} \\ \text{places}}} \times \underbrace{0.4567}_{\substack{\text{4 decimal} \\ \text{places}}}$$

produces $5 + 4 = 9$ decimal places. An eight-digit display calculator only has the ability to show eight digits, and an approximation results. The way to recognize a possible approximation is illustrated in problem 3 of the next sample set.

☆ **SAMPLE SET C**

Find each quotient using a calculator. If the result is an approximation, round to five decimal places.

1. $12.596 \div 4.7$

		Display Reads
Type	12.596	12.596
Press	$\div$	12.596
Type	4.7	4.7
Press	$=$	2.68

Since the display is not filled, we expect this to be an accurate result.

2. $0.5696376 \div 0.00123$

		Display Reads
Type	.5696376	0.5696376
Press	$\div$	0.5696376
Type	.00123	0.00123
Press	$=$	463.12

Since the display is not filled, we expect this result to be accurate.

3. $0.8215199 \div 4.113$

		Display Reads
Type	.8215199	0.8215199
Press	$\div$	0.8215199
Type	4.113	4.113
Press	$=$	0.1997373

There are EIGHT DIGITS—DISPLAY FILLED! BE AWARE OF POSSIBLE APPROXIMATIONS.

We can check for a possible approximation in the following way. Since the division $4\overline{)12}$ (with 3 above) can be checked by multiplying 4 and 3, we can check our division by performing the multiplication

$$\underbrace{4.113}_{\substack{3 \text{ decimal} \\ \text{places}}} \times \underbrace{0.1997373}_{\substack{7 \text{ decimal} \\ \text{places}}}$$

This multiplication produces $3 + 7 = 10$ decimal digits. But our suspected quotient contains only 8 decimal digits. We conclude that the answer is an approximation. Then, rounding to five decimal places, we get 0.19974.

★ **PRACTICE SET C**

Find each quotient using a calculator. If the result is an approximation, round to four decimal places.

1. $42.49778 \div 14.261$ **2.** $0.001455 \div 0.291$ **3.** $7.459085 \div 2.1192$

☐ DIVIDING DECIMALS BY POWERS OF 10

In problems 4 and 5 of Practice Set B, we found the decimal representations of 8,162.41 ÷ 10 and 8,162.41 ÷ 100. Let's look at each of these again and then, from these observations, make a general statement regarding division of a decimal number by a power of 10.

1. 8,162 ÷ 10

```
        816.241
   10) 8162.410
        80
        ‾‾
        16
        10
        ‾‾
        62
        60
        ‾‾
        24
        20
        ‾‾
        41
        40
        ‾‾
        10
        10
        ‾‾
         0
```

Thus, 8,162.41 ÷ 10 = 816.241.

Notice that the divisor 10 is composed of one 0 and that the quotient 816.241 can be obtained from the dividend 8,162.41 by moving the decimal point one place to the left.

2.

```
         81.6241
   100) 8162.4100
        800
        ‾‾‾
        162
        100
        ‾‾‾
         62 4
         60 0
         ‾‾‾‾
          2 41
          2 00
          ‾‾‾‾
            410
            400
            ‾‾‾
            100
            100
            ‾‾‾
              0
```

Thus, 8,162.41 ÷ 100 = 81.6241.

Notice that the divisor 100 is composed of two 0's and that the quotient 81.6241 can be obtained from the dividend by moving the decimal point two places to the left.

Using these observations, we can suggest the following method for dividing decimal numbers by powers of 10.

Dividing a Decimal Fraction by a Power of 10

> To divide a decimal fraction by a power of 10, move the decimal point of the decimal fraction to the *left* as many places as there are zeros in the power of 10. Add zeros if necessary.

☆ SAMPLE SET D

> Find each quotient.
>
> **1.** 9,248.6 ÷ 100
>
> Since there are 2 zeros in this power of 10, we move the decimal point 2 places to the left.
>
> 92 48.6 ÷ 100 = 92.486
>
> **2.** 3.28 ÷ 10,000
>
> Since there are 4 zeros in this power of 10, we move the decimal point 4 places to the left. To do so, we need to add three zeros.
>
> 0003.28 ÷ 10,000 = 0.000328

★ PRACTICE SET D

Find the decimal representation of each quotient.

1. 182.5 ÷ 10

2. 182.5 ÷ 100

3. 182.5 ÷ 1,000

4. 182.5 ÷ 10,000

5. 646.18 ÷ 100

6. 21.926 ÷ 1,000

Answers to Practice Sets are on p. 284.

Section 6.6 EXERCISES

For problems 1–30, find the decimal representation of each quotient. Use a calculator to check each result.

1. 4.8 ÷ 3

2. 16.8 ÷ 8

3. 18.5 ÷ 5

4. 12.33 ÷ 3

5. 54.36 ÷ 9

6. 73.56 ÷ 12

7. 159.46 ÷ 17

8. 12.16 ÷ 64

9. 37.26 ÷ 81

10. 439.35 ÷ 435

11. 36.98 ÷ 4.3

12. 46.41 ÷ 9.1

13. 3.6 ÷ 1.5

14. 0.68 ÷ 1.7

15. 50.301 ÷ 8.1

16. 2.832 ÷ 0.4

17. 4.7524 ÷ 2.18

18. 16.2409 ÷ 4.03

19. 1.002001 ÷ 1.001

20. 25.050025 ÷ 5.005

21. 12.4 ÷ 3.1

22. 0.48 ÷ 0.08

23. 30.24 ÷ 2.16

24. 48.87 ÷ 0.87

25. 12.321 ÷ 0.111

26. 64,351.006 ÷ 10

27. 64,351.006 ÷ 100

28. 64,351.006 ÷ 1,000

29. 64,351.006 ÷ 1,000,000

30. 0.43 ÷ 100

For problems 31–35, find each quotient. Round to the specified position. A calculator may be used.

	Actual Quotient	Tenths	Hundredths	Thousandths
31. 11.2944 ÷ 6.24				
32. 45.32931 ÷ 9.01				
33. 3.18186 ÷ 0.66				
34. 4.3636 ÷ 4				
35. 0.00006318 ÷ 0.018				

For problems 36–44, find each solution.

36. Divide the product of 7.4 and 4.1 by 2.6.

37. Divide the product of 11.01 and 0.003 by 2.56 and round to two decimal places.

38. Divide the difference of the products of 2.1 and 9.3, and 4.6 and 0.8 by 0.07 and round to one decimal place.

39. A ring costing $567.08 is to be paid off in equal monthly payments of $46.84. In how many months will the ring be paid off?

40. Six cans of cola cost $2.58. What is the price of one can?

41. A family traveled 538.56 miles in their car in one day on their vacation. If their car used 19.8 gallons of gas, how many miles per gallon did it get?

42. Three college students decide to rent an apartment together. The rent is $812.50 per month. How much must each person contribute toward the rent?

43. A woman notices that on slow speed her video cassette recorder runs through 296.80 tape units in 10 minutes and at fast speed through 1098.16 tape units. How many times faster is fast speed than slow speed?

44. A class of 34 first semester business law students pay a total of $1,354.90, disregarding sales tax, for their law textbooks. What is the cost of each book?

🖩 **Calculator Problems**

For problems 45–52, use a calculator to find the quotients. If the result is approximate (see Sample Set C, problem 3), round the result to three decimal places.

45. 3.8994 ÷ 2.01

46. 0.067444 ÷ 0.052

47. $14{,}115.628 \div 484.74$

48. $219{,}709.36 \div 9941.6$

49. $0.0852092 \div 0.49271$

50. $2.4858225 \div 1.11611$

51. $0.123432 \div 0.1111$

52. $2.102838 \div 1.0305$

EXERCISES FOR REVIEW

(4.2)   **53.** Convert $4\dfrac{7}{8}$ to an improper fraction.

(4.6)   **54.** $\dfrac{2}{7}$ of what number is $\dfrac{4}{5}$?

(5.2)   **55.** Find the sum. $\dfrac{4}{15} + \dfrac{7}{10} + \dfrac{3}{5}$.

(6.3)   **56.** Round 0.01628 to the nearest ten-thousandths.

(6.5)   **57.** Find the product. $(2.06)(1.39)$.

★ **Answers to Practice Sets (6.6)**

A.   **1.** 61.5   **2.** 1.884   **3.** 0.0741   **4.** 0.000062

B.   **1.** 2.96   **2.** 4.58   **3.** 40,000   **4.** 816.241   **5.** 81.6241   **6.** 8.16241   **7.** 0.816241

C.   **1.** 2.98   **2.** 0.005
    3. 3.5197645 is an approximate result. Rounding to four decimal places, we get 3.5198.

D.   **1.** 18.25   **2.** 1.825   **3.** 0.1825   **4.** 0.01825   **5.** 6.4618   **6.** 0.021926

6.7 Nonterminating Divisions

Section Overview	☐ NONTERMINATING DIVISIONS ☐ DENOTING NONTERMINATING QUOTIENTS

☐ NONTERMINATING DIVISIONS

Let's consider two divisions:

(1) $9.8 \div 3.5$ (2) $4 \div 3$

Previously, we have considered divisions like example 1, which is an example of a

Terminating Divisions

terminating division. A **terminating division** is a division in which the quotient terminates after several divisions (the *remainder is zero*).

$$
\begin{array}{r}
2.8 \\
3.5\overline{)9.8\,0} \\
\underline{7\,0} \\
2\,8\,0 \\
\underline{2\,8\,0} \\
0
\end{array}
$$

Exact Divisions,
Nonterminating Division

The quotient in this problem terminates in the tenths position. Terminating divisions are also called **exact divisions.**

The division in example 2 is an example of a nonterminating division. A **nonterminating division** is a division that, regardless of how far we carry it out, *always has a remainder.*

$$
\begin{array}{r}
1.333 \\
3\overline{)4.00000} \\
\underline{3} \\
1\,0 \\
\underline{9} \\
10 \\
\underline{9} \\
10 \\
\underline{9} \\
10 \\
\underline{9} \\
10
\end{array}
$$

Repeating Decimal

We can see that the pattern in the brace is repeated endlessly. Such a decimal quotient is called a **repeating decimal.**

☐ DENOTING NONTERMINATING QUOTIENTS

We use three dots at the end of a number to indicate that a pattern repeats itself endlessly.

$$4 \div 3 = 1.333 \ldots$$

Another way, aside from using three dots, of denoting an endlessly repeating pattern is to write a bar (⁻) above the repeating sequence of digits.

$$4 \div 3 = 1.\overline{3}$$

The bar indicates the repeated pattern of 3.

Repeating patterns in a division can be discovered in two ways:

1. As the division process progresses, should the remainder ever be the same as the dividend, it can be concluded that the division is nonterminating and that the pattern in the quotient repeats. This fact is illustrated in problem 1 of Sample Set A.

2. As the division process progresses, should the "product, difference" pattern ever repeat two consecutive times, it can be concluded that the division is nonterminating and that the pattern in the quotient repeats. This fact is illustrated in problems 2 and 4 of Sample Set A.

☆ **SAMPLE SET A**

Carry out each division until the repeating pattern can be determined.

1. $100 \div 27$

```
       3.70370
27) 100.00000
    81
    ───
    19 0
    18 9
    ─────
       100
        81
       ───
       190
       189
```

When the remainder is identical to the dividend, the division is nonterminating. This implies that the pattern in the quotient repeats.

$100 \div 27 = 3.70370370 \ldots$ The repeating block is 703.

$100 \div 27 = 3.\overline{703}$

2. $1 \div 9$

```
    .111
9) 1.000
   9
   ──
   10
    9
   ──
   10
    9
   ──
    1
```

We see that this "product, difference" pattern repeats. We can conclude that the division is nonterminating and that the quotient repeats.

$1 \div 9 = 0.111 \ldots$ The repeating block is 1.

$1 \div 9 = 0.\overline{1}$

3. Divide 2 by 11 and round to 3 decimal places.

Since we wish to round the quotient to three decimal places, we'll carry out the division so that the quotient has four decimal places.

```
     .1818
11) 2.0000
    1 1
    ───
      90
      88
    ───
      20
      11
    ───
      90
```

The number .1818 rounded to three decimal places is .182. Thus, correct to three decimal places,

$2 \div 11 = 0.182$

4. Divide 1 by 6.

```
    .166
6) 1.000
   6
   ──
   40
   36
   ──
   40
   36
   ──
    4
```

We see that this "product, difference" pattern repeats. We can conclude that the division is nonterminating and that the quotient repeats at the 6.

$1 \div 6 = 0.1\overline{6}$

★ **PRACTICE SET A**

Carry out the following divisions until the repeating pattern can be determined.

1. $1 \div 3$ **2.** $5 \div 6$ **3.** $11 \div 9$

4. $17 \div 9$ **5.** Divide 7 by 6 and round to 2 decimal places.

6. Divide 400 by 11 and round to 4 decimal places.

Answers to the Practice Set are on p. 288.

Section 6.7 EXERCISES

For problems 1–20, carry out each division until the repeating pattern is determined. If a repeating pattern is not apparent, round the quotient to three decimal places.

1. $4 \div 9$ **2.** $8 \div 11$

3. $4 \div 25$ **4.** $5 \div 6$

5. $1 \div 7$ **6.** $3 \div 1.1$

7. $20 \div 1.9$ **8.** $10 \div 2.7$

9. $1.11 \div 9.9$ **10.** $8.08 \div 3.1$

11. $51 \div 8.2$ **12.** $0.213 \div 0.31$

13. $0.009 \div 1.1$ **14.** $6.03 \div 1.9$

15. $0.518 \div 0.62$ **16.** $1.55 \div 0.27$ **22.** $8 \div 11$

23. $14 \div 27$

17. $0.333 \div 0.999$ **18.** $0.444 \div 0.999$ **24.** $1 \div 44$

25. $2 \div 44$

19. $0.555 \div 0.27$ **20.** $3.8 \div 0.99$ **26.** $0.7 \div 0.9$ (Compare this with problem 21.)

27. $80 \div 110$ (Compare this with problem 22.)

28. $0.0707 \div 0.7070$

 Calculator Problems

For problems 21–30, use a calculator to perform each division.

29. $0.1414 \div 0.2020$

21. $7 \div 9$

30. $1 \div 0.9999999$

EXERCISES FOR REVIEW

(1.1)   **31.** In the number 411,105, how many ten thousands are there?

(2.2)   **32.** Find the quotient, if it exists. $17 \div 0$.

(3.5)   **33.** Find the least common multiple of 45, 63, and 98.

(6.4)   **34.** Subtract 8.01629 from 9.00187 and round the result to three decimal places.

(6.6)   **35.** Find the quotient. $104.06 \div 12.1$.

★ **Answers to Practice Set (6.7)**

A.   **1.** $0.\overline{3}$ **2.** $0.8\overline{3}$ **3.** $1.\overline{2}$ **4.** $1.\overline{8}$ **5.** 1.17 **6.** 36.3636

6.8 Converting a Fraction to a Decimal

Now that we have studied and practiced dividing with decimals, we are also able to convert a fraction to a decimal. To do so we need only recall that a fraction bar can also be a division symbol. Thus, $\frac{3}{4}$ not only means "3 objects out of 4," but can also mean "3 divided by 4."

☆ **SAMPLE SET A**

Convert the following fractions to decimals. If the division is nonterminating, round to two decimal places.

1. $\dfrac{3}{4}$.   **Divide 3 by 4.**

$$
\begin{array}{r}
.75 \\
4\overline{)3.00} \\
\underline{2\ 8} \\
20 \\
\underline{20} \\
0
\end{array}
$$

Thus, $\dfrac{3}{4} = 0.75$.

2. $\dfrac{1}{5}$.   **Divide 1 by 5.**

$$
\begin{array}{r}
.2 \\
5\overline{)1.0} \\
\underline{1\ 0} \\
0
\end{array}
$$

Thus, $\dfrac{1}{5} = 0.2$

3. $\dfrac{5}{6}$.   **Divide 5 by 6.**

$$
\begin{array}{r}
.833 \\
6\overline{)5.000} \\
\underline{4\ 8} \\
20 \\
\underline{18} \\
20
\end{array}
$$

This recurring remainder indicates that the division is nonterminating.

$\dfrac{5}{6} = 0.833\ \cdots$ **We are to round to two decimal places.**

Thus, $\dfrac{5}{6} = 0.83$ to two decimal places.

4. $5\dfrac{1}{8}$.   **Note that** $5\dfrac{1}{8} = 5 + \dfrac{1}{8}$.

Convert $\dfrac{1}{8}$ **to a decimal.**

$$
\begin{array}{r}
.125 \\
8\overline{)1.000} \\
\underline{8} \\
20 \\
\underline{16} \\
40 \\
\underline{40} \\
0
\end{array}
$$

$\dfrac{1}{8} = .125$

Thus, $5\dfrac{1}{8} = 5 + \dfrac{1}{8} = 5 + .125 = 5.125$.

5. $0.16\frac{1}{4}$.   **This is a complex decimal.**

   Note that the 6 is in the hundredths position.

The number $0.16\frac{1}{4}$ is read as "sixteen and one-fourth hundredths."

$$0.16\frac{1}{4} = \frac{16\frac{1}{4}}{100} = \frac{\frac{16 \cdot 4 + 1}{4}}{100} = \frac{\frac{65}{4}}{\frac{100}{1}} = \frac{\overset{13}{\cancel{65}}}{4} \cdot \frac{1}{\underset{20}{\cancel{100}}} = \frac{13 \cdot 1}{4 \cdot 20} = \frac{13}{80}$$

Now, convert $\frac{13}{80}$ to a decimal.

```
        .1625
    80) 13.0000
         8 0
         5 00
         4 80
           200
           160
           400
           400
             0
```

Thus, $0.16\frac{1}{4} = 0.1625$.

★ PRACTICE SET A

Convert the following fractions and complex decimals to decimals (in which no proper fractions appear). If the divison is nonterminating, round to two decimal places.

1. $\frac{1}{4}$ **2.** $\frac{1}{25}$ **3.** $\frac{1}{6}$ **4.** $\frac{15}{16}$ **5.** $0.9\frac{1}{2}$ **6.** $8.0126\frac{3}{8}$

Answers to the Practice Set are on p. 293.

Section 6.8 EXERCISES

For problems 1–30, convert each fraction or complex decimal number to a decimal (in which no proper fractions appear).

1. $\dfrac{1}{2}$

2. $\dfrac{4}{5}$

3. $\dfrac{7}{8}$

4. $\dfrac{5}{8}$

5. $\dfrac{3}{5}$

6. $\dfrac{2}{5}$

7. $\dfrac{1}{25}$

8. $\dfrac{3}{25}$

9. $\dfrac{1}{20}$

10. $\dfrac{1}{15}$

11. $\dfrac{1}{50}$

12. $\dfrac{1}{75}$

13. $\dfrac{1}{3}$

14. $\dfrac{5}{6}$

15. $\dfrac{3}{16}$

16. $\dfrac{9}{16}$

17. $\dfrac{1}{27}$

18. $\dfrac{5}{27}$

19. $\dfrac{7}{13}$

20. $\dfrac{9}{14}$

21. $7\dfrac{2}{3}$

22. $8\dfrac{5}{16}$

23. $1\dfrac{2}{15}$

24. $65\dfrac{5}{22}$

25. $101\dfrac{6}{25}$

26. $0.1\dfrac{1}{2}$

27. $0.24\frac{1}{8}$

28. $5.66\frac{2}{3}$

39. $\frac{1}{11}$

40. $\frac{2}{11}$

29. $810.3106\frac{5}{16}$

30. $4.1\frac{1}{9}$

41. $\frac{3}{11}$

42. $\frac{4}{11}$

For problems 31–48, convert each fraction to a decimal. Round to five decimal places.

31. $\frac{1}{9}$

32. $\frac{2}{9}$

43. $\frac{5}{11}$

44. $\frac{6}{11}$

33. $\frac{3}{9}$

34. $\frac{4}{9}$

45. $\frac{7}{11}$

46. $\frac{8}{11}$

35. $\frac{5}{9}$

36. $\frac{6}{9}$

37. $\frac{7}{9}$

38. $\frac{8}{9}$

47. $\frac{9}{11}$

48. $\frac{10}{11}$

📖 **Calculator Problems**

For problems 49–55, use a calculator to convert each fraction to a decimal. If no repeating pattern seems to exist, round to four decimal places.

52. $\dfrac{1}{1469}$

49. $\dfrac{16}{125}$

53. $\dfrac{4}{21,015}$

50. $\dfrac{85}{311}$

54. $\dfrac{81,426}{106,001}$

51. $\dfrac{192}{197}$

55. $\dfrac{16,501}{426}$

EXERCISES FOR REVIEW

(1.3)   **56.** Round 2,105,106 to the nearest hundred thousand.

(4.6)   **57.** $\dfrac{8}{5}$ of what number is $\dfrac{3}{2}$?

(5.4)   **58.** Arrange $1\dfrac{9}{16}$, $1\dfrac{5}{8}$, and $1\dfrac{7}{12}$ in increasing order.

(6.2)   **59.** Convert the complex decimal $3.6\dfrac{5}{4}$ to a fraction.

(6.7)   **60.** Find the quotient. $30 \div 1.1$.

★ **Answers to Practice Set (6.8)**

A.   **1.** 0.25   **2.** 0.04   **3.** 0.17   **4.** 0.9375   **5.** 0.95   **6.** 8.0126375

6.9 Combinations of Operations with Decimals and Fractions

Having considered operations with decimals and fractions, we now consider operations that involve both decimals and fractions.

☆ SAMPLE SET A

Perform the following operations.

1. $0.38 \cdot \dfrac{1}{4}$.

Convert both numbers to decimals or both numbers to fractions. We'll convert to decimals.

$$
\begin{array}{r}
.25 \\
4\overline{)1.00} \\
\underline{8} \\
20 \\
\underline{20} \\
0
\end{array}
$$

$$
\begin{array}{r}
\overset{1}{} \\
4 \\
.38 \\
\times .25 \\
\hline
190 \\
76 \\
\hline
.0950
\end{array}
$$

To convert $\dfrac{1}{4}$ to a decimal, divide 1 by 4.

Now multiply 0.38 and .25.

Thus, $0.38 \cdot \dfrac{1}{4} = 0.095$.

In the problems that follow, the conversions from fraction to decimal, or decimal to fraction, and some of the additions, subtraction, multiplications, and divisions will be left to you.

2. $1.85 + \dfrac{3}{8} \cdot 4.1$

$1.85 + 0.375 \cdot 4.1$

$1.85 + 1.5375$

3.3875

Convert $\dfrac{3}{8}$ to a decimal.

Multiply before adding.

Now add.

3. $\dfrac{5}{13}\left(\dfrac{4}{5} - 0.28\right)$

Convert 0.28 to a fraction.

$$\dfrac{5}{13}\left(\dfrac{4}{5} - \dfrac{28}{100}\right) = \dfrac{5}{13}\left(\dfrac{4}{5} - \dfrac{7}{25}\right)$$

$$= \dfrac{5}{13}\left(\dfrac{20}{25} - \dfrac{7}{25}\right)$$

$$= \dfrac{\overset{1}{\cancel{5}}}{\cancel{13}_1} \cdot \dfrac{\overset{1}{\cancel{13}}}{\cancel{25}_5}$$

$$= \dfrac{1}{5}$$

4. $\dfrac{0.125}{1\frac{1}{3}} + \dfrac{1}{16} - 0.1211 = \dfrac{\frac{125}{1000}}{\frac{4}{3}} + \dfrac{1}{16} - 0.1211$

$= \dfrac{\frac{1}{8}}{\frac{4}{3}} + \dfrac{1}{16} - 0.1211$

$= \dfrac{1}{8} \cdot \dfrac{3}{4} + \dfrac{1}{16} - 0.1211$

$= \dfrac{3}{32} + \dfrac{1}{16} - 0.1211$

$= \dfrac{3}{32} + \dfrac{2}{32} - 0.1211 = \dfrac{5}{32} - 0.1211$

$= 0.15625 - 0.1211$

$= 0.03515$ **Convert this to fraction form.**

$= \dfrac{3515}{100,000}$

$= \dfrac{703}{20,000}$

★ PRACTICE SET A

Perform the following operations.

1. $\dfrac{3}{5} + 1.6$

2. $8.91 + \dfrac{1}{5} \cdot 1.6$

3. $1\dfrac{9}{16}\left(6.12 + \dfrac{7}{25}\right)$

4. $\dfrac{0.156}{1\frac{11}{15}} - 0.05$

Answers to the Practice Set are on p. 297.

Section 6.9 EXERCISES

For problems 1–28, perform the indicated operations and simplify.

1. $\dfrac{3}{10} + 0.7$

2. $\dfrac{1}{5} + 0.1$

3. $\dfrac{5}{8} - 0.513$

4. $0.418 - \dfrac{67}{200}$

5. $0.22 \cdot \dfrac{1}{4}$

6. $\dfrac{3}{5} \cdot 8.4$

16. $0.2 \cdot \left(\dfrac{7}{20} + 1.1143 \right)$

7. $\dfrac{1}{25} \cdot 3.19$

8. $\dfrac{3}{20} \div 0.05$

17. $\dfrac{3}{4} \cdot \left(0.875 + \dfrac{1}{8} \right)$

9. $\dfrac{7}{40} \div 0.25$

10. $1\dfrac{1}{15} \div 0.9 \cdot 0.12$

18. $5.198 - 0.26 \cdot \left(\dfrac{14}{250} + 0.119 \right)$

11. $9.26 + \dfrac{1}{4} \cdot 0.81$

12. $0.588 + \dfrac{1}{40} \cdot 0.24$

19. $0.5\dfrac{1}{4} + (0.3)^2$

13. $\dfrac{1}{20} + 3.62 \cdot \dfrac{3}{8}$

20. $(1.4)^2 - 1.6\dfrac{1}{2}$

14. $7 + 0.15 \div \dfrac{3}{30}$

21. $\left(\dfrac{3}{8} \right)^2 - 0.000625 + (1.1)^2$

15. $\dfrac{15}{16} \cdot \left(\dfrac{7}{10} - 0.5 \right)$

22. $(0.6)^2 \cdot \left(\dfrac{1}{20} - \dfrac{1}{25} \right)$

23. $\left(\dfrac{1}{2}\right)^2 - 0.125$

26. $8\dfrac{1}{3} \cdot \left(\dfrac{1\frac{1}{4}}{2.25} + \dfrac{9}{25}\right)$

24. $\dfrac{0.75}{4\frac{1}{2}} + \dfrac{5}{12}$

27. $\dfrac{\dfrac{0.32}{12}}{\dfrac{35}{0.35}}$

25. $\left(\dfrac{0.375}{2\frac{1}{16}} - \dfrac{1}{33}\right)$

28. $\dfrac{\left(\sqrt{\dfrac{49}{64}} - 5\right)\ 0.125}{1.375}$

EXERCISES FOR REVIEW

(2.4) **29.** Is 21,480 divisible by 3?

(3.1) **30.** Expand 14^4. Do not find the actual value.

(3.2) **31.** Find the prime factorization of 15,400.

(6.2) **32.** Convert 8.016 to a fraction.

(6.8) **33.** Find the quotient. $16 \div 27$.

★ **Answers to Practice Set (6.9)**

A. **1.** 2.2 or $2\dfrac{1}{5}$ **2.** 9.23 **3.** 10 **4.** $\dfrac{1}{25}$ or 0.04

Chapter 6 SUMMARY OF KEY CONCEPTS

Decimal Point (6.1)

A *decimal point* is a point that separates the units digit from the tenths digit.

Decimal or Decimal Fraction (6.1)

A *decimal fraction* is a fraction whose denominator is a power of ten.

Converting a Decimal to a Fraction (6.2)

Decimals can be converted to fractions by saying the decimal number in words, then writing what was said.

Rounding Decimals (6.3)

Decimals are rounded in much the same way whole numbers are rounded.

Addition and Subtraction of Decimals (6.4)

To add or subtract decimals,

1. Align the numbers vertically so that the decimal points line up under each other and the corresponding decimal positions are in the same column.
2. Add or subtract the numbers as if they were whole numbers.
3. Place a decimal point in the resulting sum directly under the other decimal points.

Multiplication of Decimals (6.5)

To multiply two decimals,

1. Multiply the numbers as if they were whole numbers.
2. Find the sum of the number of decimal places in the factors.
3. The number of decimal places in the product is the number found in step 2.

Multiplying Decimals by Powers of 10 (6.5)

To multiply a decimal by a power of 10, move the decimal point to the right as many places as there are zeros in the power of ten. Add zeros if necessary.

Division of a Decimal by a Decimal (6.6)

To divide a decimal by a nonzero decimal,

1. Convert the divisor to a whole number by moving the decimal point until it appears to the right of the divisor's last digit.
2. Move the decimal point of the dividend to the right the same number of digits it was moved in the divisor.
3. Proceed to divide.
4. Locate the decimal in the answer by bringing it straight up from the dividend.

Dividing Decimals by Powers of 10 (6.6)

To divide a decimal by a power of 10, move the decimal point to the left as many places as there are zeros in the power of ten. Add zeros if necessary.

Terminating Divisions (6.7)

A *terminating division* is a division in which the quotient terminates after several divisions. Terminating divisions are also called exact divisions.

Nonterminating Divisions (6.7)

A *nonterminating division* is a division that, regardless of how far it is carried out, always has a remainder. Nonterminating divisions are also called nonexact divisions.

Converting Fractions to Decimals (6.8)

A fraction can be converted to a decimal by dividing the numerator by the denominator.

EXERCISE SUPPLEMENT

Section 6.1

1. The decimal digit that appears two places to the right of the decimal point is in the _____ position.

2. The decimal digit that appears four places to the right of the decimal point is in the _____ position.

For problems 3–8, read each decimal by writing it in words.

3. 7.2

4. 8.105

5. 16.52

6. 5.9271

7. 0.005

8. 4.01701

For problems 9–13, write each decimal using digits.

9. Nine and twelve-hundredths.

10. Two and one hundred seventy-seven thousandths.

11. Fifty-six and thirty-five ten-thousandths.

12. Four tenths.

13. Four thousand eighty-one millionths.

Section 6.2

For problems 14–20, convert each decimal to a proper fraction or a mixed number.

14. 1.07

15. 85.63

16. 0.05

17. $0.14\frac{2}{3}$

18. $1.09\frac{1}{8}$

19. $4.01\frac{1}{27}$

20. $9.11\frac{1}{9}$

Section 6.3

For problems 21–25, round each decimal to the specified position.

21. 4.087 to the nearest hundredth.

22. 4.087 to the nearest tenth.

23. 16.5218 to the nearest one.

24. 817.42 to the nearest ten.

25. 0.9811602 to the nearest one.

Sections 6.4–6.7

For problems 26–45, perform each operation and simplify.

26. $7.10 + 2.98$

27. $14.007 - 5.061$

28. $1.2 \cdot 8.6$

29. $41.8 \cdot 0.19$

30. $57.51 \div 2.7$

31. $0.54003 \div 18.001$

32. $32,051.3585 \div 23,006.9999$

33. $100 \cdot 1,816.001$

34. $1,000 \cdot 1,816.001$

35. $10.000 \cdot 0.14$

36. $0.135888 \div 16.986$

37. $150.79 \div 100$

38. $4.119 \div 10,000$

39. $42.7 \div 18$

40. $6.9 \div 12$

41. $0.014 \div 47.6$. Round to three decimal places.

42. $8.8 \div 19$. Round to one decimal place.

43. $1.1 \div 9$

44. $1.1 \div 9.9$

45. $30 \div 11.1$

Section 6.8

For problems 46–55, convert each fraction to a decimal.

46. $\dfrac{3}{8}$

47. $\dfrac{43}{100}$

48. $\dfrac{82}{1000}$

49. $9\dfrac{4}{7}$

50. $8\dfrac{5}{16}$

51. $1.3\dfrac{1}{3}$

52. $25.6\dfrac{2}{3}$

53. $125.125\dfrac{1}{8}$

54. $9.11\dfrac{1}{9}$

55. $0.0\dfrac{5}{6}$

Section 6.9

For problems 56–62, perform each operation.

56. $\dfrac{5}{8} \cdot 0.25$

57. $\dfrac{3}{16} \cdot 1.36$

58. $\dfrac{3}{5} \cdot \left(\dfrac{1}{2} + 1.75\right)$

59. $\dfrac{7}{2} \cdot \left(\dfrac{5}{4} + 0.30\right)$

60. $19.375 \div \left(4.375 - 1\dfrac{1}{16}\right)$

61. $\dfrac{15}{602} \cdot \left(2.\overline{6} + 3\dfrac{1}{4}\right)$

62. $4\dfrac{13}{18} \div \left(5\dfrac{3}{14} + 3\dfrac{5}{21}\right)$

1. _____

1. (6.1) The decimal digit that appears three places to the right of the decimal point is in the _____ position.

2. (6.1) Write, using words, 15.036.

2. _____

3. (6.1) Write eighty-one and twelve hundredths using digits.

3. _____

4. (6.1) Write three thousand seventeen millionths using digits.

4. _____

5. (6.2) Convert 0.78 to a fraction. Reduce.

5. _____

6. (6.2) Convert 0.875 to a fraction. Reduce.

6. _____

7. (6.3) Round 4.8063 to the nearest tenth.

7. _____

8. (6.3) Round 187.51 to the nearest hundred.

8. _____

9. (6.3) Round 0.0652 to the nearest hundredth.

9. _____

For problems 10–20, perform each operation.

10. _____

10. (6.4) $15.026 + 5.971$

11. _____

11. (6.4) $72.15 - 26.585$

12. _____   **12. (6.5)** $16.2 \cdot 4.8$

13. _____   **13. (6.5)** $10{,}000 \cdot 0.016$

14. _____   **14. (6.6)** $44.64 \div 18.6$

15. _____   **15. (6.6)** $0.21387 \div 0.19$

16. _____   **16. (6.7)** $0.\overline{27} - \dfrac{3}{11}$

17. _____   **17. (6.8)** Convert $6\dfrac{2}{11}$ to a decimal.

18. _____   **18. (6.8)** Convert $0.5\dfrac{9}{16}$ to a decimal.

19. _____   **19. (6.9)** $3\dfrac{1}{8} + 2.325$

20. _____   **20. (6.9)** $\dfrac{3}{8} \times 0.5625$

7

Ratios and Rates

After completing this chapter, you should

Section 7.1 Ratios and Rates
- be able to distinguish between denominate and pure numbers and between ratios and rates

Section 7.2 Proportions
- be able to describe proportions and find the missing factor in a proportion
- be able to work with proportions involving rates

Section 7.3 Applications of Proportions
- solve proportion problems using the five-step method

Section 7.4 Percent
- understand the relationship between ratios and percents
- be able to make conversions between fractions, decimals, and percents

Section 7.5 Fractions of One Percent
- understand the meaning of a fraction of one percent
- be able to make conversions involving fractions of one percent

Section 7.6 Applications of Percents
- be able to distinguish between base, percent, and percentage
- be able to find the percentage, the percent, and the base

7.1 Ratios and Rates

Section
Overview
  ☐ DENOMINATE NUMBERS AND PURE NUMBERS
  ☐ RATIOS AND RATES

☐ DENOMINATE NUMBERS AND PURE NUMBERS

Denominate Numbers

Like and Unlike Denominate Numbers

It is often necessary or convenient to compare two quantities. **Denominate numbers** are numbers together with some specified unit. If the units being compared are alike, the denominate numbers are called **like denominate numbers.** If units are not alike, the numbers are called **unlike denominate numbers.** Examples of denominate numbers are shown in the diagram:

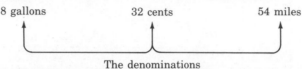

8 gallons   32 cents   54 miles

The denominations

Pure Numbers

Numbers that exist purely as numbers and do *not* represent amounts of quantities are called **pure numbers.** Examples of pure numbers are 8, 254, 0, $21\frac{5}{8}$, $\frac{2}{5}$, and 0.07.

Numbers can be *compared* in two ways: subtraction and division.

Comparing Numbers by Subtraction and Division

> **Comparison of two numbers by subtraction** indicates how *much more* one number is than another.
>
> **Comparison by division** indicates how *many times* larger or smaller one number is than another.

Comparing Pure or Like Denominate Numbers by Subtraction

> Numbers can be compared by subtraction if and only if they both are like denominate numbers or both pure numbers.

☆ SAMPLE SET A

1. Compare 8 miles and 3 miles by subtraction.

   8 miles − 3 miles = 5 miles

   This means that 8 miles is 5 miles more than 3 miles.

   Examples of use: I can now jog 8 miles whereas I used to jog only 3 miles. So, I can now jog 5 miles more than I used to.

2. Compare 12 and 5 by subtraction.

   12 − 5 = 7

   This means that 12 is 7 more than 5.

3. Comparing 8 miles and 5 gallons by subtraction makes no sense.

   8 miles − 5 gallons = ?

4. Compare 36 and 4 by division.

   36 ÷ 4 = 9

   This means that 36 is 9 times as large as 4. Recall that 36 ÷ 4 = 9 can be expressed as $\frac{36}{4} = 9$.

5. Compare 8 miles and 2 miles by division.

$$\frac{8 \text{ miles}}{2 \text{ miles}} = 4$$

This means that 8 miles is 4 times as large as 2 miles.

Example of use: I can jog 8 miles to your 2 miles. Or, for every 2 miles that you jog, I jog 8. So, I jog 4 times as many miles as you jog.

Notice that when like quantities are being compared by division, we drop the units. Another way of looking at this is that the units divide out (cancel).

6. Compare 30 miles and 2 gallons by division.

$$\frac{30 \text{ miles}}{2 \text{ gallons}} = \frac{15 \text{ miles}}{1 \text{ gallon}}$$

Example of use: A particular car goes 30 miles on 2 gallons of gasoline. This is the same as getting 15 miles to 1 gallon of gasoline.

Notice that when the quantities being compared by division are unlike quantities, we do not drop the units.

★ PRACTICE SET A

Make the following comparisons and interpret each one.

1. Compare 10 diskettes to 2 diskettes by
   (a) subtraction:

   (b) division:

2. Compare, if possible, 16 bananas and 2 bags by
   (a) subtraction:

   (b) division.

☐ RATIOS AND RATES

Ratio

> A comparison, by division, of two pure numbers or two like denominate numbers is a **ratio**.

The comparison by division of the pure numbers $\frac{36}{4}$ and the like denominate numbers $\frac{8 \text{ miles}}{2 \text{ miles}}$ are examples of ratios.

Rate

> A comparison, by division, of two unlike denominate numbers is a **rate**.

The comparison by division of two unlike denominate numbers, such as

$$\frac{55 \text{ miles}}{1 \text{ gallon}} \quad \text{and} \quad \frac{40 \text{ dollars}}{5 \text{ tickets}}$$

are examples of rates.

Let's agree to represent two numbers (pure or denominate) with the letters a and b. This means that we're letting a represent some number and b represent some, perhaps different, number. With this agreement, we can write the ratio of the two numbers a and b as

$$\frac{a}{b} \quad \text{or} \quad \frac{b}{a}$$

The ratio $\frac{a}{b}$ is read as "a to b."

The ratio $\frac{b}{a}$ is read as "b to a."

Since a ratio or a rate can be expressed as a fraction, it may be reducible.

☆ SAMPLE SET B

1. The ratio 30 to 2 can be expressed as $\frac{30}{2}$. Reducing, we get $\frac{15}{1}$.

   The ratio 30 to 2 is *equivalent* to the ratio 15 to 1.

2. The rate "4 televisions to 12 people" can be expressed as $\frac{4 \text{ televisions}}{12 \text{ people}}$. The meaning of this rate is that "for every 4 televisions, there are 12 people."

   Reducing, we get $\frac{1 \text{ television}}{3 \text{ people}}$. The meaning of this rate is that "for every 1 television, there are 3 people."

   Thus, the rate of "4 televisions to 12 people" is the *same* as the rate of "1 television to 3 people."

★ PRACTICE SET B

Write the following ratios and rates as fractions.

1. 3 to 2   **2.** 1 to 9   **3.** 5 books to 4 people   **4.** 120 miles to 2 hours

5. 8 liters to 3 liters

Write the following ratios and rates in the form "*a* to *b*." Reduce when necessary.

6. $\dfrac{9}{5}$ **7.** $\dfrac{1}{3}$ **8.** $\dfrac{25 \text{ miles}}{2 \text{ gallons}}$ **9.** $\dfrac{2 \text{ mechanics}}{4 \text{ wrenches}}$ **10.** $\dfrac{15 \text{ video tapes}}{18 \text{ video tapes}}$

Answers to Practice Sets are on p. 309.

Section 7.1 EXERCISES

For problems 1–9, complete the statements.

1. Two numbers can be compared by subtraction if and only if

_____ .

2. A comparison, by division, of two pure numbers or two like denominate numbers is called a

_____ .

3. A comparison, by division, of two unlike denominate numbers is called a

_____ .

4. $\dfrac{6}{11}$ is an example of a _____ .
    (ratio/rate)

5. $\dfrac{5}{12}$ is an example of a _____ .
    (ratio/rate)

6. $\dfrac{7 \text{ erasers}}{12 \text{ pencils}}$ is an example of a _____ .
    (ratio/rate)

7. $\dfrac{20 \text{ silver coins}}{35 \text{ gold coins}}$ is an example of a _____ .
    (ratio/rate)

8. $\dfrac{3 \text{ sprinklers}}{5 \text{ sprinklers}}$ is an example of a _____ .
    (ratio/rate)

9. $\dfrac{18 \text{ exhaust valves}}{11 \text{ exhaust valves}}$ is an example of a

_____ .
(ratio/rate)

For problems 10–16, write each ratio or rate as a verbal phrase.

10. $\dfrac{8}{3}$

11. $\dfrac{2}{5}$

12. $\dfrac{8 \text{ feet}}{3 \text{ seconds}}$

13. $\dfrac{29 \text{ miles}}{2 \text{ gallons}}$

14. $\dfrac{30,000 \text{ stars}}{300 \text{ stars}}$

15. $\dfrac{5 \text{ yards}}{2 \text{ yards}}$

16. $\dfrac{164 \text{ trees}}{28 \text{ trees}}$

For problems 17–30, write the simplified fractional form of each ratio or rate.

17. 12 to 5

18. 81 to 19

19. 42 plants to 5 homes

20. 8 books to 7 desks

21. 16 pints to 1 quart

22. 4 quarts to 1 gallon

23. 2.54 cm to 1 in

24. 80 tables to 18 tables

25. 25 cars to 10 cars

26. 37 wins to 16 losses

27. 105 hits to 315 at bats

28. 510 miles to 22 gallons

29. 1,042 characters to 1 page

30. 1,245 pages to 2 books

EXERCISES FOR REVIEW

(4.2)   **31.** Convert $\dfrac{16}{3}$ to a mixed number.

(4.6)   **32.** $1\dfrac{5}{9}$ of $2\dfrac{4}{7}$ is what number?

(5.2)   **33.** Find the difference. $\dfrac{11}{28} - \dfrac{7}{45}$.

(6.7)   **34.** Perform the division. If no repeating pattern seems to exist, round the quotient to three decimal places: $22.35 \div 17$.

(6.9)   **35.** Find the value of $1.85 + \dfrac{3}{8} \cdot 4.1$.

★ **Answers to Practice Sets (7.1)**

A.   **1. (a)** 8 diskettes; 10 diskettes is 8 diskettes more than 2 diskettes.
      (b) 5; 10 diskettes is 5 times as many diskettes as 2 diskettes.
   2. (a) Comparison by subtraction makes no sense.
      (b) $\dfrac{16 \text{ bananas}}{2 \text{ bags}} = \dfrac{8 \text{ bananas}}{\text{bag}}$, 8 bananas per bag.

B.   **1.** $\dfrac{3}{2}$ **2.** $\dfrac{1}{9}$ **3.** $\dfrac{5 \text{ books}}{4 \text{ people}}$ **4.** $\dfrac{60 \text{ miles}}{1 \text{ hour}}$ **5.** $\dfrac{8}{3}$ **6.** 9 to 5   **7.** 1 to 3
   8. 25 miles to 2 gallons   **9.** 1 mechanic to 2 wrenches   **10.** 5 to 6

7.2 Proportions

Section Overview

- ☐ **RATIOS, RATES, AND PROPORTIONS**
- ☐ **FINDING THE MISSING FACTOR IN A PROPORTION**
- ☐ **PROPORTIONS INVOLVING RATES**

☐ RATIOS, RATES, AND PROPORTIONS

Ratio
Rate

We have defined a **ratio** as a comparison, by division, of two pure numbers or two *like* denominate numbers. We have defined a **rate** as a comparison, by division, of two *unlike* denominate numbers.

Proportion

A **proportion** is a statement that two ratios or rates are equal. The following two examples show how to read proportions.

$$\frac{3}{4} = \frac{6}{8} \qquad\qquad \frac{25 \text{ miles}}{1 \text{ gallon}} = \frac{50 \text{ miles}}{2 \text{ gallons}}$$

  ↓ ↓ ↓   ↓   ↓   ↓

3 is to 4 as 6 is to 8   25 miles is to 1 gallon as 50 miles is to 2 gallons

☆ SAMPLE SET A

Write or read each proportion.

1. $\dfrac{3}{5} = \dfrac{12}{20}$

   3 is to 5 as 12 is to 20

2. $\dfrac{10 \text{ items}}{5 \text{ dollars}} = \dfrac{2 \text{ items}}{1 \text{ dollar}}$

   10 items is to 5 dollars as 2 items is to 1 dollar

3. 8 is to 12 as 16 is to 24.

   $\dfrac{8}{12} = \dfrac{16}{24}$

4. 50 milligrams of vitamin C is to 1 tablet as 300 milligrams of vitamin C is to 6 tablets.

   $\dfrac{50}{1} = \dfrac{300}{6}$

★ PRACTICE SET A

Write or read each proportion.

1. $\dfrac{3}{8} = \dfrac{6}{16}$

2. $\dfrac{2 \text{ people}}{1 \text{ window}} = \dfrac{10 \text{ people}}{5 \text{ windows}}$

3. 15 is to 4 as 75 is to 20.

4. 2 plates are to 1 tray as 20 plates are to 10 trays.

☐ FINDING THE MISSING FACTOR IN A PROPORTION

Many practical problems can be solved by writing the given information as proportions. Such proportions will be composed of three specified numbers and one unknown number. It is customary to let a letter, such as x, represent the unknown number. An example of such a proportion is

$$\frac{x}{4} = \frac{20}{16}$$

This proportion is read as "x is to 4 as 20 is to 16."

There is a method of solving these proportions that is based on the equality of fractions. Recall that two fractions are equivalent if and only if their cross products are equal. For example,

$$\frac{3}{4} = \frac{6}{8} \qquad \text{since} \qquad \frac{3}{4} \diagdown \frac{6}{8}$$

$$3 \cdot 8 = 6 \cdot 4$$
$$24 = 24$$

Notice that in a proportion that contains three specified numbers and a letter representing an unknown quantity, that regardless of where the letter appears, the following situation always occurs.

$$\underbrace{(\text{number}) \cdot (\text{letter}) = (\text{number}) \cdot (\text{number})}$$

We recognize this as a multiplication statement. Specifically, it is a missing factor statement. (See Section 4.6 for a discussion of multiplication statements.) For example,

$\dfrac{x}{4} = \dfrac{20}{16}$ means that $16 \cdot x = 4 \cdot 20$.

$\dfrac{4}{x} = \dfrac{16}{20}$ means that $4 \cdot 20 = 16 \cdot x$.

$\dfrac{5}{4} = \dfrac{x}{16}$ means that $5 \cdot 16 = 4 \cdot x$.

$\dfrac{5}{4} = \dfrac{20}{x}$ means that $5 \cdot x = 4 \cdot 20$.

Each of these statements is a multiplication statement. Specifically, each is a missing factor statement. (The letter used here is x, whereas M was used in Section 4.6.)

Finding the Missing Factor in a Proportion

> The missing factor in a missing factor statement can be determined by dividing the product by the known factor, that is, if x represents the missing factor, then
>
> $x = (\text{product}) \div (\text{known factor})$

☆ SAMPLE SET B

Find the unknown number in each proportion.

1. $\dfrac{x}{4} = \dfrac{20}{16}$.   **Find the cross product.**

$16 \cdot x = 20 \cdot 4$
$16 \cdot x = 80$ **Divide the product 80 by the known factor 16.**

$x = \dfrac{80}{16}$

$x = 5$ **The unknown number is 5.**

This means that $\dfrac{5}{4} = \dfrac{20}{16}$, or 5 is to 4 as 20 is to 16.

2. $\dfrac{5}{x} = \dfrac{20}{16}$.   **Find the cross product.**

$5 \cdot 16 = 20 \cdot x$
$80 = 20 \cdot x$ **Divide the product 80 by the known factor 20.**

$\dfrac{80}{20} = x$

$4 = x$ **The unknown number is 4.**

This means that $\dfrac{5}{4} = \dfrac{20}{16}$, or, 5 is to 4 as 20 is to 16.

Continued

3. $\dfrac{16}{3} = \dfrac{64}{x}$ **Find the cross product.**

$16 \cdot x = 64 \cdot 3$
$16 \cdot x = 192$ **Divide 192 by 16.**

$x = \dfrac{192}{16}$

$x = 12$ **The unknown number is 12.**

This means that $\dfrac{16}{3} = \dfrac{64}{12}$, or, 16 is to 3 as 64 is to 12.

4. $\dfrac{9}{8} = \dfrac{x}{40}$ **Find the cross products.**

$9 \cdot 40 = 8 \cdot x$
$360 = 8 \cdot x$ **Divide 360 by 8.**

$\dfrac{360}{8} = x$

$45 = x$ **The unknown number is 45.**

★ **PRACTICE SET B**

Find the unknown number in each proportion.

1. $\dfrac{x}{8} = \dfrac{12}{32}$ 2. $\dfrac{7}{x} = \dfrac{14}{10}$ 3. $\dfrac{9}{11} = \dfrac{x}{55}$ 4. $\dfrac{1}{6} = \dfrac{8}{x}$

Answers to Practice Sets are on p. 316.

❏ PROPORTIONS INVOLVING RATES

Recall that a rate is a comparison, by division, of unlike denominate numbers. We must be careful when setting up proportions that involve rates. The *form* is important. For example, if a rate involves two types of units, say unit type 1 and unit type 2, we can write

$$\frac{\text{unit type 1}}{\text{unit type 2}} = \frac{\text{unit type 1}}{\text{unit type 2}}$$

Same units appear on same side.
Same units appear on same side.

or

$$\frac{\text{unit type 1}}{\text{unit type 1}} = \frac{\text{unit type 2}}{\text{unit type 2}}$$

Same units appear on same side.
Same units appear on same side.

Both cross products produce a statement of the type

(unit type 1) $\cdot$ (unit type 2) $=$ (unit type 1) $\cdot$ (unit type 2)

which we take to mean the comparison

$\underbrace{\text{(unit type 1) is to (unit type 2)}}_{\substack{\text{Comparison of type 1} \\ \text{with type 2}}}$ as $\underbrace{\text{(unit type 1) is to (unit type 2)}}_{\substack{\text{Comparison of type 1} \\ \text{with type 2}}}$

Same overall type

Examples of correctly expressed proportions are the following:

1. $\dfrac{\text{mi}}{\text{hr}} = \dfrac{\text{mi}}{\text{hr}}$ Same units appear on the same side

2. $\dfrac{\text{mi}}{\text{mi}} = \dfrac{\text{hr}}{\text{hr}}$

Same units appear on the same side.

However, if we write the same types of units on different sides, such as,

$$\frac{\text{unit type 1}}{\text{unit type 2}} = \frac{\text{unit type 2}}{\text{unit type 1}}$$

the cross product produces a statement of the form

$\underbrace{\text{(unit type 1)} \cdot \text{(unit type 1)}}_{\substack{\text{Comparison of type 1} \\ \text{with type 1}}} = \underbrace{\text{(unit type 2)} \cdot \text{(unit type 2)}}_{\substack{\text{Comparison of type 2} \\ \text{with type 2}}}$

Different overall types

We can see that this is an incorrect comparison by observing the following example: It is *incorrect* to write

$$\frac{2 \text{ hooks}}{3 \text{ poles}} = \frac{6 \text{ poles}}{4 \text{ hooks}}$$

for two reasons.

1. The cross product is numerically wrong: $(2 \cdot 4 \neq 3 \cdot 6)$.
2. The cross product produces the statement "hooks are to hooks as poles are to poles," which makes no sense.

Section 7.2 EXERCISES

1. A statement that two ratios or _____ are equal is called a _____.

For problems 2–10, write each proportion in fractional form.

2. 3 is to 7 as 18 is to 42.

3. 1 is to 11 as 3 is to 33.

4. 9 is to 14 as 27 is to 42.

5. 6 is to 90 as 3 is to 45.

6. 5 liters is to 1 bottle as 20 liters is to 4 bottles.

7. 18 grams of cobalt is to 10 grams of silver as 36 grams of cobalt is to 20 grams of silver.

8. 4 cups of water is to 1 cup of sugar as 32 cups of water is to 8 cups of sugar.

9. 3 people absent is to 31 people present as 15 people absent is to 155 people present.

10. 6 dollars is to 1 hour as 90 dollars is to 15 hours.

For problems 11–20, write each proportion as a sentence.

11. $\dfrac{3}{4} = \dfrac{15}{20}$

12. $\dfrac{1}{8} = \dfrac{5}{40}$

13. $\dfrac{3 \text{ joggers}}{100 \text{ feet}} = \dfrac{6 \text{ joggers}}{200 \text{ feet}}$

14. $\dfrac{12 \text{ marshmallows}}{3 \text{ sticks}} = \dfrac{36 \text{ marshmallows}}{9 \text{ sticks}}$

15. $\dfrac{40 \text{ miles}}{80 \text{ miles}} = \dfrac{2 \text{ gallons}}{4 \text{ gallons}}$

16. $\dfrac{4 \text{ couches}}{10 \text{ couches}} = \dfrac{2 \text{ houses}}{5 \text{ houses}}$

17. $\dfrac{1 \text{ person}}{1 \text{ job}} = \dfrac{8 \text{ people}}{8 \text{ jobs}}$

18. $\dfrac{1 \text{ popsicle}}{2 \text{ children}} = \dfrac{\frac{1}{2} \text{ popsicle}}{1 \text{ child}}$

19. $\dfrac{2{,}000 \text{ pounds}}{1 \text{ ton}} = \dfrac{60{,}000 \text{ pounds}}{30 \text{ tons}}$

20. $\dfrac{1 \text{ table}}{5 \text{ tables}} = \dfrac{2 \text{ people}}{10 \text{ people}}$

For problems 21–30, solve each proportion.

21. $\dfrac{x}{5} = \dfrac{6}{15}$

22. $\dfrac{x}{10} = \dfrac{28}{40}$

23. $\dfrac{5}{x} = \dfrac{10}{16}$

24. $\dfrac{13}{x} = \dfrac{39}{60}$

25. $\dfrac{1}{3} = \dfrac{x}{24}$

26. $\dfrac{7}{12} = \dfrac{x}{60}$

27. $\dfrac{8}{3} = \dfrac{72}{x}$

28. $\dfrac{16}{1} = \dfrac{48}{x}$

29. $\dfrac{x}{25} = \dfrac{200}{125}$

30. $\dfrac{65}{30} = \dfrac{x}{60}$

For problems 31–36, express each sentence as a proportion then solve the proportion.

31. 5 hats are to 4 coats as x hats are to 24 coats.

32. x cushions are to 2 sofas as 24 cushions are to 16 sofas.

33. 1 spacecraft is to 7 astronauts as 5 spacecraft are to x astronauts.

34. 56 microchips are to x circuit boards as 168 microchips are to 3 circuit boards.

35. 18 calculators are to 90 calculators as x students are to 150 students.

36. x dollars are to $40,000 as 2 sacks are to 1 sack.

For problems 37–44, indicate whether the proportion is true or false.

37. $\dfrac{3}{16} = \dfrac{12}{64}$

38. $\dfrac{2}{15} = \dfrac{10}{75}$

39. $\dfrac{1}{9} = \dfrac{3}{30}$

40. $\dfrac{6 \text{ knives}}{7 \text{ forks}} = \dfrac{12 \text{ knives}}{15 \text{ forks}}$

41. $\dfrac{33 \text{ miles}}{1 \text{ gallon}} = \dfrac{99 \text{ miles}}{3 \text{ gallons}}$

42. $\dfrac{320 \text{ feet}}{5 \text{ seconds}} = \dfrac{65 \text{ feet}}{1 \text{ second}}$

43. $\dfrac{35 \text{ students}}{70 \text{ students}} = \dfrac{1 \text{ class}}{2 \text{ classes}}$

44. $\dfrac{9 \text{ ml chloride}}{45 \text{ ml chloride}} = \dfrac{1 \text{ test tube}}{7 \text{ test tubes}}$

EXERCISES FOR REVIEW

(1.6) **45.** Use the numbers 5 and 7 to illustrate the commutative property of addition.

(2.6) **46.** Use the numbers 5 and 7 to illustrate the commutative property of multiplication.

(5.2) **47.** Find the difference. $\dfrac{5}{14} - \dfrac{3}{22}$.

(6.5) **48.** Find the product. $8.06129 \cdot 1{,}000$.

(7.1) **49.** Write the simplified fractional form of the rate "sixteen sentences to two paragraphs."

★ **Answers to Practice Sets (7.2)**

A. **1.** 3 is to 8 as 6 is to 16   **2.** 2 people are to 1 window as 10 people are to 5 windows   **3.** $\dfrac{15}{4} = \dfrac{75}{20}$

   4. $\dfrac{2 \text{ plates}}{1 \text{ tray}} = \dfrac{20 \text{ plates}}{10 \text{ trays}}$

B. **1.** $x = 3$ **2.** $x = 5$ **3.** $x = 45$ **4.** $x = 48$

7.3 Applications of Proportions

Section Overview

☐ **THE FIVE-STEP METHOD**
☐ **PROBLEM SOLVING**

☐ THE FIVE-STEP METHOD

In Section 7.2 we noted that many practical problems can be solved by writing the given information as proportions. Such proportions will be composed of three specified numbers and one unknown number represented by a letter.

The first and most important part of solving a proportion problem is to determine, by careful reading, what the unknown quantity is and to represent it with some letter.

The Five-Step Method

The five-step method for solving proportion problems:

1. By careful reading, determine what the unknown quantity is and represent it with some letter. There will be only one unknown in a problem.
2. Identify the three specified numbers.
3. Determine which comparisons are to be made and set up the proportion.
4. Solve the proportion (using the methods of Section 7.2).
5. Interpret and write a conclusion in a sentence with the appropriate units of measure.

Step 1 is extremely important. Many problems go unsolved because time is not taken to establish what quantity is to be found.

When solving an applied problem, **always** begin by determining the unknown quantity and representing it with a letter.

☐ PROBLEM SOLVING

☆ **SAMPLE SET A**

1. On a map, 2 inches represents 25 miles. How many miles are represented by 8 inches?

Step 1:   The unknown quantity is miles.

Let x = number of miles represented by 8 inches
Step 2:   The three specified numbers are

2 inches
25 miles
8 inches

Step 3: The comparisons are

$$2 \text{ inches to } 25 \text{ miles} \longrightarrow \frac{2 \text{ inches}}{25 \text{ miles}}$$

$$8 \text{ inches to } x \text{ miles} \longrightarrow \frac{8 \text{ inches}}{x \text{ miles}}$$

Proportions involving ratios and rates are more readily solved by suspending the units while doing the computations.

$$\frac{2}{25} = \frac{8}{x}$$

Step 4: $\frac{2}{25} = \frac{8}{x}$. **Perform the cross multiplication.**

$$2 \cdot x = 8 \cdot 25$$
$$2 \cdot x = 200 \qquad \textbf{Divide 200 by 2.}$$
$$x = \frac{200}{2}$$
$$x = 100$$

In step 1, we let x represent the number of miles. So, x represents 100 miles.

Step 5: If 2 inches represents 25 miles, then 8 inches represents 100 miles.

Try problem 1 in Practice Set A.

2. An acid solution is composed of 7 parts water to 2 parts acid. How many parts of water are there in a solution composed of 20 parts acid?

Step 1: The unknown quantity is the number of parts of water.

Let $n =$ number of parts of water.

Step 2: The three specified numbers are

7 parts water
2 parts acid
20 parts acid

Step 3: The comparisons are

$$7 \text{ parts water to } 2 \text{ parts acid} \longrightarrow \frac{7}{2}$$

$$n \text{ parts water to } 20 \text{ parts acid} \longrightarrow \frac{n}{20}$$

$$\frac{7}{2} = \frac{n}{20}$$

Step 4: $\frac{7}{2} = \frac{n}{20}$. **Perform the cross multiplication.**

$$7 \cdot 20 = 2 \cdot n$$
$$140 = 2 \cdot n \qquad \textbf{Divide 140 by 2.}$$
$$\frac{140}{2} = n$$
$$70 = n$$

In step 1 we let n represent the number of parts of water. So, n represents 70 parts of water.

Step 5: 7 parts water to 2 parts acid indicates 70 parts water to 20 parts acid.

Try problem 2 in Practice Set A.

Continued

3. A 5-foot girl casts a $3\frac{1}{3}$-foot shadow at a particular time of the day. How tall is a person who casts a 3-foot shadow at the same time of the day?

Step 1: The unknown quantity is the height of the person.

Let h = height of the person.

Step 2: The three specified numbers are

5 feet (height of girl)

$3\frac{1}{3}$ feet (length of shadow)

3 feet (length of shadow)

Step 3: The comparisons are

5-foot girl is to $3\frac{1}{3}$-foot shadow $\longrightarrow \dfrac{5}{3\frac{1}{3}}$

h-foot person is to 3-foot shadow $\longrightarrow \dfrac{h}{3}$

$$\frac{5}{3\frac{1}{3}} = \frac{h}{3}$$

Step 4: $\dfrac{5}{3\frac{1}{3}} = \dfrac{h}{3}$

$$5 \cdot 3 = 3\frac{1}{3} \cdot h$$

$$15 = \frac{10}{3} \cdot h \qquad \textbf{Divide 15 by } \frac{10}{3}.$$

$$\frac{15}{\frac{10}{3}} = h$$

$$\frac{\overset{3}{\cancel{15}}}{1} \cdot \frac{3}{\underset{2}{\cancel{10}}} = h$$

$$\frac{9}{2} = h$$

$$h = 4\frac{1}{2}$$

Step 5: A person who casts a 3-foot shadow at this particular time of the day is $4\frac{1}{2}$ feet tall.

Try problem 3 in Practice Set A.

4. The ratio of men to women in a particular town is 3 to 5. How many women are there in the town if there are 19,200 men in town?

Step 1: The unknown quantity is the number of women in town.

Let x = number of women in town.

Step 2: The three specified numbers are

3
5
19,200

Step 3: The comparisons are 3 men to 5 women $\longrightarrow \dfrac{3}{5}$

19,200 men to x women $\longrightarrow \dfrac{19,200}{x}$

$$\dfrac{3}{5} = \dfrac{19,200}{x}$$

Step 4: $\dfrac{3}{5} = \dfrac{19,200}{x}$

$3 \cdot x = 19,200 \cdot 5$
$3 \cdot x = 96,000$

$x = \dfrac{96,000}{3}$

$x = 32,000$

Step 5: There are 32,000 women in town.

5. The rate of wins to losses of a particular baseball team is $\dfrac{9}{2}$. How many games did this team lose if they won 63 games?

Step 1: The unknown quantity is the number of games lost.
Let $n =$ number of games lost.

Step 2: Since $\dfrac{9}{2} \longrightarrow$ means 9 wins to 2 losses, the three specified numbers are

9 (wins)
2 (losses)
63 (wins)

Step 3: The comparisons are

9 wins to 2 losses $\longrightarrow \dfrac{9}{2}$

63 wins to n losses $\longrightarrow \dfrac{63}{n}$

$$\dfrac{9}{2} = \dfrac{63}{n}$$

Step 4: $\dfrac{9}{2} = \dfrac{63}{n}$

$9 \cdot n = 2 \cdot 63$
$9 \cdot n = 126$

$n = \dfrac{126}{9}$

$n = 14$

Step 5: This team had 14 losses.
Try problem 4 in Practice Set A.

★ **PRACTICE SET A**

Solve each problem.

1. On a map, 3 inches represents 100 miles. How many miles are represented by 15 inches?

Step 1:

Step 2:

Step 3:

Step 4:

Step 5:

2. An alcohol solution is composed of 14 parts water to 3 parts alcohol. How many parts of alcohol are in a solution that is composed of 112 parts water?

Step 1:

Step 2:

Step 3:

Step 4:

Step 5:

3. A $5\frac{1}{2}$-foot woman casts a 7-foot shadow at a particular time of the day. How long of a shadow does a 3-foot boy cast at that same time of day?

Step 1:

Step 2:

Step 3:

Step 4:

Step 5:

4. The rate of houseplants to outside plants at a nursery is 4 to 9. If there are 384 houseplants in the nursery, how many outside plants are there?

Step 1:

Step 2:

Step 3:

Step 4:

Step 5:

5. The odds for a particular event occurring are 11 to 2. (For every 11 times the event does occur, it will not occur 2 times.) How many times does the event occur if it does not occur 18 times?

   Step 1:

   Step 2:

   Step 3:

   Step 4:

   Step 5:

6. The rate of passing grades to failing grades in a particular chemistry class is $\frac{7}{2}$. If there are 21 passing grades, how many failing grades are there?

   Step 1:

   Step 2:

   Step 3:

   Step 4:

   Step 5:

Answers to the Practice Set are on p. 325.

Section 7.3 EXERCISES

For problems 1–20, use the five-step method to solve each problem.

1. On a map, 4 inches represents 50 miles. How many inches represent 300 miles?

2. On a blueprint for a house, 2 inches represents 3 feet. How many inches represent 10 feet?

3. A model is built to $\dfrac{2}{15}$ scale. If a particular part of the model measures 6 inches, how long is the actual structure?

4. An acid solution is composed of 5 parts acid to 9 parts of water. How many parts of acid are there in a solution that contains 108 parts of water?

5. An alloy contains 3 parts of nickel to 4 parts of silver. How much nickel is in an alloy that contains 44 parts of silver?

6. The ratio of water to salt in a test tube is 5 to 2. How much salt is in a test tube that contains 35 ml of water?

7. The ratio of sulfur to air in a container is $\dfrac{4}{45}$. How many ml of air are there in a container that contains 207 ml of sulfur?

8. A 6-foot man casts a 4-foot shadow at a particular time of the day. How tall is a person that casts a 3-foot shadow at that same time of the day?

9. A $5\dfrac{1}{2}$-foot woman casts a $1\dfrac{1}{2}$-foot shadow at a particular time of the day. How long a shadow does her $3\dfrac{1}{2}$-foot niece cast at the same time of the day?

10. A man, who is 6 feet tall, casts a 7-foot shadow at a particular time of the day. How tall is a tree that casts an 84-foot shadow at that same time of the day?

11. The ratio of books to shelves in a bookstore is 350 to 3. How many books are there in a store that has 105 shelves?

12. The ratio of algebra classes to geometry classes at a particular community college is 13 to 2. How many geometry classes does this college offer if it offers 13 algebra classes?

13. The odds for a particular event to occur are 16 to 3. If this event occurs 64 times, how many times would you predict it does not occur?

14. The odds against a particular event occurring are 8 to 3. If this event does occur 64 times, how many times would you predict it does not occur?

15. The owner of a stationery store knows that a 1-inch stack of paper contains 300 sheets. The owner wishes to stack the paper in units of 550 sheets. How many inches tall should each stack be?

16. A recipe that requires 6 cups of sugar for 15 servings is to be used to make 45 servings. How much sugar will be needed?

17. A pond loses $7\frac{1}{2}$ gallons of water every 2 days due to evaporation. How many gallons of water are lost, due to evaporation, in $\frac{1}{2}$ day?

18. A photograph that measures 3 inches wide and $4\frac{1}{2}$ inches high is to be enlarged so that it is 5 inches wide. How high will it be?

19. If 25 pounds of fertilizer covers 400 square feet of grass, how many pounds will it take to cover 500 square feet of grass?

20. Every $1\frac{1}{2}$ teaspoons of a particular multiple vitamin, in granular form, contains 0.65 the minimum daily requirement of vitamin C. How many teaspoons of this vitamin are required to supply 1.25 the minimum daily requirement?

EXERCISES FOR REVIEW

(2.1) **21.** Find the product. $818 \cdot 0$.

(4.3) **22.** Determine the missing numerator: $\dfrac{8}{15} = \dfrac{N}{90}$.

(5.5) **23.** Find the value of $\dfrac{\dfrac{3}{10} + \dfrac{4}{12}}{\dfrac{19}{20}}$.

(6.4) **24.** Subtract 0.249 from the sum of 0.344 and 0.612.

(7.2) **25.** Solve the proportion: $\dfrac{6}{x} = \dfrac{36}{30}$.

★ **Answers to Practice Set (7.3)**

A. **1.** 500 miles **2.** 24 parts of alcohol **3.** $3\dfrac{9}{11}$ feet **4.** 864 outside plants

5. The event occurs 99 times. **6.** 6 failing grades

7.4 Percent

Section
Overview

☐ **RATIOS AND PERCENTS**
☐ **THE RELATIONSHIP BETWEEN FRACTIONS, DECIMALS, AND PERCENTS — MAKING CONVERSIONS**

☐ RATIOS AND PERCENTS

Ratio

Percent

We defined a **ratio** as a comparison, by division, of two pure numbers or two like denominate numbers. A most convenient number to compare numbers to is 100. Ratios in which one number is compared to 100 are called **percents.** The word *percent* comes from the Latin word "per centum." The word "per" means "for each" or "for every," and the word "centum" means "hundred." Thus, we have the following definition.

Percent means "for each hundred," or "for every hundred."

%

The symbol % is used to represent the word percent.

☆ **SAMPLE SET A**

1. The ratio 26 to 100 can be written as 26%.

We read 26% as "twenty-six percent."

2. The ratio $\dfrac{165}{100}$ can be written as 165%.

We read 165% as "one hundred sixty-five percent."

Continued

3. The percent 38% can be written as the fraction $\frac{38}{100}$.

4. The percent 210% can be written as the fraction $\frac{210}{100}$ or the mixed number $2\frac{10}{100}$ or 2.1.

5. Since one dollar is 100 cents, 25 cents is $\frac{25}{100}$ of a dollar. This implies that 25 cents is 25% of one dollar.

★ **PRACTICE SET A**

1. Write the ratio 16 to 100 as a percent.

2. Write the ratio 195 to 100 as a percent.

3. Write the percent 83% as a ratio in fractional form.

4. Write the percent 362% as a ratio in fractional form.

❑ THE RELATIONSHIP BETWEEN FRACTIONS, DECIMALS, AND PERCENTS — MAKING CONVERSIONS

Since a percent is a ratio, and a ratio can be written as a fraction, and a fraction can be written as a decimal, any of these forms can be converted to any other.

Before we proceed to the problems in Sample Set B and Practice Set B, let's summarize the conversion techniques.

CONVERSION TECHNIQUES — FRACTIONS, DECIMALS, PERCENTS

To Convert a Fraction	To Convert a Decimal	To Convert a Percent
To a decimal: Divide the numerator by the denominator.	**To a fraction:** Read the decimal and reduce the resulting fraction.	**To a decimal:** Move the decimal point 2 places to the left and drop the % symbol.
To a percent: Convert the fraction first to a decimal, then move the decimal point 2 places to the right and affix the % symbol.	**To a percent:** Move the decimal point 2 places to the right and affix the % symbol.	**To a fraction:** Drop the % sign and write the number "over" 100. Reduce, if possible.

☆ **SAMPLE SET B**

1. Convert 12% to a decimal.

$$12\% = \frac{12}{100} = 0.12$$

Note that $12\% = 12.\% = 0.12$

The % symbol is dropped, and the decimal point moves 2 places to the left.

2. Convert 0.75 to a percent.

$$0.75 = \frac{75}{100} = 75\%$$

Note that $0.75 = 75\% = 75.\%$

The % symbol is affixed, and the decimal point moves 2 units to the right.

3. Convert $\frac{3}{5}$ to a percent.

We see in problem 2 that we can convert a decimal to a percent. We also know that we can convert a fraction to a decimal. Thus, we can see that if we first convert the fraction to a decimal, we can then convert the decimal to a percent.

$$\frac{3}{5} \longrightarrow 5\overline{)3.0} \qquad \text{or} \qquad \frac{3}{5} = 0.6 = \frac{6}{10} = \frac{60}{100} = 60\%$$
$$\underline{.6}$$
$$\underline{3\,0}$$
$$0$$

4. Convert 42% to a fraction.

$$42\% = \frac{42}{100} = \frac{21}{50}$$

or

$$42\% = 0.42 = \frac{42}{100} = \frac{21}{50}$$

★ **PRACTICE SET B**

1. Convert 21% to a decimal.

2. Convert 461% to a decimal.

3. Convert 0.55 to a percent.

4. Convert 5.64 to a percent.

5. Convert $\frac{3}{20}$ to a percent.

6. Convert $\frac{11}{8}$ to a percent.

7. Convert $\frac{3}{11}$ to a percent.

Answers to Practice Sets are on p. 329.

Section 7.4 EXERCISES

For problems 1–12, convert each decimal to a percent.

1. 0.25

2. 0.36

3. 0.48

4. 0.343

5. 0.771

6. 1.42

7. 2.58

8. 4.976

9. 16.1814

10. 533.01

11. 2

12. 14

For problems 13–22, convert each percent to a decimal.

13. 15%

14. 43%

15. 16.2%

16. 53.8%

17. 5.05%

18. 6.11%

19. 0.78%

20. 0.88%

21. 0.09%

22. 0.001%

For problems 23–36, convert each fraction to a percent.

23. $\dfrac{1}{5}$

24. $\dfrac{3}{5}$

25. $\dfrac{5}{8}$

26. $\dfrac{1}{16}$

27. $\dfrac{7}{25}$

28. $\dfrac{16}{45}$

29. $\dfrac{27}{55}$

30. $\dfrac{15}{8}$

31. $\frac{41}{25}$ **32.** $6\frac{4}{5}$ **41.** 65%   **42.** 18%

33. $9\frac{9}{20}$ **34.** $\frac{1}{200}$ **43.** 12.5%   **44.** 37.5%

35. $\frac{6}{11}$ **36.** $\frac{35}{27}$ **45.** 512.5%   **46.** 937.5%

For problems 37–50, convert each percent to a fraction.

47. $9.\bar{9}$%   **48.** $55.\bar{5}$%

37. 80%   **38.** 60%

39. 25%   **40.** 75%   **49.** $22.\bar{2}$%   **50.** $63.\bar{6}$%

EXERCISES
FOR REVIEW

(4.5)   **51.** Find the quotient. $\frac{40}{54} \div 8\frac{7}{21}$.

(4.6)   **52.** $\frac{3}{8}$ of what number is $2\frac{2}{3}$?

(5.2)   **53.** Find the value of $\frac{28}{15} + \frac{7}{10} - \frac{5}{12}$.

(6.3)   **54.** Round 6.99997 to the nearest ten thousandths.

(7.3)   **55.** On a map, 3 inches represent 40 miles. How many inches represent 480 miles?

★ **Answers to Practice Sets (7.4)**

A. **1.** 16%   **2.** 195%   **3.** $\frac{83}{100}$ **4.** $\frac{362}{100}$ or $\frac{181}{50}$

B. **1.** 0.21   **2.** 4.61   **3.** 55%   **4.** 564%   **5.** 15%   **6.** 137.5%   **7.** $27.\overline{27}$%

7.5 Fractions of One Percent

❑ **CONVERSIONS INVOLVING FRACTIONS OF ONE PERCENT**
❑ **CONVERSIONS INVOLVING NONTERMINATING FRACTIONS**

❑ CONVERSIONS INVOLVING FRACTIONS OF ONE PERCENT

Percents such as $\frac{1}{2}\%$, $\frac{3}{5}\%$, $\frac{5}{8}\%$, and $\frac{7}{11}\%$, where 1% has not been attained, are fractions of 1%. This implies that

$$\frac{1}{2}\% = \frac{1}{2} \text{ of } 1\%$$

$$\frac{3}{5}\% = \frac{3}{5} \text{ of } 1\%$$

$$\frac{5}{8}\% = \frac{5}{8} \text{ of } 1\%$$

$$\frac{7}{11}\% = \frac{7}{11} \text{ of } 1\%$$

Since "percent" means "for each hundred," and "of" means "times," we have

$$\frac{1}{2}\% = \frac{1}{2} \text{ of } 1\% = \frac{1}{2} \cdot \frac{1}{100} = \frac{1}{200}$$

$$\frac{3}{5}\% = \frac{3}{5} \text{ of } 1\% = \frac{3}{5} \cdot \frac{1}{100} = \frac{3}{500}$$

$$\frac{5}{8}\% = \frac{5}{8} \text{ of } 1\% = \frac{5}{8} \cdot \frac{1}{100} = \frac{5}{800}$$

$$\frac{7}{11}\% = \frac{7}{11} \text{ of } 1\% = \frac{7}{11} \cdot \frac{1}{100} = \frac{7}{1100}$$

☆ **SAMPLE SET A**

1. Convert $\frac{2}{3}\%$ to a fraction.

$$\frac{2}{3}\% = \frac{2}{3} \text{ of } 1\% = \frac{\overset{1}{\cancel{2}}}{3} \cdot \frac{1}{\underset{50}{\cancel{100}}}$$

$$= \frac{1 \cdot 1}{3 \cdot 50}$$

$$= \frac{1}{150}$$

2. Convert $\frac{5}{8}\%$ to a decimal.

$$\frac{5}{8}\% = \frac{5}{8} \text{ of } 1\% = \frac{5}{8} \cdot \frac{1}{100}$$

$$= 0.625 \cdot 0.01$$

$$= 0.00625$$

★ **PRACTICE SET A**

1. Convert $\frac{1}{4}$% to a fraction.

2. Convert $\frac{3}{8}$% to a fraction.

3. Convert $3\frac{1}{3}$% to a fraction.

☐ CONVERSIONS INVOLVING NONTERMINATING FRACTIONS

We must be careful when changing a fraction of 1% to a decimal. The number $\frac{2}{3}$, as we know, has a nonterminating decimal representation. Therefore, it cannot be expressed exactly as a decimal.

When converting nonterminating fractions of 1% to decimals, it is customary to express the fraction as a rounded decimal with at least three decimal places.

Converting a Nonterminating Fraction to a Decimal

> To convert a nonterminating fraction of 1% to a decimal:
>
> 1. Convert the fraction as a rounded decimal.
> 2. Move the decimal point two digits to the left and remove the percent sign.

☆ **SAMPLE SET B**

1. Convert $\frac{2}{3}$% to a three-place decimal.

1. Convert $\frac{2}{3}$ to a decimal.

Since we wish the resulting decimal to have three decimal digits, and removing the percent sign will account for two of them, we need to round $\frac{2}{3}$ to one place ($2 + 1 = 3$).

$\frac{2}{3}$% = 0.7% to one decimal place. $\left(\frac{2}{3} = 0.6666\ldots\right)$

2. Move the decimal point two digits to the left and remove the % sign. We'll need to add zeros to locate the decimal point in the correct location.

$\frac{2}{3}$% = 0.007 to 3 decimal places

2. Convert $5\frac{4}{11}$% to a four-place decimal.

1. Since we wish the resulting decimal to have four decimal places, and removing the percent sign will account for two, we to round $\frac{4}{11}$ to two places.

$5\frac{4}{11}$% = 5.36% to two decimal places. $\left(\frac{4}{11} = 0.3636\ldots\right)$

2. Move the decimal point two places to the left and drop the percent sign.

$5\frac{4}{11}$% = 0.0536 to four decimal places.

Continued

3. Convert $28\frac{5}{9}\%$ to a decimal rounded to ten thousandths.

1. Since we wish the resulting decimal to be rounded to ten thousandths (four decimal places), and removing the percent sign will account for two, we need to round $\frac{5}{9}$ to two places.

$$28\frac{5}{9}\% = 28.56\% \text{ to two decimal places.} \qquad \left(\frac{5}{9} = 0.5555 \ldots\right)$$

2. Move the decimal point to the left two places and drop the percent sign.

$$28\frac{5}{9}\% = 0.2856 \text{ correct to ten thousandths.}$$

★ **PRACTICE SET B**

1. Convert $\frac{7}{9}\%$ to a three-place decimal.

2. Convert $51\frac{5}{11}\%$ to a decimal rounded to ten thousandths.

Answers to Practice Sets are on p. 334.

Section 7.5 EXERCISES

For problems 1–26, make the conversions as indicated.

1. Convert $\frac{3}{4}\%$ to a fraction.

2. Convert $\frac{5}{6}\%$ to a fraction.

3. Convert $\frac{1}{9}\%$ to a fraction.

4. Convert $\frac{15}{19}\%$ to a fraction.

5. Convert $\frac{5}{4}\%$ to a fraction.

6. Convert $\frac{7}{3}\%$ to a fraction.

7. Convert $1\frac{6}{7}\%$ to a fraction.

8. Convert $2\frac{5}{16}\%$ to a fraction.

9. Convert $25\frac{1}{4}\%$ to a fraction.

10. Convert $50\frac{1}{2}\%$ to a fraction.

11. Convert $72\frac{3}{5}\%$ to a fraction.

12. Convert $99\frac{1}{8}\%$ to a fraction.

13. Convert $136\frac{2}{3}\%$ to a fraction.

14. Convert $521\frac{3}{4}\%$ to a fraction.

15. Convert $10\frac{1}{5}\%$ to a decimal.

16. Convert $12\frac{3}{4}\%$ to a decimal.

17. Convert $3\frac{7}{8}\%$ to a decimal.

18. Convert $7\frac{1}{16}\%$ to a decimal.

19. Convert $\frac{3}{7}\%$ to a three-place decimal.

20. Convert $\frac{1}{9}\%$ to a three-place decimal.

21. Convert $6\frac{3}{11}\%$ to a four-place decimal.

22. Convert $9\frac{2}{7}\%$ to a four-place decimal.

23. Convert $24\frac{5}{21}\%$ to a three-place decimal.

24. Convert $45\frac{8}{27}\%$ to a three-place decimal.

25. Convert $11\frac{16}{17}\%$ to a four-place decimal.

26. Convert $5\frac{1}{7}\%$ to a three-place decimal.

EXERCISES FOR REVIEW

(3.1) **27.** Write $8 \cdot 8 \cdot 8 \cdot 8 \cdot 8$ using exponents.

(4.2) **28.** Convert $4\frac{7}{8}$ to an improper fraction.

(5.3) **29.** Find the sum. $\frac{7}{10} + \frac{2}{21} + \frac{1}{7}$.

(6.5) **30.** Find the product. $(4.21)(0.006)$.

(7.4) **31.** Convert 8.062 to a percent.

★ **Answers to Practice Sets (7.5)**

A. **1.** $\frac{1}{400}$ **2.** $\frac{3}{800}$ **3.** $\frac{1}{30}$

B. **1.** 0.008 **2.** 0.5145

7.6 Applications of Percents

Section Overview	☐ **BASE, PERCENT, AND PERCENTAGE** ☐ **FINDING THE PERCENTAGE** ☐ **FINDING THE PERCENT** ☐ **FINDING THE BASE**

☐ BASE, PERCENT, AND PERCENTAGE

There are three basic types of percent problems. Each type involves a base, a percent, and a percentage, and when they are translated from words to mathematical symbols *each becomes a multiplication statement*. Examples of these types of problems are the following:

1. What number is 30% of 50? (Missing product statement.)
2. 15 is what percent of 50? (Missing factor statement.)
3. 15 is 30% of what number? (Missing factor statement.)

In problem 1, the product is missing. To solve the problem, we represent the missing product with P.

$P = 30\% \cdot 50$

Percentage

The missing product P is called the **percentage**. Percentage means *part*, or *portion*. In $P = 30\% \cdot 50$, P represents a particular *part* of 50.

In problem 2, one of the factors is missing. Here we represent the missing factor with Q.

$$15 = Q \cdot 50$$

Percent

The missing factor is the **percent.** Percent, we know, means *per* 100, or *part of* 100. In $15 = Q \cdot 50$, Q indicates what part of 50 is being taken or considered. Specifically, $15 = Q \cdot 50$ means that if 50 was to be divided into 100 equal parts, then Q indicates 15 are being considered.

In problem 3, one of the factors is missing. Represent the missing factor with B.

$$15 = 30\% \cdot B$$

Base

The missing factor is the **base.** Some meanings of base are a *source of supply,* or a *starting place.* In $15 = 30\% \cdot B$, B indicates the amount of supply. Specifically, $15 = 30\% \cdot B$ indicates that 15 represents 30% of the total supply.

Each of these three types of problems is of the form

$$(\text{percentage}) = (\text{percent}) \cdot (\text{base})$$

We can determine any one of the three values given the other two using the methods discussed in Section 4.6.

☐ FINDING THE PERCENTAGE

☆ **SAMPLE SET A**

1. What number is 30% of 50? **Missing product statement.**

   $(\text{percentage}) = (\text{percent}) \cdot (\text{base})$

   | P | $=$ | 30% | $\cdot$ | 50 | **Convert 30% to a decimal.** |

   $P = .30 \cdot 50$ **Multiply.**

   $P = 15$

   Thus, 15 is 30% of 50.

   Do Practice Set A, problem 1.

2. What number is 36% of 95? **Missing product statement.**

   $(\text{percentage}) = (\text{percent}) \cdot (\text{base})$

   $P = 36\% \cdot 95$ **Convert 36% to a decimal.**

   $P = .36 \cdot 95$ **Multiply.**

   $P = 34.2$

   Thus, 34.2 is 36% of 95.

   Do Practice Set A, problem 1.

3. A salesperson, who gets a commission of 12% of each sale she makes, makes a sale of \$8,400.00. How much is her commission?

   We need to determine what part of \$8,400.00 is to be taken. What *part* indicates *percentage.*

Continued

$\underbrace{\text{What number}}$ is 12% of 8,400.00?   **Missing product statement.**

$$\underset{\downarrow}{\text{(percentage)}} = \underset{\downarrow}{\text{(percent)}} \cdot \underset{\downarrow}{\text{(base)}}$$

$$P = 12\% \cdot 8,400.00 \qquad \textbf{Convert to decimals.}$$

$$P = .12 \cdot 8,400.00 \qquad \textbf{Multiply.}$$

$$P = 1008.00$$

Thus, the salesperson's commission is $1,008.00.

Do Practice Set A, problem 2.

4. A girl, by practicing typing on her home computer, has been able to increase her typing speed by 110%. If she originally typed 16 words per minute, by how many words per minute was she able to increase her speed?

We need to determine what part of 16 has been taken. What *part* indicates *percentage*.

$\underbrace{\text{What number}}$ is 110% of 16?   **Missing product statement.**

$$\underset{\downarrow}{\text{(percentage)}} = \underset{\downarrow}{\text{(percent)}} \cdot \underset{\downarrow}{\text{(base)}}$$

$$P = 110\% \cdot 16 \qquad \textbf{Convert to decimals.}$$

$$P = 1.10 \cdot 16 \qquad \textbf{Multiply.}$$

$$P = 17.6$$

Thus, the girl has increased her typing speed by 17.6 words per minute. Her new speed is 16 + 17.6 = 33.6 words per minute.

Do Practice Set A, problem 3.

5. A student who makes $125 a month working part-time receives a 4% salary raise. What is the student's new monthly salary?

With a 4% raise, this student will make 100% of the original salary + 4% of the original salary. This means the new salary will be 104% of the original salary. We need to determine what part of $125 is to be taken. What *part* indicates *percentage*.

$\underbrace{\text{What number}}$ is 104% of 125   **Missing product statement.**

$$\underset{\downarrow}{\text{(percentage)}} = \underset{\downarrow}{\text{(percent)}} \cdot \underset{\downarrow}{\text{(base)}}$$

$$P = 104\% \cdot 125 \qquad \textbf{Convert to decimals.}$$

$$P = 1.04 \cdot 125 \qquad \textbf{Multiply.}$$

$$P = 130$$

Thus, this student's new monthly salary is $130.

Do Practice Set A, problem 4.

6. An article of clothing is on sale at 15% off the marked price. If the marked price is $24.95, what is the sale price?

Since the item is discounted 15%, the new price will be 100% − 15% = 85% of the marked price. We need to determine what part of 24.95 is to be taken. What *part* indicates *percentage*.

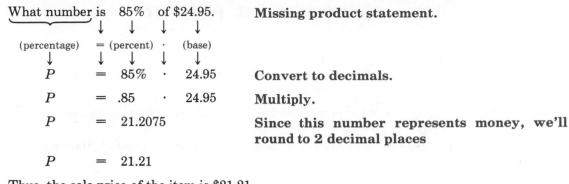

What number is 85% of $24.95.   **Missing product statement.**

$$(\text{percentage}) = (\text{percent}) \cdot (\text{base})$$

$P = 85\% \cdot 24.95$ **Convert to decimals.**

$P = .85 \cdot 24.95$ **Multiply.**

$P = 21.2075$ **Since this number represents money, we'll round to 2 decimal places**

$P = 21.21$

Thus, the sale price of the item is $21.21.

★ PRACTICE SET A

1. What number is 42% of 85?

2. A sales person makes a commission of 16% on each sale he makes. How much is his commission if he makes a sale of $8,500?

3. An assembly line worker can assemble 14 parts of a product in one hour. If he can increase his assembly speed by 35%, by how many parts per hour would he increase his assembly of products?

4. A computer scientist in the Silicon Valley makes $42,000 annually. What would this scientist's new annual salary be if she were to receive an 8% raise?

☐ FINDING THE PERCENT

☆ SAMPLE SET B

1.   15   is what percent of 50?   **Missing factor statement.**

(percentage) = (percent) · (base)   [(product) = (factor) · (factor)]

    15   = Q ·   50

Recall that (missing factor) = (product) ÷ (known factor).

$Q = 15 \div 50$ **Divide.**

$Q = 0.3$ **Convert to a percent.**

$Q = 30\%$

Thus, 15 is 30% of 50.

Do Practice Set B, problem 1.

2.   4.32   is what percent of 72?   **Missing factor statement.**

(percentage) = (percent) · (base)   [(product) = (factor) · (factor)]

    4.32   = Q ·   72

$Q = 4.32 \div 72$ **Divide.**

$Q = 0.06$ **Convert to a percent.**

$Q = 6\%$

Thus, 4.32 is 6% of 72.

Do Practice Set B, problem 1.

3. On a 160 question exam, a student got 125 correct answers. What percent is this? Round the result to two decimal places.

We need to determine the percent.

    125   is what percent of 160?   **Missing factor statement.**

(percentage) = (percent) · (base)   [(product) = (factor) · (factor)]

    125   = Q ·   160

$Q = 125 \div 160$ **Divide.**

$Q = 0.78125$ **Round to two decimal places.**

$Q = 0.78$

Thus, this student received a 78% on the exam.

Do Practice Set B, problem 2.

4. A bottle contains 80 milliliters of hydrochloric acid (HCl) and 30 milliliters of water. What percent of HCl does the bottle contain? Round the result to two decimal places.

We need to determine the percent. The total amount of liquid in the bottle is

80 milliliters + 30 milliliters = 110 milliliters.

80 is what percent of 110?   **Missing factor statement.**

(percentage) = (percent) · (base)   **[(product) = (factor) · (factor)]**

80 = Q · 110

$Q = 80 \div 110$ **Divide.**

$Q = 0.727272 \ldots$ **Round to two decimal places.**

$Q \approx 73\%$ **The symbol "≈" is read as "approximately."**

Thus, this bottle contains approximately 73% HCl.

Do Practice Set B, problem 3.

5. Five years ago a woman had an annual income of $19,200. She presently earns $42,000 annually. By what percent has her salary increased? Round the result to two decimal places.

We need to determine the percent.

42,000 is what percent of 19,200?   **Missing factor statement.**

(percentage) = (percent) · (base)

42,000 = Q · 19,200

$Q = 42,000 \div 19,200$ **Divide.**

$Q = 2.1875$ **Round to two decimal places.**

$Q = 2.19$ **Convert to a percent.**

$Q = 219\%$

Thus, this woman's annual salary has increased 219%.

★ PRACTICE SET B

1. 99.13 is what percent of 431?

2. On an 80 question exam, a student got 72 correct answers. What percent did the student get on the exam?

3. A bottle contains 45 milliliters of sugar and 67 milliliters of water. What fraction of sugar does the bottle contain? Round the result to two decimal places (then express as a percent).

☐ FINDING THE BASE

1.

$$15 \quad \text{is} \quad 30\% \quad \text{of} \underbrace{\text{what number?}}$$

(percentage) = (percent) · (base)

$$15 = 30\% \cdot B$$

$$15 = .30 \cdot B$$

$B = 15 \div .30$
$B = 50$

Thus, 15 is 30% of 50.

Try problem 1 in Practice Set C.

Missing factor statement.

[(percentage) = (factor) · (factor)]

Convert to decimals.

[(missing factor) = (product) ÷ (known factor)]

2.

$$56.43 \quad \text{is} \quad 33\% \quad \text{of} \underbrace{\text{what number?}}$$

(percentage) = (percent) · (base)

$$56.43 = 33\% \cdot B$$

$$56.43 = .33 \cdot B$$

$B = 56.43 \div .33$
$B = 171$

Thus, 56.43 is 33% of 171.

Try problem 1 in Practice Set C.

Missing factor statement.

Convert to decimals.

Divide.

3. Fifteen milliliters of water represents 2% of a hydrochloric acid (HCl) solution. How many milliliters of solution are there?

We need to determine the total supply. The word *supply* indicates *base*.

$$15 \quad \text{is} \quad 2\% \quad \text{of} \underbrace{\text{what number?}}$$

(percentage) = (percent) · (base)

$$15 = 2\% \cdot B$$

$$15 = .02 \cdot B$$

$B = 15 \div .02$
$B = 750$

Thus, there are 750 milliliters of solution in the bottle.

Try problem 2 in Practice Set C.

Missing factor statement.

Convert to decimals.

Divide.

4. In a particular city, a sales tax of $6\frac{1}{2}\%$ is charged on items purchased in local stores. If the tax on an item is $2.99, what is the price of the item?

We need to determine the price of the item. We can think of *price* as the *starting place*. *Starting place* indicates *base*. We need to determine the base.

2.99   is $6\frac{1}{2}\%$ of what number?   Missing factor statement.

$\downarrow$ $\downarrow$ $\downarrow$ $\downarrow$

(percentage) = (percent) · (base)

$\downarrow$ $\downarrow$ $\downarrow$ $\downarrow$ $\downarrow$

2.99   = $6\frac{1}{2}\%$ · B Convert to decimals.

2.99   =   6.5%   · B

2.99   =   .065   · B [(missing factor) = (product) ÷ (known factor)]

$B = 2.99 \div .065$ Divide.

$B = 46$

Thus, the price of the item is $46.00.

Try problem 3 in Practice Set C.

5. A clothing item is priced at $20.40. This marked price includes a 15% discount. What is the original price?

We need to determine the original price. We can think of the original price as the *starting place*. *Starting place* indicates *base*. We need to determine the base. The new price, $20.40, represents $100\% - 15\% = 85\%$ of the original price.

20.40   is   85%   of what number?   Missing factor statement.

$\downarrow$ $\downarrow$ $\downarrow$ $\downarrow$

(percentage) = (percent) · (base)

$\downarrow$ $\downarrow$ $\downarrow$ $\downarrow$ $\downarrow$

20.40   =   85%   · B Convert to decimals.

20.40   =   .85   · B [(missing factor) = (product) ÷ (known factor)]

$B = 20.40 \div .85$ Divide.

$B = 24$

Thus, the original price of the item is $24.00.

Try problem 4 in Practice Set C.

★ PRACTICE SET C

1. 1.98 is 2% of what number?

2. 3.3 milliliters of HCl represents 25% of an HCl solution. How many milliliters of solution are there?

3. A salesman, who makes a commission of $18\frac{1}{4}$% on each sale, makes a commission of $152.39 on a particular sale. Rounded to the nearest dollar, what is the amount of the sale?

4. At "super-long play," $2\frac{1}{2}$ hours of play of a video cassette recorder represents 31.25% of the total playing time. What is the total playing time?

Answers to Practice Sets are on p. 346.

Section 7.6 EXERCISES

For problems 1–25, find each indicated quantity.

1. What is 21% of 104?

2. What is 8% of 36?

3. What is 98% of 545?

4. What is 143% of 33?

5. What is $10\frac{1}{2}$% of 20?

6. 3.25 is what percent of 88?

7. 22.44 is what percent of 44?

8. 0.0036 is what percent of 0.03?

9. 31.2 is what percent of 26?

10. 266.4 is what percent of 74?

11. 0.0101 is what percent of 0.0505?

12. 2.4 is 24% of what number?

13. 24.19 is 41% of what number?

14. 61.12 is 16% of what number?

15. 82.81 is 91% of what number?

16. 115.5 is 20% of what number?

17. 43.92 is 480% of what number?

18. What is 85% of 62?

19. 29.14 is what percent of 5.13?

20. 0.6156 is what percent of 5.13?

21. What is 0.41% of 291.1?

22. 26.136 is 121% of what number?

23. 1,937.5 is what percent of 775?

24. 1 is what percent of 2,000?

25. 0 is what percent of 59?

26. An item of clothing is on sale for 10% off the marked price. If the marked price is $14.95, what is the sale price? (Round to two decimal places.)

27. A grocery clerk, who makes $365 per month, receives a 7% raise. How much is her new monthly salary?

28. An item of clothing which originally sells for $55.00 is marked down to $46.75. What percent has it been marked down?

29. On a 25 question exam, a student gets 21 correct. What percent is this?

30. On a 45 question exam, a student gets 40%. How many questions did this student get correct?

31. A vitamin tablet, which weighs 250 milligrams, contains 35 milligrams of vitamin C. What percent of the weight of this tablet is vitamin C?

32. Five years ago a secretary made $11,200 annually. The secretary now makes $17,920 annually. By what percent has this secretary's salary been increased?

33. A baseball team wins $48\frac{3}{4}$% of all their games. If they won 78 games, how many games did they play?

34. A typist was able to increase his speed by 120% to 42 words per minute. What was his original typing speed?

35. A salesperson makes a commission of 12% on the total amount of each sale. If, in one month, she makes a total of $8,520 in sales, how much has she made in commission?

36. A salesperson receives a salary of $850 per month plus a commission of $8\frac{1}{2}$% of her sales. If, in a particular month, she sells $22,800 worth of merchandise, what will be her monthly earnings?

37. A man borrows $1150.00 from a loan company. If he makes 12 equal monthly payments of $130.60, what percent of the loan is he paying in interest?

38. The distance from the sun to the earth is approximately 93,000,000 miles. The distance from the sun to Pluto is approximately 860.2% of the distance from the sun to the Earth. Approximately, how many miles is Pluto from the sun?

39. The number of people on food stamps in Maine in 1975 was 151,000. By 1980, the number had decreased to 59,200. By what percent did the number of people on food stamps decrease? (Round the result to the nearest percent.)

40. In Nebraska, in 1960, there were 734,000 motor-vehicle registrations. By 1979, the total had increased by about 165.6%. About how many motor-vehicle registrations were there in Nebraska in 1979?

41. From 1973 to 1979, in the United States, there was an increase of 166.6% of Ph.D. social scientists to 52,000. How many were there in 1973?

42. In 1950, in the United States, there were 1,894 daily newspapers. That number decreased to 1,747 by 1981. What percent did the number of daily newspapers decrease?

43. A particular alloy is 27% copper. How many pounds of copper are there in 55 pounds of the alloy?

44. A bottle containing a solution of hydrochloric acid (HCl) is marked 15% (meaning that 15% of the HCl solution is acid). If a bottle contains 65 milliliters of solution, how many milliliters of water does it contain?

45. A bottle containing a solution of HCl is marked 45%. A test shows that 36 of the 80 milliliters contained in the bottle are hydrochloric acid. Is the bottle marked correctly? If not, how should it be remarked?

EXERCISES FOR REVIEW

(2.5)   **46.** Use the numbers 4 and 7 to illustrate the commutative property of multiplication.

(4.2)   **47.** Convert $\dfrac{14}{5}$ to a mixed number.

(5.4)   **48.** Arrange the numbers $\dfrac{7}{12}, \dfrac{5}{9}$, and $\dfrac{4}{7}$ in increasing order.

(6.2)   **49.** Convert 4.006 to a mixed number.

(7.5)   **50.** Convert $\dfrac{7}{8}\%$ to a fraction.

★ **Answers to Practice Sets (7.6)**

A. **1.** 35.7   **2.** $1,360   **3.** 4.9   **4.** $45,360

B. **1.** 23%   **2.** 90%   **3.** 40%

C. **1.** 99   **2.** 13.2 ml   **3.** $835   **4.** 8 hours

Denominate Numbers (7.1)

Numbers that appear along with units are *denominate numbers*. The amounts 6 dollars and 4 pints are examples of denominate numbers.

Like and Unlike Denominate Numbers (7.1)

Like denominate numbers are denominate numbers with like units. If the units are not the same, the numbers are *unlike denominate numbers*.

Pure Numbers (7.1)

Numbers appearing without a unit are *pure numbers*.

Comparing Numbers by Subtraction and Division (7.1)

Comparison of two numbers by subtraction indicates how much more one number is than another. Comparison by division indicates how many times larger or smaller one number is than another.

Comparing Pure or Like Denominate Numbers by Subtraction (7.1)

Numbers can be compared by subtraction if and only if they are pure numbers or like denominate numbers.

Ratio Rate (7.1)

A comparison, by division, of two pure or two like denominate numbers is a *ratio*. A comparison, by division, of two unlike denominate numbers is a *rate*.

Proportion (7.2)

A *proportion* is a statement that two ratios or rates are equal.

$$\frac{3 \text{ people}}{2 \text{ jobs}} = \frac{6 \text{ people}}{4 \text{ jobs}}$$

is a proportion.

Solving a Proportion (7.2)

To *solve a proportion* that contains three known numbers and a letter that represents an unknown quantity, perform the cross multiplication, then divide the product of the two numbers by the number that multiplies the letter.

Proportions Involving Rates (7.2)

When writing a proportion involving rates it is very important to write it so that the same type of units appears on the same side of either the equal sign or the fraction bar.

$$\frac{\text{unit type 1}}{\text{unit type 2}} = \frac{\text{unit type 1}}{\text{unit type 2}} \qquad \text{or} \qquad \frac{\text{unit type 1}}{\text{unit type 1}} = \frac{\text{unit type 2}}{\text{unit type 2}}$$

Five-Step Method for Solving Proportions (7.3)

Five-Step Method

1. By careful reading, determine what the unknown quantity is and represent it with some letter. There will be only one unknown in a problem.
2. Identify the three specified numbers.
3. Determine which comparisons are to be made and set up the proportion.
4. Solve the proportion.
5. Interpret and write a conclusion.

When solving applied problems, ALWAYS begin by determining the unknown quantity and representing it with a letter.

Percents (7.4)

A ratio in which one number is compared to 100 is a *percent*. Percent means "for each hundred."

Conversions of Fractions, Decimals, and Percents (7.4)

It is possible to convert decimals to percents, fractions to percents, percents to decimals, and percents to fractions.

Applications of Percents:

The three basic types of percent problems involve a *base*, a *percentage*, and a *percent*.

Base
Percentage
Percent (7.6)

The *base* is the number used for comparison.
The *percentage* is the number being compared to the base.
By its definition, *percent* means *part of*.

Solving Problems (7.6)

$$\text{Percentage} = (\text{percent}) \times (\text{base})$$

$$\text{Percent} = \frac{\text{percentage}}{\text{base}}$$

$$\text{Base} = \frac{\text{percentage}}{\text{percent}}$$

347

EXERCISE SUPPLEMENT

Section 7.1

1. Compare 250 watts to 100 watts by subtraction.

2. Compare 126 and 48 by subtraction.

3. Compare 98 radishes to 41 radishes by division.

4. Compare 144 to 9 by division.

5. Compare 100 tents to 5 tents by division.

6. Compare 28 feet to 7 feet by division.

7. Comparison, by division, of two pure numbers or two like denominate numbers is called a _____.

8. A comparison, by division, of two unlike denominate numbers is called a _____.

For problems 9–12, express each ratio or rate as a fraction.

9. 15 to 5

10. 72 to 12

11. 8 millimeters to 5 milliliters

12. 106 tablets to 52 tablets

For problems 13–16, write each ratio in the form "a to b."

13. $\dfrac{9}{16}$

14. $\dfrac{5}{11}$

15. $\dfrac{1 \text{ diskette}}{8 \text{ diskettes}}$

16. $\dfrac{5 \text{ papers}}{3 \text{ pens}}$

For problems 17–21, write each ratio or rate using words.

17. $\dfrac{9}{16} = \dfrac{18}{32}$

18. $\dfrac{1}{4} = \dfrac{12}{48}$

19. $\dfrac{8 \text{ items}}{4 \text{ dollars}} = \dfrac{2 \text{ items}}{1 \text{ dollar}}$

20. 150 milligrams of niacin is to 2 tablets as 300 milligrams of niacin is to 4 tablets.

21. 20 people is to 4 seats as 5 people is to 1 seat.

Section 7.2

For problems 22–27, determine the missing number in each proportion.

22. $\dfrac{x}{3} = \dfrac{24}{9}$

23. $\dfrac{15}{7} = \dfrac{60}{x}$

24. $\dfrac{1}{1} = \dfrac{x}{44}$

25. $\dfrac{3}{x} = \dfrac{15}{50}$

26. $\dfrac{15 \text{ bats}}{16 \text{ balls}} = \dfrac{x \text{ bats}}{128 \text{ balls}}$

27. $\dfrac{36 \text{ rooms}}{29 \text{ fans}} = \dfrac{504 \text{ rooms}}{x \text{ fans}}$

Section 7.3

28. On a map, 3 inches represents 20 miles. How many miles does 27 inches represent?

29. A salt solution is composed of 8 parts of salt to 5 parts of water. How many parts of salt are there in a solution that contains 50 parts of water?

30. A model is built to $\dfrac{4}{15}$ scale. If a particular part of the model measures 8 inches in length, how long is the actual structure?

31. The ratio of ammonia to air in a container is $\frac{3}{40}$. How many milliliters of air should be in a container that contains 8 milliliters of ammonia?

32. A 4-foot girl casts a 9-foot shadow at a particular time of the day. How tall is a pole that casts a 144-foot shadow at the same time of the day?

33. The odds that a particular event will occur are 11 to 2. If this event occurs 55 times, how many times would you predict it does not occur?

34. Every $1\frac{3}{4}$ teaspoon of a multiple vitamin, in granular form, contains 0.85 the minimum daily requirement of vitamin A. How many teaspoons of this vitamin are required to supply 2.25 the minimum daily requirement?

Section 7.4 and 7.5

For problems 35–39, convert each decimal to a percent.

35. 0.16

36. 0.818

37. 5.3536

38. 0.50

39. 3

For problems 40–48, convert each percent to a decimal.

40. 62%

41. 1.58%

42. 9.15%

43. 0.06%

44. 0.003%

45. $5\frac{3}{11}$% to a three-place decimal

46. $\frac{9}{13}$% to a three-place decimal

47. $82\frac{25}{29}$% to a four-place decimal

48. $18\frac{1}{7}$% to a four-place decimal

For problems 49–55, convert each fraction or mixed number to a percent.

49. $\frac{3}{5}$

50. $\frac{2}{10}$

51. $\frac{5}{16}$

52. $\frac{35}{8}$

53. $\frac{105}{16}$

54. $45\frac{1}{11}$

55. $6\frac{278}{9}$

For problems 56–64, convert each percent to a fraction or mixed number.

56. 95%

57. 12%

58. 83%

59. 38.125%

60. $61.\overline{2}$%

61. $\frac{5}{8}$%

62. $6\frac{9}{20}$%

63. $15\frac{3}{22}$%

64. $106\frac{19}{45}$%

Section 7.6

For problems 65–72, find each solution.

65. What is 16% of 40?

66. 29.4 is what percent of 105?

67. $3\dfrac{21}{50}$ is 547.2% of what number?

68. 0.09378 is what percent of 52.1?

69. What is 680% of 1.41?

70. A kitchen knife is on sale for 15% off the marked price. If the marked price is $39.50, what is the sale price?

71. On an 80 question geology exam, a student gets 68 correct. What percent is correct?

72. A salesperson makes a commission of 18% of her monthly sales total. She also receives a monthly salary of $1,600.00. If, in a particular month, she sells $4,000.00 worth of merchandise, how much will she make that month?

1. _____

1. (7.1) Compare 4 cassette tapes to 7 dollars.

2. _____

2. (7.1) What do we call a comparison, by division, of two unlike denominate numbers?

For problems 3 and 4, express each ratio or rate as a fraction.

3. _____

3. (7.1) 11 to 9

4. (7.1) 5 televisions to 2 radios

4. _____

For problems 5 and 6, write each ratio or rate in the form "a to b."

5. _____

5. (7.1) $\dfrac{8 \text{ maps}}{3 \text{ people}}$

6. (7.1) $\dfrac{2 \text{ psychologists}}{75 \text{ people}}$

6. _____

For problems 7–9, solve each proportion.

7. _____

7. (7.2) $\dfrac{8}{x} = \dfrac{48}{90}$

8. _____

8. (7.2) $\dfrac{x}{7} = \dfrac{4}{28}$

9. _____

9. (7.2) $\dfrac{3 \text{ computers}}{8 \text{ students}} = \dfrac{24 \text{ computers}}{x \text{ students}}$

10. _____

10. (7.3) On a map, 4 inches represents 50 miles. How many miles does 3 inches represent?

11. _____

11. (7.3) An acid solution is composed of 6 milliliters of acid to 10 milliliters of water. How many milliliters of acid are there in an acid solution that is composed of 3 milliliters of water?

12. _____

12. (7.3) The odds that a particular event will occur are 9 to 7. If the event occurs 27 times, how many times would you predict it will it not occur?

13. _____

For problems 13 and 14, convert each decimal to a percent.

13. (7.4) 0.82

14. _____

14. (7.4) $3.\overline{7}$

15. _____

For problems 15 and 16, convert each percent to a decimal.

15. (7.4) 2.813%

16. _____

16. (7.4) 0.006%

For problems 17–19, convert each fraction to a percent.

17. _____

17. (7.4) $\dfrac{42}{5}$

18. _____

18. (7.4) $\dfrac{1}{8}$

19. _____

19. (7.4) $\dfrac{800}{80}$

For problems 20 and 21, convert each percent to a fraction.

20. _(7.4)_ 15%

20. _____

21. _(7.4)_ $\dfrac{4}{27}$%

21. _____

For problems 22–25, find each indicated quantity.

22. _(7.6)_ What is 18% of 26?

22. _____

23. _(7.6)_ 0.618 is what percent of 0.3?

23. _____

24. _(7.6)_ 0.1 is 1.1% of what number?

24. _____

25. _(7.6)_ A salesperson makes a monthly salary of $1,000.00. He also gets a commission of 12% of his total monthly sales. If, in a particular month, he sells $5,500.00 worth of merchandise, what is his income that month?

25. _____

8

Techniques of Estimation

After completing this chapter, you should

Section 8.1 Estimation by Rounding
- understand the reason for estimation
- be able to estimate the result of an addition, multiplication, subtraction, or division using the rounding technique

Section 8.2 Estimation by Clustering
- understand the concept of clustering
- be able to estimate the result of adding more than two numbers when clustering occurs using the clustering technique

Section 8.3 Mental Arithmetic—Using the Distributive Property
- understand the distributive property
- be able to obtain the exact result of a multiplication using the distributive property

Section 8.4 Estimation by Rounding Fractions
- be able to estimate the sum of two or more fractions using the technique of rounding fractions

8.1 Estimation by Rounding

☐ **ESTIMATION BY ROUNDING**

When beginning a computation, it is valuable to have an idea of what value to expect for the result. When a computation is completed, it is valuable to know if the result is reasonable.

In the rounding process, it is important to note two facts:

1. The rounding that is done in estimation does not always follow the rules of rounding discussed in Section 1.3 (Rounding Whole Numbers). Since estimation is concerned with the expected value of a computation, rounding is done using *convenience* as the guide rather than using hard-and-fast rounding rules. For example, if we wish to estimate the result of the division 80 ÷ 26, we might round 26 to 20 rather than to 30 since 80 is more *conveniently* divided by 20 than by 30.

2. Since rounding may occur out of convenience, and different people have different ideas of what may be convenient, results of an estimation done by rounding may vary. For a particular computation, different people may get different estimated results. *Results may vary.*

Estimation

> **Estimation** is the process of determining an expected value of a computation.

Common words used in estimation are *about, near,* and *between.*

☐ ESTIMATION BY ROUNDING

The rounding technique estimates the result of a computation by rounding the numbers involved in the computation to one or two nonzero digits.

☆ **SAMPLE SET A**

Estimate the sum: 2,357 + 6,106.

Notice that 2,357 is near 2,400, and that 6,106 is near 6,100.
two nonzero digits   two nonzero digits

The sum can be estimated by 2,400 + 6,100 = 8,500. (It is quick and easy to add 24 and 61.)

Thus, 2,357 + 6,106 is *about* 8,400. In *fact,* 2,357 + 6,106 = 8,463.

★ **PRACTICE SET A**

1. Estimate the sum: 4,216 + 3,942.   **2.** Estimate the sum: 812 + 514.

3. Estimate the sum: 43,892 + 92,106.

☆ **SAMPLE SET B**

Estimate the difference: 5,203 — 3,015.

Notice that 5,203 is near 5,200, and that 3,015 is near 3,000.

two nonzero digits — one nonzero digit

The difference can be estimated by 5,200 — 3,000 = 2,200.

Thus, 5,203 — 3,015 is *about* 2,200. In *fact*, 5,203 — 3,015 = 2,188.

We could make a less accurate estimation by observing that 5,203 is near 5,000. The number 5,000 has only one nonzero digit rather than two (as does 5,200). This fact makes the estimation quicker (but a little less accurate). We then estimate the difference by 5,000 — 3,000 = 2,000, and conclude that 5,203 — 3,015 is about 2,000. This is why we say "answers may vary."

★ **PRACTICE SET B**

1. Estimate the difference: 628 — 413.

2. Estimate the difference: 7,842 — 5,209.

3. Estimate the difference: 73,812 — 28,492.

☆ **SAMPLE SET C**

1. Estimate the product: 73 · 46.

Notice that 73 is near 70, and that 46 is near 50.

one nonzero digit — one nonzero digit

The product can be estimated by 70 · 50 = 3,500. (Recall that to multiply numbers ending in zeros, we multiply the nonzero digits and affix to this product the total number of ending zeros in the factors. See Section 2.1 for a review of this technique.)

Thus, 73 · 46 is about 3,500. In fact, 73 · 46 = 3,358.

2. Estimate the product: 87 · 4,316.

Notice that 87 is close to 90, and that 4,316 is close to 4,000.

one nonzero digit — one nonzero digit

The product can be estimated by 90 · 4,000 = 360,000.

Thus, 87 · 4,316 is about 360,000. In fact, 87 · 4,316 = 375,492.

★ **PRACTICE SET C**

1. Estimate the product: 31 · 87.

2. Estimate the product: 18 · 42.

3. Estimate the product: 16 · 94.

☆ SAMPLE SET D

1. Estimate the quotient: $153 \div 17$.

Notice that 153 is close to 150, and that 17 is close to 15.
(underbraced: two nonzero digits)   (underbraced: two nonzero digits)

The quotient can be estimated by $150 \div 15 = 10$.

Thus, $153 \div 17$ is about 10. In fact, $153 \div 17 = 9$.

2. Estimate the quotient: $742,000 \div 2,400$.

Notice that 742,000 is close to 700,000, and that 2,400 is close to 2,000.
(underbraced: one nonzero digit)   (underbraced: one nonzero digit)

The quotient can be estimated by $700,000 \div 2,000 = 350$.

Thus, $742,000 \div 2,400$ is about 350. In fact, $742,000 \div 2,400 = 309.1\overline{6}$.

★ PRACTICE SET D

1. Estimate the quotient: $221 \div 18$.

2. Estimate the quotient: $4,079 \div 381$.

3. Estimate the quotient: $609,000 \div 16,000$.

☆ SAMPLE SET E

Estimate the sum: $53.82 + 41.6$.

Notice that 53.82 is close to 54, and that 41.6 is close to 42.
(underbraced: two nonzero digits)   (underbraced: two nonzero digits)

The sum can be estimated by $54 + 42 = 96$.

Thus, $53.82 + 41.6$ is about 96. In fact, $53.82 + 41.6 = 95.42$.

★ PRACTICE SET E

1. Estimate the sum: $61.02 + 26.8$.

2. Estimate the sum: $109.12 + 137.88$.

⭐ **SAMPLE SET F**

1. Estimate the product: (31.28)(14.2).

   Notice that 31.28 is close to 30, and that 14.2 is close to 15.

   one nonzero   two nonzero
   digit   digits

   The product can be estimated by $30 \cdot 15 = 450$. ($3 \cdot 15 = 45$, then affix one zero.)

   Thus, (31.28)(14.2) is about 450. In fact, (31.28)(14.2) = 444.176.

2. Estimate 21% of 5.42.

   Notice that 21% = .21 as a decimal, and that .21 is close to .2.

                                                     one nonzero
                                                     digit

   Notice also that 5.42 is close to 5.

                          one nonzero
                          digit

   Then, 21% of 5.42 can be estimated by $(.2)(5) = 1$.

   Thus, 21% of 5.42 is about 1. In fact, 21% of 5.42 is 1.1382.

★ **PRACTICE SET F**

1. Estimate the product: (47.8)(21.1).   2. Estimate 32% of 14.88.

Answers to Practice Sets are on p. 362.

Section 8.1 EXERCISES

For problems 1–45, estimate each calculation using the method of rounding. After you have made an estimate, find the exact value and compare this to the estimated result to see if your estimated value is reasonable. Results may vary.

1. $1{,}402 + 2{,}198$

2. $3{,}481 + 4{,}216$

3. $921 + 796$

4. $611 + 806$

5. $4{,}681 + 9{,}325$

6. $6{,}476 + 7{,}814$

7. $7{,}805 - 4{,}266$

8. $8,427 - 5,342$

9. $14,106 - 8,412$

10. $26,486 - 18,931$

11. $32 \cdot 53$

12. $67 \cdot 42$

13. $628 \cdot 891$

14. $426 \cdot 741$

15. $18,012 \cdot 32,416$

16. $22,481 \cdot 51,076$

17. $287 \div 19$

18. $884 \div 33$

19. $1,254 \div 57$

20. $2,189 \div 42$

21. $8,092 \div 239$

22. $2,688 \div 48$

23. $72.14 + 21.08$

24. $43.016 + 47.58$

25. $96.53 - 26.91$

26. $115.0012 - 25.018$

34. $(5{,}137.118)(263.56)$

27. $206.19 + 142.38$

35. $(6.92)(0.88)$

28. $592.131 + 211.6$

36. $(83.04)(1.03)$

29. $(32.12)(48.7)$

37. $(17.31)(.003)$

30. $(87.013)(21.07)$

38. $(14.016)(.016)$

31. $(3.003)(16.52)$

39. 93% of 7.01

32. $(6.032)(14.091)$

40. 107% of 12.6

33. $(114.06)(384.3)$

41. 32% of 15.3

42. 74% of 21.93

44. 4% of .863

43. 18% of 4.118

45. 2% of .0039

EXERCISES FOR REVIEW

(5.2) **46.** Find the difference: $\dfrac{7}{10} - \dfrac{5}{16}$.

(5.5) **47.** Find the value of $\dfrac{6 - \dfrac{1}{4}}{6 + \dfrac{1}{4}}$.

(6.2) **48.** Convert the complex decimal $1.11\frac{1}{4}$ to a decimal.

(7.3) **49.** A woman 5 foot tall casts an 8-foot shadow at a particular time of the day. How tall is a tree that casts a 96-foot shadow at the same time of the day?

(7.6) **50.** 11.62 is 83% of what number?

★ Answers to Practice Sets (8.1)

A. Results may vary. **1.** $4,216 + 3,942$: $4,200 + 3,900$. About 8,100. In fact, 8,158.
2. $812 + 514$: $800 + 500$. About 1,300. In fact, 1,326.
3. $43,892 + 92,106$: $44,000 + 92,000$. About 136,000. In fact, 135,998.

B. Results may vary. **1.** $628 - 413$: $600 - 400$. About 200. In fact, 215.
2. $7,842 - 5,209$: $7,800 - 5,200$. About 2,600. In fact, 2,633.
3. $73,812 - 28,492$: $74,000 - 28,000$. About 46,000. In fact, 45,320.

C. Results may vary. **1.** $31 \cdot 87$: $30 \cdot 90$. About 2,700. In fact, 2,697.
2. $18 \cdot 42$: $20 \cdot 40$. About 800. In fact, 756. **3.** $16 \cdot 94$: $15 \cdot 100$. About 1,500. In fact, 1,504.

D. Results may vary. **1.** $221 \div 18$: $200 \div 20$. About 10. In fact, $12.2\overline{7}$.
2. $4,079 \div 381$: $4,000 \div 400$. About 10. In fact, 10.70603675 . . .
3. $609,000 \div 16,000$: $600,000 \div 15,000$. About 40. In fact, 38.0625.

E. Results may vary. **1.** $61.02 + 26.8$: $61 + 27$. About 88. In fact, 87.82. **2.** $109.12 + 137.88$: $110 + 138$. About 248. In fact, 247. We could have estimated 137.88 with 140. Then $110 + 140$ is an easy mental addition. We would conclude then that $109.12 + 137.88$ is about 250.

F. Results may vary. **1.** $(47.8)(21.1)$: $(50)(20)$. About 1,000. In fact, 1,008.58. **2.** 32% of 14.88: $(.3)(15)$. About 4.5. In fact, 4.7616.

8.2 Estimation by Clustering

☐ **ESTIMATION BY CLUSTERING**

Cluster

When more than two numbers are to be added, the sum may be estimated using the clustering technique. The rounding technique could also be used, but if several of the numbers are seen to **cluster** (are seen to be close to) one particular number, the clustering technique provides a quicker estimate. Consider a sum such as

$32 + 68 + 29 + 73$

Notice two things:

1. There are more than two numbers to be added.
2. Clustering occurs.

    (a) Both 68 and 73 cluster around 70, so $68 + 73$ is close to $70 + 70 = 2(70) = 140$.

$$32 + 68 + 29 + 71$$

    (b) Both 32 and 29 cluster around 30, so $32 + 29$ is close to $30 + 30 = 2(30) = 60$.

The sum may be estimated by

$$(2 \cdot 30) + (2 \cdot 70) = 60 + 140$$
$$= 200$$

In fact, $32 + 68 + 29 + 73 = 202$.

☆ **SAMPLE SET A**

Estimate each sum. Results may vary.

1. $27 + 48 + 31 + 52$.

27 and 31 cluster near 30. Their sum is about $2 \cdot 30 = 60$.
48 and 52 cluster near 50. Their sum is about $2 \cdot 50 = 100$.

Thus, $27 + 48 + 31 + 52$ is about $(2 \cdot 30) + (2 \cdot 50) = 60 + 100$
$$= 160$$

In fact, $27 + 48 + 31 + 52 = 158$.

2. $88 + 21 + 19 + 91$.

88 and 91 cluster near 90. Their sum is about $2 \cdot 90 = 180$.
21 and 19 cluster near 20. Their sum is about $2 \cdot 20 = 40$.

Thus, $88 + 21 + 19 + 91$ is about $(2 \cdot 90) + (2 \cdot 20) = 180 + 40$
$$= 220$$

In fact, $88 + 21 + 19 + 91 = 219$.

3. $17 + 21 + 48 + 18$.

17, 21, and 18 cluster near 20. Their sum is about $3 \cdot 20 = 60$.
48 is about 50.

Thus, $17 + 21 + 48 + 18$ is about $(3 \cdot 20) + 50 = 60 + 50$
$$= 110$$

In fact, $17 + 21 + 48 + 18 = 104$.

Continued

4. $61 + 48 + 49 + 57 + 52$.

61 and 57 cluster near 60. Their sum is about $2 \cdot 60 = 120$.
48, 49, and 52 cluster near 50. Their sum is about $3 \cdot 50 = 150$.

Thus, $61 + 48 + 49 + 57 + 52$ is about $(2 \cdot 60) + (3 \cdot 50) = 120 + 150$
$= 270$

In fact, $61 + 48 + 49 + 57 + 52 = 267$.

5. $706 + 321 + 293 + 684$.

706 and 684 cluster near 700. Their sum is about $2 \cdot 700 = 1,400$.
321 and 293 cluster near 300. Their sum is about $2 \cdot 300 = 600$.

Thus, $706 + 321 + 293 + 684$ is about $(2 \cdot 700) + (2 \cdot 300) = 1,400 + 600$
$= 2,000$

In fact, $706 + 321 + 293 + 684 = 2,004$.

★ PRACTICE SET A

Use the clustering method to estimate each sum.

1. $28 + 51 + 31 + 47$ **2.** $42 + 39 + 68 + 41$ **3.** $37 + 39 + 83 + 42 + 79$

4. $612 + 585 + 830 + 794$

Answers to the Practice Set are on p. 365.

Section 8.2 EXERCISES

For problems 1–20, use the clustering method to estimate each sum. Results may vary.

1. $28 + 51 + 31 + 47$

2. $42 + 19 + 39 + 23$

3. $88 + 62 + 59 + 90$

4. $76 + 29 + 33 + 82$

5. $19 + 23 + 87 + 21$

6. $41 + 28 + 42 + 37$

7. $89 + 32 + 89 + 93$

8. $73 + 72 + 27 + 71$

9. $43 + 62 + 61 + 55$

15. $19 + 24 + 87 + 23 + 91 + 93$

10. $31 + 77 + 31 + 27$

16. $108 + 61 + 63 + 96 + 57 + 99$

11. $57 + 34 + 28 + 61 + 62$

17. $518 + 721 + 493 + 689$

12. $94 + 18 + 23 + 91 + 19$

18. $981 + 1208 + 1214 + 1006$

13. $103 + 72 + 66 + 97 + 99$

19. $23 + 81 + 77 + 79 + 19 + 81$

14. $42 + 121 + 119 + 124 + 41$

20. $94 + 68 + 66 + 101 + 106 + 71 + 110$

EXERCISES FOR REVIEW

(1.1)   **21.** Specify all the digits greater than 6.

(4.4)   **22.** Find the product: $\dfrac{2}{3} \cdot \dfrac{9}{14} \cdot \dfrac{7}{12}$.

(6.2)   **23.** Convert 0.06 to a fraction.

(7.2)   **24.** Write the proportion in fractional form: "5 is to 8 as 25 is to 40."

(8.1)   **25.** Estimate the sum using the method of rounding: $4,882 + 2,704$.

★ **Answers to Practice Set (8.2)**

A.   Results may vary.   **1.** $(2 \cdot 30) + (2 \cdot 50) = 60 + 100 = 160$ **2.** $(3 \cdot 40) + 70 = 120 + 70 = 190$
    3. $(3 \cdot 40) + (2 \cdot 80) = 120 + 160 = 280$ **4.** $(2 \cdot 600) + (2 \cdot 800) = 1,200 + 1,600 = 2,800$

8.3 Mental Arithmetic — Using the Distributive Property

☐ **THE DISTRIBUTIVE PROPERTY**
☐ **ESTIMATION USING THE DISTRIBUTIVE PROPERTY**

☐ THE DISTRIBUTIVE PROPERTY

Distributive Property

The **distributive property** is a characteristic of numbers that involves both addition and multiplication. It is used often in algebra, and we can use it now to obtain exact results for a multiplication.

Suppose we wish to compute $3(2 + 5)$. We can proceed in either of two ways, one way which is known to us already (the order of operations), and a new way (the distributive property).

1. Compute $3(2 + 5)$ using the order of operations.

   $3(2 + 5)$

   Operate inside the parentheses first: $2 + 5 = 7$.

   $3(2 + 5) = 3 \cdot 7$

   Now multiply 3 and 7.

   $3(2 + 5) = 3 \cdot 7 = 21$

   Thus, $3(2 + 5) = 21$.

2. Compute $3(2 + 5)$ using the distributive property.

   We know that multiplication describes repeated addition. Thus,

   $3(2 + 5) = \underbrace{2 + 5 + 2 + 5 + 2 + 5}_{2 + 5 \text{ appears 3 times}}$

   $= 2 + 2 + 2 + 5 + 5 + 5$ (by the commutative property of addition)
   $= 3 \cdot 2 + 3 \cdot 5$ (since multiplication describes repeated addition)
   $= 6 + 15$
   $= 21$

Thus, $3(2 + 5) = 21$.
Let's look again at this use of the distributive property.

$3(2 + 5) = \underbrace{2 + 5 + 2 + 5 + 2 + 5}_{2 + 5 \text{ appears 3 times}}$

$3(2 + 5) = \underbrace{2 + 2 + 2}_{\substack{2 \text{ appears} \\ 3 \text{ times}}} + \underbrace{5 + 5 + 5}_{\substack{5 \text{ appears} \\ 3 \text{ times}}}$

$3(2 + 5) = \underset{3 \text{ times 2}}{3 \cdot 2} + \underset{3 \text{ times 5}}{3 \cdot 5}$

The 3 has been *distributed* to the 2 and 5.

This is the distributive property. We distribute the *factor* to each *addend* in the parentheses. The distributive property works for both sums and differences.

☆ **SAMPLE SET A**

1. $4(6 + 2) = 4 \cdot 6 + 4 \cdot 2$
$ = 24 + 8$
$ = 32$

Using the order of operations, we get

$4(6 + 2) = 4 \cdot 8$
$ = 32$

2. $8(9 + 6) = 8 \cdot 9 + 8 \cdot 6$
$ = 72 + 48$
$ = 120$

Using the order of operations, we get

$8(9 + 6) = 8 \cdot 15$
$ = 120$

3. $4(9 - 5) = 4 \cdot 9 - 4 \cdot 5$
$ = 36 - 20$
$ = 16$

4. $25(20 - 3) = 25 \cdot 20 - 25 \cdot 3$
$ = 500 - 75$
$ = 425$

★ **PRACTICE SET A**

Use the distributive property to compute each value.

1. $6(8 + 4)$ **2.** $4(4 + 7)$ **3.** $8(2 + 9)$ **4.** $12(10 + 3)$ **5.** $6(11 - 3)$

6. $8(9 - 7)$ **7.** $15(30 - 8)$

☐ **ESTIMATION USING THE DISTRIBUTIVE PROPERTY**

We can use the distributive property to obtain exact results for products such as $25 \cdot 23$. The distributive property works best for products when one of the factors ends in 0 or 5. We shall restrict our attention to only such products.

☆ **SAMPLE SET B**

Use the distributive property to compute each value.

1. $25 \cdot 23$

   Notice that $23 = 20 + 3$. We now write

   $25 \cdot 23 = 25(20 + 3)$

   $= 25 \cdot 20 + 25 \cdot 3$
   $= 500 + 75$
   $= 575$

   Thus, $25 \cdot 23 = 575$

   We could have proceeded by writing 23 as $30 - 7$.

   $25 \cdot 23 = 25(30 - 7)$

   $= 25 \cdot 30 - 25 \cdot 7$
   $= 750 - 175$
   $= 575$

2. $15 \cdot 37$

   Notice that $37 = 30 + 7$. We now write

   $15 \cdot 37 = 15(30 + 7)$

   $= 15 \cdot 30 + 15 \cdot 7$
   $= 450 + 105$
   $= 555$

   Thus, $15 \cdot 37 = 555$

   We could have proceeded by writing 37 as $40 - 3$.

   $15 \cdot 37 = 15(40 - 3)$

   $= 15 \cdot 40 - 15 \cdot 3$
   $= 600 - 45$
   $= 555$

3. $15 \cdot 86$

   Notice that $86 = 80 + 6$. We now write

   $15 \cdot 86 = 15(80 + 6)$

   $= 15 \cdot 80 + 15 \cdot 6$
   $= 1,200 + 90$
   $= 1,290$

   We could have proceeded by writing 86 as $90 - 4$.

   $15 \cdot 86 = 15(90 - 4)$

   $= 15 \cdot 90 - 15 \cdot 4$
   $= 1,350 - 60$
   $= 1,290$

★ **PRACTICE SET B**

Use the distributive property to compute each value.

1. 25 · 12   **2.** 35 · 14   **3.** 80 · 58   **4.** 65 · 62

Answers to Practice Sets are on p. 370.

Section 8.3 EXERCISES

For problems 1–25, use the distributive property to compute each product.

1. 15 · 13   **2.** 15 · 14

3. 25 · 11   **4.** 25 · 16

5. 15 · 16   **6.** 35 · 12

7. 45 · 83   **8.** 45 · 38

9. 25 · 38   **10.** 25 · 96

11. 75 · 14   **12.** 85 · 34

13. 65 · 26   **14.** 55 · 51

15. 15 · 107   **16.** 25 · 208

17. 35 · 402   **18.** 85 · 110

19. 95 · 12   **20.** 65 · 40

21. 80 · 32   **22.** 30 · 47

23. 50 · 63   **24.** 90 · 78

25. 40 · 89

EXERCISES FOR REVIEW

(3.4)   **26.** Find the greatest common factor of 360 and 3,780.

(4.4)   **27.** Reduce $\dfrac{594}{5,148}$ to lowest terms.

(4.6)   **28.** $1\dfrac{5}{9}$ of $2\dfrac{4}{7}$ is what number?

(7.2)   **29.** Solve the proportion: $\dfrac{7}{15} = \dfrac{x}{90}$.

(8.2)   **30.** Use the clustering method to estimate the sum: $88 + 106 + 91 + 114$.

★ **Answers to Practice Sets (8.3)**

A.   **1.** $6 \cdot 8 + 6 \cdot 4 = 48 + 24 = 72$ **2.** $4 \cdot 4 + 4 \cdot 7 = 16 + 28 = 44$ **3.** $8 \cdot 2 + 8 \cdot 9 = 16 + 72 = 88$
   4. $12 \cdot 10 + 12 \cdot 3 = 120 + 36 = 156$ **5.** $6 \cdot 11 - 6 \cdot 3 = 66 - 18 = 48$
   6. $8 \cdot 9 - 8 \cdot 7 = 72 - 56 = 16$ **7.** $15 \cdot 30 - 15 \cdot 8 = 450 - 120 = 330$

B.   **1.** $25(10 + 2) = 25 \cdot 10 + 25 \cdot 2 = 250 + 50 = 300$
   2. $35(10 + 4) = 35 \cdot 10 + 35 \cdot 4 = 350 + 140 = 490$
   3. $80(50 + 8) = 80 \cdot 50 + 80 \cdot 8 = 4,000 + 640 = 4,640$
   4. $65(60 + 2) = 65 \cdot 60 + 65 \cdot 2 = 3,900 + 130 = 4,030$

8.4 Estimation by Rounding Fractions

Section Overview

☐ **ESTIMATION BY ROUNDING FRACTIONS**

Estimation by rounding fractions is a useful technique for estimating the result of a computation involving fractions. Fractions are commonly rounded to $\dfrac{1}{4}, \dfrac{1}{2}, \dfrac{3}{4}, 0$, and 1. Remember that rounding may cause estimates to vary.

☆ **SAMPLE SET A**

Make each estimate remembering that results may vary.

1. Estimate $\dfrac{3}{5} + \dfrac{5}{12}$.

Notice that $\dfrac{3}{5}$ is about $\dfrac{1}{2}$, and that $\dfrac{5}{12}$ is about $\dfrac{1}{2}$.

Thus, $\dfrac{3}{5} + \dfrac{5}{12}$ is about $\dfrac{1}{2} + \dfrac{1}{2} = 1$. In fact, $\dfrac{3}{5} + \dfrac{5}{12} = \dfrac{61}{60}$, a little more than 1.

2. Estimate $5\frac{3}{8} + 4\frac{9}{10} + 11\frac{1}{5}$.

Adding the whole number parts, we get 20. Notice that $\frac{3}{8}$ is close to $\frac{1}{4}$, $\frac{9}{10}$ is close to 1, and $\frac{1}{5}$ is close to $\frac{1}{4}$. Then $\frac{3}{8} + \frac{9}{10} + \frac{1}{5}$ is close to $\frac{1}{4} + 1 + \frac{1}{4} = 1\frac{1}{2}$.

Thus, $5\frac{3}{8} + 4\frac{9}{10} + 11\frac{1}{5}$ is close to $20 + 1\frac{1}{2} = 21\frac{1}{2}$.

In fact, $5\frac{3}{8} + 4\frac{9}{10} + 11\frac{1}{5} = 21\frac{19}{40}$, a little less than $21\frac{1}{2}$.

★ **PRACTICE SET A**

Use the method of rounding fractions to estimate the result of each computation. Results may vary.

1. $\frac{5}{8} + \frac{5}{12}$ **2.** $\frac{7}{9} + \frac{3}{5}$ **3.** $8\frac{4}{15} + 3\frac{7}{10}$ **4.** $16\frac{1}{20} + 4\frac{7}{8}$

Answers to the Practice Set are on p. 373.

Section 8.4 EXERCISES

For problems 1–20, estimate each sum or difference using the method of rounding. After you have made an estimate, find the exact value of the sum or difference and compare this result to the estimated value. Results may vary.

1. $\frac{5}{6} + \frac{7}{8}$

2. $\frac{3}{8} + \frac{11}{12}$

3. $\frac{9}{10} + \frac{3}{5}$

4. $\frac{13}{15} + \frac{1}{20}$

5. $\frac{3}{20} + \frac{6}{25}$

6. $\frac{1}{12} + \frac{4}{5}$

7. $\dfrac{15}{16} + \dfrac{1}{12}$

14. $5\dfrac{3}{20} + 2\dfrac{8}{15}$

8. $\dfrac{29}{30} + \dfrac{11}{20}$

15. $9\dfrac{1}{15} + 6\dfrac{4}{5}$

9. $\dfrac{5}{12} + 6\dfrac{4}{11}$

16. $7\dfrac{5}{12} + 10\dfrac{1}{16}$

10. $\dfrac{3}{7} + 8\dfrac{4}{15}$

17. $3\dfrac{11}{20} + 2\dfrac{13}{25} + 1\dfrac{7}{8}$

11. $\dfrac{9}{10} + 2\dfrac{3}{8}$

18. $6\dfrac{1}{12} + 1\dfrac{1}{10} + 5\dfrac{5}{6}$

12. $\dfrac{19}{20} + 15\dfrac{5}{9}$

19. $\dfrac{15}{16} - \dfrac{7}{8}$

13. $8\dfrac{3}{5} + 4\dfrac{1}{20}$

20. $\dfrac{12}{25} - \dfrac{9}{20}$

EXERCISES
FOR REVIEW

(2.5) **21.** The fact that

(a first number · a second number) · a third number = a first number · (a second number · a third number)

is an example of which property of multiplication?

(4.5) **22.** Find the quotient: $\dfrac{14}{15} \div \dfrac{4}{45}$.

(5.3) **23.** Find the difference: $3\dfrac{5}{9} - 2\dfrac{2}{3}$.

(6.7) **24.** Find the quotient: $4.6 \div 0.11$.

(8.3) **25.** Use the distributive property to compute the product: $25 \cdot 37$.

★ **Answers to Practice Set (8.4)**

A. Results may vary. **1.** $\dfrac{1}{2} + \dfrac{1}{2} = 1$. In fact, $\dfrac{5}{8} + \dfrac{5}{12} = \dfrac{25}{24} = 1\dfrac{1}{24}$

2. $1 + \dfrac{1}{2} = 1\dfrac{1}{2}$. In fact, $\dfrac{7}{9} + \dfrac{3}{5} = 1\dfrac{17}{45}$ **3.** $8\dfrac{1}{4} + 3\dfrac{3}{4} = 11 + 1 = 12$. In fact, $8\dfrac{4}{15} + 3\dfrac{7}{10} = 11\dfrac{29}{30}$

4. $(16 + 0) + (4 + 1) = 16 + 5 = 21$. In fact, $16\dfrac{1}{20} + 4\dfrac{7}{8} = 20\dfrac{37}{40}$

Estimation (8.1)

Estimation is the process of determining an expected value of a computation.

Estimation By Rounding (8.1)

The *rounding technique* estimates the result of a computation by rounding the numbers involved in the computation to one or two nonzero digits. For example, $512 + 896$ can be estimated by $500 + 900 = 1,400$.

Cluster (8.2)

When several numbers are close to one particular number, they are said to *cluster* near that particular number.

Estimation By Clustering (8.2)

The *clustering technique of estimation* can be used when

1. there are more than two numbers to be added, and
2. clustering occurs.

For example, $31 + 62 + 28 + 59$ can be estimated by

$$(2 \cdot 30) + (2 \cdot 60) = 60 + 120 = 180$$

Distributive Property (8.3)

The *distributive property* is a characteristic of numbers that involves both addition and multiplication. For example,

$$3(4 + 6) = 3 \cdot 4 + 3 \cdot 6 = 12 + 18 = 30$$

Estimation Using the Distributive Property (8.3)

The *distributive property* can be used to obtain exact results for a multiplication. For example,

$$15 \cdot 23 = 15 \cdot (20 + 3) = 15 \cdot 20 + 15 \cdot 3 = 300 + 45 = 345$$

Estimation by Rounding Fractions (8.4)

Estimation by rounding fractions commonly rounds fractions to $\frac{1}{4}, \frac{1}{2}, \frac{3}{4}, 0$, and 1.

For example,

$$\frac{5}{12} + \frac{5}{16} \text{ can be estimated by } \frac{1}{2} + \frac{1}{4} = \frac{3}{4}$$

EXERCISE SUPPLEMENT

Section 8.1

For problems 1–70, estimate each value using the method of rounding. After you have made an estimate, find the exact value. Compare the exact and estimated values. Results may vary.

1. 286 + 312

2. 419 + 582

3. 689 + 511

4. 926 + 1,105

5. 1,927 + 3,017

6. 5,026 + 2,814

7. 1,408 + 2,352

8. 1,186 + 4,228

9. 5,771 + 246

10. 8,305 + 484

11. 3,812 + 2,906

12. 5,293 + 8,007

13. 28,481 + 32,856

14. 92,512 + 26,071

15. 87,612 + 2,106

16. 42,612 + 4,861

17. 212,413 + 609

18. 487,235 + 494

19. 2,409 + 1,526

20. 3,704 + 4,704

21. 41 · 63

22. 38 · 81

23. 18 · 28

24. 52 · 21

25. 307 · 489

26. 412 · 807

27. 77 · 614

28. 62 · 596

29. 27 · 473

30. 92 · 336

31. 12 · 814

32. 8 · 2,106

33. 192 · 452

34. 374 · 816

35. 88 · 4,392

36. 126 · 2,834

37. 3,896 · 413

38. 5,794 · 837

39. 6,311 · 3,512

40. 7,471 · 5,782

41. 180 ÷ 12

42. 309 ÷ 16

43. 286 ÷ 22

44. 527 ÷ 17

45. 1,007 ÷ 19

46. 1,728 ÷ 36

47. 2,703 ÷ 53

48. 2,562 ÷ 61

49. 1,260 ÷ 12

50. 3,618 ÷ 18

51. 3,344 ÷ 76

52. 7,476 ÷ 356

53. 20,984 ÷ 488

54. 43,776 ÷ 608

55. 7,196 ÷ 514

56. 51,492 ÷ 514

57. 26,962 ÷ 442

58. 33,712 ÷ 112

59. 105,152 ÷ 106

60. 176,978 ÷ 214

61. 48.06 + 23.11

62. 73.73 + 72.9

63. 62.91 + 56.4

64. 87.865 + 46.772

65. 174.6 + 97.2

66. (48.3)(29.6)

67. (87.11)(23.2)

68. (107.02)(48.7)

69. (0.76)(5.21)

70. (1.07)(13.89)

Section 8.2

For problems 71–90, estimate each value using the method of clustering. After you have made an estimate, find the exact value. Compare the exact and estimated values. Results may vary.

71. $38 + 51 + 41 + 48$

72. $19 + 73 + 23 + 71$

73. $27 + 62 + 59 + 31$

74. $18 + 73 + 69 + 19$

75. $83 + 49 + 79 + 52$

76. $67 + 71 + 84 + 81$

77. $16 + 13 + 24 + 26$

78. $34 + 56 + 36 + 55$

79. $14 + 17 + 83 + 87$

80. $93 + 108 + 96 + 111$

81. $18 + 20 + 31 + 29 + 24 + 38$

82. $32 + 27 + 48 + 51 + 72 + 69$

83. $64 + 17 + 27 + 59 + 31 + 21$

84. $81 + 41 + 92 + 38 + 88 + 80$

85. $87 + 22 + 91$

86. $44 + 38 + 87$

87. $19 + 18 + 39 + 22 + 42$

88. $31 + 28 + 49 + 29$

89. $88 + 86 + 27 + 91 + 29$

90. $57 + 62 + 18 + 23 + 61 + 21$

Section 8.3

For problems 91–110, compute each product using the distributive property.

91. $15 \cdot 33$

92. $15 \cdot 42$

93. $35 \cdot 36$

94. $35 \cdot 28$

95. $85 \cdot 23$

96. $95 \cdot 11$

97. $30 \cdot 14$

98. $60 \cdot 18$

99. $75 \cdot 23$

100. $65 \cdot 31$

101. $17 \cdot 15$

102. $38 \cdot 25$

103. $14 \cdot 65$

104. $19 \cdot 85$

105. $42 \cdot 60$

106. $81 \cdot 40$

107. $15 \cdot 105$

108. $35 \cdot 202$

109. $45 \cdot 306$

110. $85 \cdot 97$

Section 8.4

For problems 111–125, estimate each sum using the method of rounding fractions. After you have made an estimate, find the exact value. Compare the exact and estimated values. Results may vary.

111. $\dfrac{3}{8} + \dfrac{5}{6}$

112. $\dfrac{7}{16} + \dfrac{1}{24}$

113. $\dfrac{7}{15} + \dfrac{13}{30}$

114. $\dfrac{14}{15} + \dfrac{19}{20}$

121. $11\dfrac{5}{18} + 7\dfrac{22}{45}$

122. $14\dfrac{19}{36} + 2\dfrac{7}{18}$

115. $\dfrac{13}{25} + \dfrac{7}{30}$

116. $\dfrac{11}{12} + \dfrac{7}{8}$

123. $6\dfrac{1}{20} + 2\dfrac{1}{10} + 8\dfrac{13}{60}$

117. $\dfrac{9}{32} + \dfrac{15}{16}$

118. $\dfrac{5}{8} + \dfrac{1}{32}$

124. $5\dfrac{7}{8} + 1\dfrac{1}{4} + 12\dfrac{5}{12}$

119. $2\dfrac{3}{4} + 6\dfrac{3}{5}$

120. $4\dfrac{5}{9} + 8\dfrac{1}{27}$

125. $10\dfrac{1}{2} + 6\dfrac{15}{16} + 8\dfrac{19}{80}$

For problems 1–16, estimate each value. After you have made an estimate, find the exact value. Results may vary.

1. _____

1. (8.1) $3,716 + 6,789$

2. _____

2. (8.1) $8,821 + 9,217$

3. _____

3. (8.1) $7,316 - 2,305$

4. _____

4. (8.1) $110,812 - 83,406$

5. _____

5. (8.1) $82 \cdot 38$

6. _____

6. (8.1) $51 \cdot 92$

7. _____

7. (8.1) $48 \cdot 6,012$

8. _____

8. (8.1) $238 \div 17$

9. _____

9. (8.1) $2,660 : 28$

10. _____

10. (8.1) $43.06 + 37.94$

11. _____

11. (8.1) $307.006 + 198.0005$

12. _____   **12. (8.1)** $(47.2)(92.8)$

13. _____   **13. (8.2)** $58 + 91 + 61 + 88$

14. _____   **14. (8.2)** $43 + 39 + 89 + 92$

15. _____   **15. (8.2)** $81 + 78 + 27 + 79$

16. _____   **16. (8.2)** $804 + 612 + 801 + 795 + 606$

17. _____   For problems 17–21, use the distributive property to obtain the exact result.
                       17. (8.3) $25 \cdot 14$

18. _____   **18. (8.3)** $15 \cdot 83$

19. _____   **19. (8.3)** $65 \cdot 98$

20. _____   **20. (8.3)** $80 \cdot 107$

380

21. _____

21. (8.3) $400 \cdot 215$

For problems 22–25, estimate each value. After you have made an estimate, find the exact value. Results may vary.

22. _____

22. (8.4) $\dfrac{15}{16} + \dfrac{5}{8}$

23. _____

23. (8.4) $\dfrac{1}{25} + \dfrac{11}{20} + \dfrac{17}{30}$

24. _____

24. (8.4) $8\dfrac{9}{16} + 14\dfrac{1}{12}$

25. _____

25. (8.4) $5\dfrac{4}{9} + 1\dfrac{17}{36} + 6\dfrac{5}{12}$

9 Measurement and Geometry

After completing this chapter, you should

Section 9.1 Measurement and the United States System
- know what the word measurement means
- be familiar with United States system of measurement
- be able to convert from one unit of measure in the United States system to another unit of measure

Section 9.2 The Metric System of Measurement
- be more familiar with some of the advantages of the base ten number system
- know the prefixes of the metric measures
- be familiar with the metric system of measurement
- be able to convert from one unit of measure in the metric system to another unit of measure

Section 9.3 Simplification of Denominate Numbers
- be able to convert an unsimplified unit of measure to a simplified unit of measure
- be able to add and subtract denominate numbers
- be able to multiply and divide a denominate number by a whole number

Section 9.4 Perimeter and Circumference of Geometric Figures
- know what a polygon is
- know what perimeter is and how to find it
- know what the circumference, diameter, and radius of a circle is and how to find each one
- know the meaning of the symbol π and its approximating value
- know what a formula is and four versions of the circumference formula of a circle

Section 9.5 Area and Volume of Geometric Figures and Objects
- know the meaning and notation for area
- know the area formulas for some common geometric figures
- be able to find the areas of some common geometric figures
- know the meaning and notation for volume
- know the volume formulas for some common geometric objects
- be able to find the volume of some common geometric objects

9.1 Measurement and the United States System

Section
Overview

☐ MEASUREMENT
☐ THE UNITED STATES SYSTEM OF MEASUREMENT
☐ CONVERSIONS IN THE UNITED STATES SYSTEM

☐ MEASUREMENT

There are two major systems of measurement in use today. They are the *United States system* and the *metric system*. Before we describe these systems, let's gain a clear understanding of the concept of measurement.

Measurement

Measurement is comparison to some standard.

The concept of measurement is based on the idea of direct comparison. This means that measurement is the result of the comparison of two quantities. The quantity that is used for comparison is called the **standard unit of measure.**

Standard Unit of Measure

Over the years, standards have changed. Quite some time in the past, the standard unit of measure was determined by a king. For example,

1 inch was the distance between the tip of the thumb and the knuckle of the king. 1 inch was also the length of 16 barley grains placed end to end.

Today, standard units of measure rarely change. Standard units of measure are the responsibility of the Bureau of Standards in Washington D.C.

Some desirable properties of a standard are the following:

1. *Accessibility.* We should have access to the standard so we can make comparisons.
2. *Invariance.* We should be confident that the standard is not subject to change.
3. *Reproducibility.* We should be able to reproduce the standard so that measurements are convenient and accessible to many people.

☐ THE UNITED STATES SYSTEM OF MEASUREMENT

Some of the common units (along with their abbreviations) for the United States system of measurement are listed in the following table.

UNIT CONVERSION TABLE

Length	1 foot (ft) = 12 inches (in.)
	1 yard (yd) = 3 feet (ft)
	1 mile (mi) = 5,280 feet
Weight	1 pound (lb) = 16 ounces (oz)
	1 ton (T) = 2,000 pounds
Liquid Volume	1 tablespoon (tbsp) = 3 teaspoons (tsp)
	1 fluid ounce (fl oz) = 2 tablespoons
	1 cup (c) = 8 fluid ounces
	1 pint (pt) = 2 cups
	1 quart (qt) = 2 pints
	1 gallon (gal) = 4 quarts
Time	1 minute (min) = 60 seconds (sec)
	1 hour (hr) = 60 minutes
	1 day (da) = 24 hours
	1 week (wk) = 7 days

☐ CONVERSIONS IN THE UNITED STATES SYSTEM

It is often convenient or necessary to convert from one unit of measure to another. For example, it may be convenient to convert a measurement of length that is given in feet to one that is given in inches. Such conversions can be made using *unit fractions*.

Unit Fraction

A **unit fraction** is a fraction with a value of 1.

Unit fractions are formed by using two equal measurements. One measurement is placed in the numerator of the fraction, and the other in the denominator. **Placement depends on the desired conversion.**

Placement of Units

Place the unit being converted *to* in the **numerator**.
Place the unit being converted *from* in the **denominator**.

For example,

Equal Measurements	Unit Fraction
1 ft = 12 in.	$\dfrac{1 \text{ ft}}{12 \text{ in.}}$ or $\dfrac{12 \text{ in.}}{1 \text{ ft}}$
1 pt = 16 fl oz	$\dfrac{1 \text{ pt}}{16 \text{ fl oz}}$ or $\dfrac{16 \text{ fl oz}}{1 \text{ pt}}$
1 wk = 7 da	$\dfrac{7 \text{ da}}{1 \text{ wk}}$ or $\dfrac{1 \text{ wk}}{7 \text{ da}}$

☆ SAMPLE SET A

Make the following conversions. If a fraction occurs, convert it to a decimal rounded to two decimal places.

1. Convert 11 yards to feet.

Looking in the unit conversion table under *length,* we see that 1 yd = 3 ft. There are two corresponding unit fractions, $\dfrac{1 \text{ yd}}{3 \text{ ft}}$ and $\dfrac{3 \text{ ft}}{1 \text{ yd}}$. Which one should we use? Look to see which unit we wish to convert to. Choose the unit fraction with this unit in the *numerator*. We will choose $\dfrac{3 \text{ ft}}{1 \text{ yd}}$ since this unit fraction has feet in the numerator. Now, multiply 11 yd by the unit fraction. Notice that since the unit fraction has the value of 1, multiplying by it does not change the value of 11 yd.

$11 \text{ yd} = \dfrac{11 \text{ yd}}{1} \cdot \dfrac{3 \text{ ft}}{1 \text{ yd}}$

Divide out common units. (Units can be added, subtracted, multiplied, and divided, just as numbers can.)

$= \dfrac{11 \text{ yd}}{1} \cdot \dfrac{3 \text{ ft}}{1 \text{ yd}}$

$= \dfrac{11 \cdot 3 \text{ ft}}{1}$

$= 33 \text{ ft}$

Thus, 11 yd = 33 ft.

Continued

2. Convert 36 fl oz to pints.

Looking in the unit conversion table under *liquid volume,* we see that 1 pt = 16 fl oz. Since we are to convert to pints, we will construct a unit fraction with pints in the numerator.

$$36 \text{ fl oz} = \frac{36 \text{ fl oz}}{1} \cdot \frac{1 \text{ pt}}{16 \text{ fl oz}}$$ **Divide out common units.**

$$= \frac{36 \text{ fl oz}}{1} \cdot \frac{1 \text{ pt}}{16 \text{ fl oz}}$$

$$= \frac{36 \cdot 1 \text{ pt}}{16}$$

$$= \frac{36 \text{ pt}}{16}$$ **Reduce.**

$$= \frac{9}{4} \text{ pt}$$ **Convert to decimals:** $\frac{9}{4} = 2.25$.

Thus, 36 fl oz = 2.25 pt.

3. Convert 2,016 hr to weeks.

Looking in the unit conversion table under *time,* we see that 1 wk = 7 da and that 1 da = 24 hr. To convert from hours to weeks, we must first convert from hours to days and then from days to weeks. We need two unit fractions.

The unit fraction needed for converting from hours to days is $\frac{1 \text{ da}}{24 \text{ hr}}$. The unit fraction needed for converting from days to weeks is $\frac{1 \text{ wk}}{7 \text{ da}}$.

$$2,016 \text{ hr} = \frac{2,016 \text{ hr}}{1} \cdot \frac{1 \text{ da}}{24 \text{ hr}} \cdot \frac{1 \text{ wk}}{7 \text{ da}}$$ **Divide out common units.**

$$= \frac{2,016 \text{ hr}}{1} \cdot \frac{1 \text{ da}}{24 \text{ hr}} \cdot \frac{1 \text{ wk}}{7 \text{ da}}$$

$$= \frac{2,016 \cdot 1 \text{ wk}}{24 \cdot 7}$$ **Reduce.**

$$= 12 \text{ wk}$$

Thus, 2,016 hr = 12 wk.

★ PRACTICE SET A

Make the following conversions. If a fraction occurs, convert it to a decimal rounded to two decimal places.

1. Convert 18 ft to yards.

2. Convert 2 mi to feet.

3. Convert 26 ft to yards.

4. Convert 9 qt to pints

5. Convert 52 min to hours.

6. Convert 412 hr to weeks.

Answers to the Practice Set are on p. 389.

Section 9.1 EXERCISES

For problems 1–30, make each conversion using unit fractions. If fractions occur, convert them to decimals rounded to two decimal places.

1. 14 yd to feet

2. 3 mi to yards

3. 8 mi to inches

4. 2 mi to inches

5. 18 in. to feet

6. 84 in. to yards

7. 5 in. to yards

8. 106 ft to miles

9. 62 in. to miles

10. 0.4 in. to yards

11. 3 qt to pints

12. 5 lb to ounces

13. 6 T to ounces

14. 4 oz to pounds

15. 15,000 oz to pounds

16. 15,000 oz to tons

17. 9 tbsp to teaspoons

18. 3 c to tablespoons

19. 5 pt to fluid ounces

20. 16 tsp to cups

21. 5 fl oz to quarts

22. 3 qt to gallons

23. 5 pt to teaspoons

24. 3 qt to tablespoons

25. 18 min to seconds

26. 4 da to hours

27. 3 hr to days

28. $\frac{1}{2}$ hr to days

29. $\frac{1}{2}$ da to weeks

30. $3\frac{1}{7}$ wk to seconds

EXERCISES FOR REVIEW

(2.4) **31.** Specify the digits by which 23,840 is divisible.

(4.4) **32.** Find $2\frac{4}{5}$ of $5\frac{5}{6}$ of $7\frac{5}{7}$.

(6.2) **33.** Convert $0.3\frac{2}{3}$ to a fraction.

(8.2) **34.** Use the clustering method to estimate the sum: $53 + 82 + 79 + 49$.

(8.3) **35.** Use the distributive property to compute the product: $60 \cdot 46$.

★ **Answers to Practice Set (9.1)**

A. **1.** 6 yd   **2.** 10,560 ft   **3.** 8.67 yd   **4.** 18 pt   **5.** 0.87 hr   **6.** 2.45 wk

9.2 The Metric System of Measurement

Section Overview

- ☐ **THE ADVANTAGES OF THE BASE TEN NUMBER SYSTEM**
- ☐ **PREFIXES**
- ☐ **CONVERSION FROM ONE UNIT TO ANOTHER UNIT**
- ☐ **CONVERSION TABLE**

☐ THE ADVANTAGES OF THE BASE TEN NUMBER SYSTEM

The metric system of measurement takes advantage of our base ten number system. The advantage of the metric system over the United States system is that in the metric system it is possible to convert from one unit of measure to another simply by multiplying or dividing the given number by a power of 10. This means we can make a conversion simply by moving the decimal point to the right or the left.

☐ PREFIXES

Common units of measure in the metric system are the meter (for length), the liter (for volume), and the gram (for mass). To each of the units can be attached a prefix. The **metric prefixes** along with their meaning are listed below.

Metric Prefixes

kilo	— thousand	**deci**	— tenth
hecto	— hundred	**centi**	— hundredth
deka	— ten	**milli**	— thousandth

For example, if length is being measured,

1 kilometer is equivalent to 1000 meters.
1 centimeter is equivalent to one hundredth of a meter.
1 millimeter is equivalent to one thousandth of a meter.

❑ CONVERSION FROM ONE UNIT TO ANOTHER UNIT

Let's note three characteristics of the metric system that occur in the metric table of measurements.

1. In each category, the prefixes are the same.
2. We can move from a *larger to a smaller* unit of measure by moving the decimal point to the *right*.
3. We can move from a *smaller to a larger* unit of measure by moving the decimal point to the *left*.

The following table provides a summary of the relationship between the basic unit of measure (meter, gram, liter) and each prefix, and how many places the decimal point is moved and in what direction.

kilo   hecto   deka   unit   deci   centi   milli

Basic Unit to Prefix		Move the Decimal Point
unit to deka	1 to 10	1 place to the left
unit to hecto	1 to 100	2 places to the left
unit to kilo	1 to 1,000	3 places to the left
unit to deci	1 to 0.1	1 place to the right
unit to centi	1 to 0.01	2 places to the right
unit to milli	1 to 0.001	3 places to the right

❑ CONVERSION TABLE

Listed below, in the unit conversion table, are some of the common metric units of measure.

Unit Conversion Table

Length		
1 kilometer (km) = 1,000 meters (m)		$1{,}000 \times 1$ m
1 hectometer (hm) = 100 meters		100×1 m
1 dekameter (dam) = 10 meters		10×1 m
1 meter (m)		1×1 m
1 decimeter (dm) = $\dfrac{1}{10}$ meter		$.1 \times 1$ m
1 centimeter (cm) = $\dfrac{1}{100}$ meter		$.01 \times 1$ m
1 millimeter (mm) = $\dfrac{1}{1{,}000}$ meter		$.001 \times 1$ m

Mass		
1 kilogram (kg) = 1,000 grams (g)		$1{,}000 \times 1$ g
1 hectogram (hg) = 100 grams		100×1 g
1 dekagram (dag) = 10 grams		10×1 g
1 gram (g)		1×1 g
1 decigram (dg) = $\dfrac{1}{10}$ gram		$.1 \times 1$ g
1 centigram (cg) = $\dfrac{1}{100}$ gram		$.01 \times 1$ g
1 milligram (mg) = $\dfrac{1}{1{,}000}$ gram		$.001 \times 1$ g

Volume	1 kiloliter (kL) = 1,000 liters (L)	$1,000 \times 1$ L
	1 hectoliter (hL) = 100 liters	100×1 L
	1 dekaliter (daL) = 10 liters	10×1 L
	1 liter (L)	1×1 L
	1 deciliter (dL) = $\frac{1}{10}$ liter	$.1 \times 1$ L
	1 centiliter (cL) = $\frac{1}{100}$ liter	$.01 \times 1$ L
	1 milliliter (mL) = $\frac{1}{1,000}$ liter	$.001 \times 1$ L

| **Time** | Same as the United States system |

Distinction Between Mass and Weight

There is a distinction between mass and weight. The **weight** of a body is related to gravity whereas the mass of a body is not. For example, your weight on the earth is different than it is on the moon, but your mass is the same in both places. **Mass** is a measure of a body's resistance to motion. The more massive a body, the more resistant it is to motion. Also, more massive bodies weigh more than less massive bodies.

Converting Metric Units

> To convert from one metric unit to another metric unit:
>
> 1. Determine the location of the original number on the metric scale (pictured in each of the following examples).
> 2. Move the decimal point of the original number in the same direction and same number of places as is necessary to move to the metric unit you wish to go to.

We can also convert from one metric unit to another using unit fractions. Both methods are shown in problem 1 of Sample Set A.

☆ SAMPLE SET A

1. Convert 3 kilograms to grams.

   (a) 3 kg can be written as 3.0 kg. Then,

kg	hg	dag	g	dg	cg	mg

   1 2 3
   places to the right

   3.0 kg = 3 000. g
           1 2 3

   Thus, 3 kg = 3,000 g.

   (b) We can also use unit fractions to make this conversion.

   Since we are converting to grams, and 1,000 g = 1 kg, we choose the unit fraction $\frac{1,000 \text{ g}}{1 \text{ kg}}$ since grams is in the numerator.

   $3 \text{ kg} = 3 \text{ kg} \cdot \dfrac{1,000 \text{ g}}{1 \text{ kg}}$

   $= 3 \, \cancel{\text{kg}} \cdot \dfrac{1,000 \text{ g}}{1 \, \cancel{\text{kg}}}$

   $= 3 \cdot 1,000 \text{ g}$

   $= 3,000 \text{ g}$

Continued

2. Convert 67.2 hectoliters to milliliters.

kL	hL	daL	L	dL	cL	mL

places to the right
1 2 3 4 5

67.2 hL = 67 20000 mL
1 2 3 4 5

Thus, 67.2 hL = 6,720,000 mL.

3. Convert 100.07 centimeters to meters.

km	hm	dam	m	dm	cm	mm

places to the left
2 1

100.07 cm = 1.0007 m
2 1

Thus, 100.07 cm = 1.0007 m.

4. Convert 0.16 milligrams to grams.

kg	hg	dg	g	dg	cg	mg

places to the left
3 2 1

0.16 mg = 0.000 16 g
3 2 1

Thus, 0.16 mg = 0.00016 g.

★ **PRACTICE SET A**

1. Convert 411 kilograms to grams.

2. Convert 5.626 liters to centiliters.

3. Convert 80 milliliters to kiloliters.

4. Convert 150 milligrams to centigrams.

5. Convert 2.5 centimeters to meters.

Answers to the Practice Set are on p. 394.

Section 9.2 EXERCISES

For problems 1–20, make each conversion.

1. 87 m to cm

2. 905 L to mL

3. 16,005 mg to g

4. 48.66 L to dL

5. 11.161 kL to L

6. 521.85 cm to mm

7. 1.26 dag to dg

8. 99.04 dam to cm

9. 0.51 kL to daL

10. 0.17 kL to daL

11. 0.05 m to dm

12. 0.001 km to mm

13. 8.106 hg to cg

14. 17.0186 kL to mL

15. 3 cm to m

16. 9 mm to m

17. 4 g to mg

18. 2 L to kL

19. 6 kg to mg

20. 7 daL to mL

EXERCISES FOR REVIEW

(5.2)   **21.** Find the value of $\dfrac{5}{8} - \dfrac{1}{3} + \dfrac{3}{4}$.

(7.2)   **22.** Solve the proportion: $\dfrac{9}{x} = \dfrac{27}{60}$.

(8.1)   **23.** Use the method of rounding to estimate the sum: $8{,}226 + 4{,}118$.

(8.2)   **24.** Use the clustering method to estimate the sum: $87 + 121 + 118 + 91 + 92$.

(9.1)   **25.** Convert 3 in. to yd.

★ **Answers to Practice Set (9.2)**

A.   **1.** 411,000 g   **2.** 562.6 cL   **3.** 0.00008 kL   **4.** 15 cg   **5.** 0.025 m

9.3 Simplification of Denominate Numbers

Section Overview	☐ **CONVERTING TO MULTIPLE UNITS** ☐ **ADDING AND SUBTRACTING DENOMINATE NUMBERS** ☐ **MULTIPLYING A DENOMINATE NUMBER BY A WHOLE NUMBER** ☐ **DIVIDING A DENOMINATE NUMBER BY A WHOLE NUMBER**

☐ CONVERTING TO MULTIPLE UNITS

Denominate Numbers

Numbers that have units of measure associated with them are called **denominate numbers.** It is often convenient, or even necessary, to simplify a denominate number.

Simplified Denominate Number

> A denominate number is **simplified** when the number of standard units of measure associated with it does not exceed the next higher type of unit.

The denominate number 55 min is simplified since it is smaller than the next higher type of unit, 1 hr. The denominate number 65 min is *not* simplified since it is not smaller than the next higher type of unit, 1 hr. The denominate number 65 min can be simplified to 1 hr 5 min. The denominate number 1 hr 5 min is simplified since the next higher type of unit is day, and 1 hr does not exceed 1 day.

☆ **SAMPLE SET A**

1. Simplify 19 in.

Since 12 in. = 1 ft, and 19 = 12 + 7,

19 in. = 12 in. + 7 in.
      = 1 ft + 7 in.
      = 1 ft 7 in.

2. Simplify 4 gal 5 qt.

Since 4 qt = 1 gal, and 5 = 4 + 1,

$$4 \text{ gal } 5 \text{ qt} = 4 \text{ gal} + 4 \text{ qt} + 1 \text{ qt}$$
$$= 4 \text{ gal} + 1 \text{ gal} + 1 \text{ qt}$$
$$= 5 \text{ gal} + 1 \text{ qt}$$
$$= 5 \text{ gal } 1 \text{ qt}$$

3. Simplify 2 hr 75 min.

Since 60 min = 1 hr, and 75 = 60 + 15,

$$2 \text{ hr } 75 \text{ min} = 2 \text{ hr} + 60 \text{ min} + 15 \text{ min}$$
$$= 2 \text{ hr} + 1 \text{ hr} + 15 \text{ min}$$
$$= 3 \text{ hr} + 15 \text{ min}$$
$$= 3 \text{ hr } 15 \text{ min}$$

4. Simplify 43 fl oz.

Since 8 fl oz = 1 c (1 cup), and 43 ÷ 8 = 5 R3,

$$43 \text{ fl oz} = 40 \text{ fl oz} + 3 \text{ fl oz}$$
$$= 5 \cdot 8 \text{ fl oz} + 3 \text{ fl oz}$$
$$= 5 \cdot 1 \text{ c} + 3 \text{ fl oz}$$
$$= 5 \text{ c} + 3 \text{ fl oz}$$

But, 2 c = 1 pt and 5 ÷ 2 = 2 R1. So,

$$5 \text{ c} + 3 \text{ fl oz} = 2 \cdot 2 \text{ c} + 1 \text{ c} + 3 \text{ fl oz}$$
$$= 2 \cdot 1 \text{ pt} + 1 \text{ c} + 3 \text{ fl oz}$$
$$= 2 \text{ pt} + 1 \text{ c} + 3 \text{ fl oz}$$

But, 2 pt = 1 qt, so

$$2 \text{ pt} + 1 \text{ c} + 3 \text{ fl oz} = 1 \text{ qt } 1 \text{ c } 3 \text{ fl oz}$$

★ **PRACTICE SET A**

Simplify each denominate number. Refer to the conversion tables given in Section 9.1, if necessary.

1. 18 in. **2.** 8 gal 9 qt **3.** 5 hr 80 min **4.** 8 wk 11 da **5.** 86 da

☐ ADDING AND SUBTRACTING DENOMINATE NUMBERS

Adding and Subtracting Denominate Numbers

Denominate numbers can be added or subtracted by:

1. writing the numbers vertically so that the like units appear in the same column.
2. adding or subtracting the number parts, carrying along the unit.
3. simplifying the sum or difference.

☆ **SAMPLE SET B**

1. Add 6 ft 8 in. to 2 ft 9 in.

$$\begin{array}{r} 6 \text{ ft} \quad 8 \text{ in.} \\ +2 \text{ ft} \quad 9 \text{ in.} \\ \hline 8 \text{ ft } 17 \text{ in.} \end{array}$$ **Simplify this denominate number.**

Since 12 in. = 1 ft,

$$8 \text{ ft} + 12 \text{ in.} + 5 \text{ in.} = 8 \text{ ft} + 1 \text{ ft} + 5 \text{ in.}$$
$$= 9 \text{ ft} + 5 \text{ in.}$$
$$= 9 \text{ ft } 5 \text{ in.}$$

2. Subtract 5 da 3 hr from 8 da 11 hr.

$$\begin{array}{r} 8 \text{ da } 11 \text{ hr} \\ -5 \text{ da} \quad 3 \text{ hr} \\ \hline 3 \text{ da} \quad 8 \text{ hr} \end{array}$$

3. Subtract 3 lb 14 oz from 5 lb 3 oz.

$$\begin{array}{r} 5 \text{ lb} \quad 3 \text{ oz} \\ -3 \text{ lb } 14 \text{ oz} \end{array}$$

We cannot directly subtract 14 oz from 3 oz, so we must borrow 16 oz from the pounds.

$$5 \text{ lb } 3 \text{ oz} = 5 \text{ lb} + 3 \text{ oz}$$
$$= 4 \text{ lb} + 1 \text{ lb} + 3 \text{ oz}$$
$$= 4 \text{ lb} + 16 \text{ oz} + 3 \text{ oz} \qquad \text{(Since 1 lb} = 16 \text{ oz.)}$$
$$= 4 \text{ lb} + 19 \text{ oz}$$
$$= 4 \text{ lb } 19 \text{ oz}$$

$$\begin{array}{r} 4 \text{ lb } 19 \text{ oz} \\ -3 \text{ lb } 14 \text{ oz} \\ \hline 1 \text{ lb} \quad 5 \text{ oz} \end{array}$$

4. Subtract 4 da 9 hr 21 min from 7 da 10 min.

$$\begin{array}{r} 7 \text{ da } 0 \text{ hr } 10 \text{ min} \\ -4 \text{ da } 9 \text{ hr } 21 \text{ min} \end{array}$$ **Borrow 1 da from the 7 da.**

$$\begin{array}{r} 6 \text{ da } 24 \text{ hr } 10 \text{ min} \\ -4 \text{ da} \quad 9 \text{ hr } 21 \text{ min} \end{array}$$ **Borrow 1 hr from the 24 hr.**

$$\begin{array}{r} 6 \text{ da } 23 \text{ hr } 70 \text{ min} \\ -4 \text{ da} \quad 9 \text{ hr } 21 \text{ min} \\ \hline 2 \text{ da } 14 \text{ hr } 49 \text{ min} \end{array}$$

★ **PRACTICE SET B**

Perform each operation. Simplify when possible.

1. Add 4 gal 3 qt to 1 gal 2 qt.   **2.** Add 9 hr 48 min to 4 hr 26 min.

3. Subtract 2 ft 5 in. from 8 ft 7 in.   **4.** Subtract 15 km 460 m from 27 km 800 m.

5. Subtract 8 min 35 sec from 12 min 10 sec.　　　**6.** Add 4 yd 2 ft 7 in. to 9 yd 2 ft 8 in.

7. Subtract 11 min 55 sec from 25 min 8 sec.

☐ MULTIPLYING A DENOMINATE NUMBER BY A WHOLE NUMBER

Let's examine the repeated sum

$$\underbrace{4 \text{ ft } 9 \text{ in.} + 4 \text{ ft } 9 \text{ in.} + 4 \text{ ft } 9 \text{ in.}}_{3 \text{ times}} = 12 \text{ ft } 27 \text{ in.}$$

Recalling that multiplication is a description of repeated addition, by the distributive property we have

$$
\begin{aligned}
3(4 \text{ ft } 9 \text{ in.}) &= 3(4 \text{ ft} + 9 \text{ in.}) \\
&= 3 \cdot 4 \text{ ft} + 3 \cdot 9 \text{ in.} \\
&= 12 \text{ ft} + 27 \text{ in.} \qquad \text{Now, } 27 \text{ in.} = 2 \text{ ft } 3 \text{ in.} \\
&= 12 \text{ ft} + 2 \text{ ft} + 3 \text{ in.} \\
&= 14 \text{ ft} + 3 \text{ in.} \\
&= 14 \text{ ft } 3 \text{ in.}
\end{aligned}
$$

From these observations, we can suggest the following rule.

Multiplying a Denominate Number by a Whole Number

> To multiply a denominate number by a whole number, multiply the number part of each unit by the whole number and affix the unit to this product.

☆ SAMPLE SET C

Perform the following multiplications. Simplify if necessary.

1. $6 \cdot (2 \text{ ft } 4 \text{ in.}) = 6 \cdot 2 \text{ ft} + 6 \cdot 4 \text{ in.}$
$$= 12 \text{ ft} + 24 \text{ in.}$$

Since 3 ft = 1 yd and 12 in. = 1 ft,

$$
\begin{aligned}
12 \text{ ft} + 24 \text{ in.} &= 4 \text{ yd} + 2 \text{ ft} \\
&= 4 \text{ yd } 2 \text{ ft}
\end{aligned}
$$

2. $8 \cdot (5 \text{ hr } 21 \text{ min } 55 \text{ sec}) = 8 \cdot 5 \text{ hr} + 8 \cdot 21 \text{ min} + 8 \cdot 55 \text{ sec}$
$$
\begin{aligned}
&= 40 \text{ hr} + 168 \text{ min} + 440 \text{ sec} \\
&= 40 \text{ hr} + 168 \text{ min} + 7 \text{ min} + 20 \text{ sec} \\
&= 40 \text{ hr} + 175 \text{ min} + 20 \text{ sec} \\
&= 40 \text{ hr} + 2 \text{ hr} + 55 \text{ min} + 20 \text{ sec} \\
&= 42 \text{ hr} + 55 \text{ min} + 20 \text{ sec} \\
&= 24 \text{ hr} + 18 \text{ hr} + 55 \text{ min} + 20 \text{ sec} \\
&= 1 \text{ da} + 18 \text{ hr} + 55 \text{ min} + 20 \text{ sec} \\
&= 1 \text{ da } 18 \text{ hr } 55 \text{ min } 20 \text{ sec}
\end{aligned}
$$

★ **PRACTICE SET C**

Perform the following multiplications. Simplify.

1. 2 · (10 min)　　　**2.** 5 · (3 qt)　　　**3.** 4 · (5 ft 8 in.)　　　**4.** 10 · (2 hr 15 min 40 sec)

☐ DIVIDING A DENOMINATE NUMBER BY A WHOLE NUMBER

Dividing a Denominate
Number by a Whole Number

> To divide a denominate number by a whole number, divide the number part of each unit by the whole number beginning with the largest unit. Affix the unit to this quotient. Carry any remainder to the next unit.

☆ **SAMPLE SET D**

Perform the following divisions. Simplify if necessary.

1. (12 min 40 sec) ÷ 4

```
       3 min 10 sec
  4) 12 min 40 sec
    ↳12 min
            ↘40 sec
             40 sec
                  0
```

Thus (12 min 40 sec) ÷ 4 = 3 min 10 sec

2. (5 yd 2 ft 9 in.) ÷ 3

```
     1 yd 2 ft 11 in.
  3) 5 yd 2 ft  9 in.
    ↰3 yd
     2 yd 2 ft              Convert to feet:   2 yd 2 ft = 8 ft.
         ↘8 ft
          6 ft
          2 ft  9 in.       Convert to inches:   2 ft 9 in. = 33 in.
              ↘33 in.
               33 in.
                    0
```

Thus (5 yd 2 ft 9 in.) ÷ 3 = 1 yd 2 ft 11 in.

★ **PRACTICE SET D**

Perform the following divisions. Simplify if necessary.

1. (18 hr 36 min) ÷ 9   **2.** (34 hr 8 min) ÷ 8   **3.** (13 yd 7 in.) ÷ 5

4. (47 gal 2 qt 1 pt) ÷ 3

Answers to Practice Sets are on p. 401.

Section 9.3 EXERCISES

For problems 1–15, simplify the denominate numbers.

1. 16 in.

2. 19 ft

3. 85 min

4. 90 min

5. 17 da

6. 25 oz

7. 240 oz

8. 3,500 lb

9. 26 qt

10. 300 sec

11. 135 oz

12. 14 tsp

13. 18 pt

14. 3,500 m

15. 16,300 mL

For problems 16–30, perform the indicated operations and simplify the answers if possible.

16. Add 6 min 12 sec to 5 min 15 sec.

17. Add 14 da 6 hr to 1 da 5 hr.

18. Add 9 gal 3 qt to 2 gal 3 qt.

19. Add 16 lb 10 oz to 42 lb 15 oz.

20. Subtract 3 gal 1 qt from 8 gal 3 qt.

21. Subtract 3 ft 10 in. from 5 ft 8 in.

22. Subtract 5 lb 9 oz from 12 lb 5 oz.

23. Subtract 10 hr 10 min from 11 hr 28 min.

24. Add 3 fl oz 1 tbsp 2 tsp
to 5 fl oz 1 tbsp 2 tsp.

25. Add 4 da 7 hr 12 min
to   1 da 8 hr 53 min.

26. Subtract 5 hr 21 sec
from   11 hr 2 min 14 sec.

27. Subtract 6 T 1,300 lb 10 oz
from   8 T   400 lb 10 oz.

28. Subtract 15 mi 10 in.
from   27 mi 800 ft 7 in.

29. Subtract 3 wk 5 da 50 min 12 sec
from   5 wk 6 da 20 min 5 sec.

30. Subtract   3 gal 3 qt 1 pt 1 oz
from   10 gal 2 qt 2 oz.

EXERCISES FOR REVIEW

(4.5)   **31.** Find the value: $\left(\dfrac{5}{8}\right)^2 + \dfrac{39}{64}$.

(5.3)   **32.** Find the sum: $8 + 6\dfrac{3}{5}$.

(6.2)   **33.** Convert $2.05\dfrac{1}{11}$ to a fraction.

(7.3)   **34.** An acid solution is composed of 3 parts acid to 7 parts water. How many parts of acid are there in a solution that contains 126 parts water?

(9.2)   **35.** Convert 126 kg to grams.

★ **Answers to Practice Sets (9.3)**

A. **1.** 1 ft 6 in.   **2.** 10 gal 1 qt   **3.** 6 hr 20 min   **4.** 9 wk 4 da   **5.** 12 wk 2 da

B. **1.** 6 gal 1 qt   **2.** 14 hr 14 min   **3.** 6 ft 2 in.   **4.** 12 km 340 m   **5.** 3 min 35 sec
6. 14 yd 2 ft 3 in.   **7.** 13 min 13 sec

C. **1.** 20 min   **2.** 15 qt = 3 gal 3 qt   **3.** 20 ft 32 in. = 7 yd 1 ft 8 in.
4. 20 hr 150 min 400 sec = 22 hr 36 min 40 sec

D. **1.** 2 hr 4 min   **2.** 4 hr 16 min   **3.** 2 yd 1 ft 11 in.   **4.** 15 gal 3 qt 1 pt

9.4 Perimeter and Circumference of Geometric Figures

Section Overview

- ☐ **POLYGONS**
- ☐ **PERIMETER**
- ☐ **CIRCUMFERENCE/DIAMETER/RADIUS**
- ☐ **THE NUMBER** π
- ☐ **FORMULAS**

☐ POLYGONS

We can make use of conversion skills with denominate numbers to make measurements of geometric figures such as rectangles, triangles, and circles. To make these measurements we need to be familiar with several definitions.

Polygon

> A **polygon** is a closed plane (flat) figure whose sides are line segments (portions of straight lines).

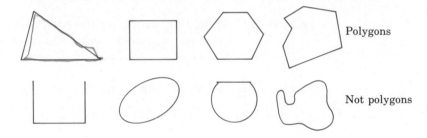

Polygons

Not polygons

☐ PERIMETER

Perimeter

> The **perimeter** of a polygon is the distance around the polygon.

To find the perimeter of a polygon, we simply add up the lengths of all the sides.

☆ SAMPLE SET A

Find the perimeter of each polygon.

1.

```
       5 cm
   ┌─────────┐
2 cm│    V    │2 cm
   └─────────┘
       5 cm
```

Perimeter = 2 cm + 5 cm + 2 cm + 5 cm
= 14 cm

2.

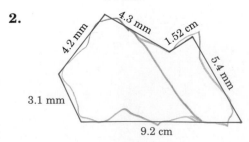

$$
\begin{array}{r}
\text{Perimeter} = \quad 3.1 \ \ \text{mm} \\
4.2 \ \ \text{mm} \\
4.3 \ \ \text{mm} \\
1.52 \ \text{mm} \\
5.4 \ \ \text{mm} \\
+9.2 \ \ \text{mm} \\
\hline
27.72 \ \text{mm}
\end{array}
$$

3.

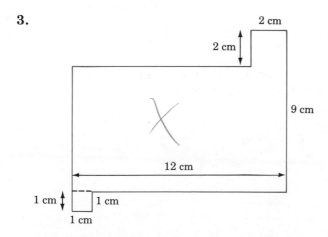

Our first observation is that three of the dimensions are missing. However, we can determine the missing measurements using the following process. Let A, B, and C represent the missing measurements. Visualize

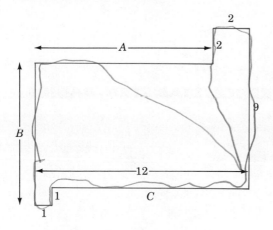

$A = 12 \text{ m} - 2 \text{ m} = 10 \text{ m}$
$B = \ \ 9 \text{ m} + 1 \text{ m} - \ \ 2 \text{ m} = 8 \text{ m}$
$C = 12 \text{ m} - 1 \text{ m} = 11 \text{ m}$

$$
\begin{array}{r}
\text{Perimeter} = \quad 8 \ \text{m} \\
10 \ \text{m} \\
2 \ \text{m} \\
2 \ \text{m} \\
9 \ \text{m} \\
11 \ \text{m} \\
1 \ \text{m} \\
+ \ 1 \ \text{m} \\
\hline
44 \ \text{m}
\end{array}
$$

★ PRACTICE SET A

Find the perimeter of each polygon.

1.
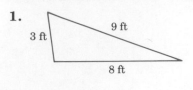
3 ft

9 ft

8 ft

2.
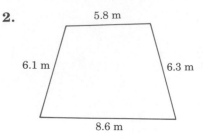
5.8 m

6.1 m

6.3 m

8.6 m

3.
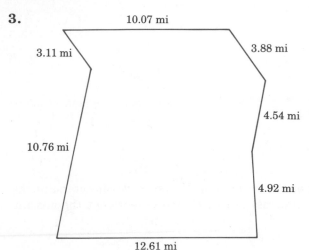
10.07 mi

3.11 mi

3.88 mi

4.54 mi

10.76 mi

4.92 mi

12.61 mi

□ CIRCUMFERENCE/DIAMETER/RADIUS

Circumference
Diameter

Radius

> The **circumference** of a circle is the distance around the circle.
> A **diameter** of a circle is any line segment that passes through the center of the circle and has its endpoints on the circle.
> A **radius** of a circle is any line segment having as its endpoints the center of the circle and a point on the circle.
> The radius is one half the diameter.

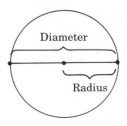
Diameter

Radius

□ THE NUMBER π

π

The symbol π, read "pi," represents the nonterminating, nonrepeating decimal number 3.14159 . . . . This number has been computed to millions of decimal places without the appearance of a repeating block of digits.

For computational purposes, π is often approximated as 3.14. We will write π ≈ 3.14 to denote that π is approximately equal to 3.14. The symbol "≈" means "approximately equal to."

☐ FORMULAS

To find the circumference of a circle, we need only know its diameter or radius. We then use a formula for computing the circumference of the circle.

Formula

> A **formula** is a rule or method for performing a task. In mathematics, a formula is a rule that directs us in computations.

Formulas are usually composed of letters that represent important, but possibly unknown, quantities.

If C, d, and r represent, respectively, the circumference, diameter, and radius of a circle, then the following two formulas give us directions for computing the circumference of the circle.

Circumference Formulas

> 1. $C = \pi d$ or $C \approx (3.14)d$
> 2. $C = 2\pi r$ or $C \approx 2(3.14)r$

☆ SAMPLE SET B

1. Find the exact circumference of the circle.

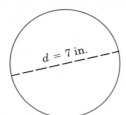

Use the formula $C = \pi d$.

$C = \pi \cdot 7$ in.

By commutativity of multiplication,

$C = 7$ in. $\cdot \pi$
$C = 7\pi$ in., exactly

This result is exact since π has not been approximated.

2. Find the approximate circumference of the circle.

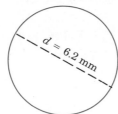

Use the formula $C \approx \pi d$.

$C \approx (3.14)(6.2)$
$C \approx 19.648$ mm

This result is approximate since π has been approximated by 3.14.

3. Find the approximate circumference of a circle with radius 18 inches.

Since we're given that the radius, r, is 18 in., we'll use the formula $C = 2\pi r$.

$C \approx (2)(3.14)(18$ in.$)$
$C \approx 113.04$ in.

4. Find the approximate perimeter of the figure.

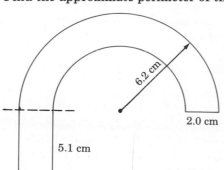

We notice that we have two semicircles (half circles).

The larger radius is 6.2 cm.
The smaller radius is 6.2 cm − 2.0 cm = 4.2 cm.
The width of the bottom part of the rectangle is 2.0 cm.

Continued

Perimeter = 2.0 cm
5.1 cm
2.0 cm
5.1 cm
(0.5) · (2) · (3.14) · (6.2 cm)   **Circumference of outer semicircle.**
+(0.5) · (2) · (3.14) · (4.2 cm)   **Circumference of inner semicircle**

6.2 cm − 2.0 cm = 4.2 cm

The 0.5 appears because we want the perimeter of only *half* a circle.

Perimeter ≈   2.0   cm
5.1   cm
2.0   cm
5.1   cm
19.468 cm
+13.188 cm
46.856 cm

★ PRACTICE SET B

1. Find the exact circumference of the circle.

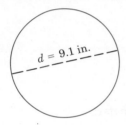

d = 9.1 in.

2. Find the approximate circumference of the circle.

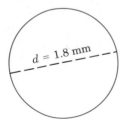

d = 1.8 mm

3. Find the approximate circumference of the circle with radius 20.1 m.

4. Find the approximate outside perimeter of

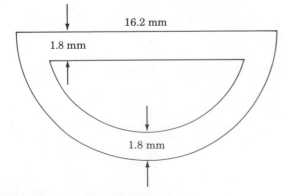

16.2 mm

1.8 mm

1.8 mm

Answers to Practice Sets are on p. 409.

Section 9.4 EXERCISES

For problems 1–20, find each perimeter or approximate circumference. Use $\pi \approx 3.14$.

1.

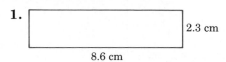

2.3 cm

8.6 cm

2.

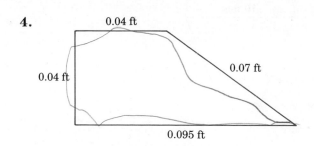

3.8 mm

9.3 mm

8 mm

3.

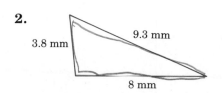

4.8 in.

17.23 in.

16.11 in.

4.

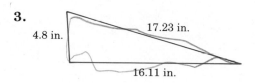

0.04 ft

0.04 ft

0.07 ft

0.095 ft

5.

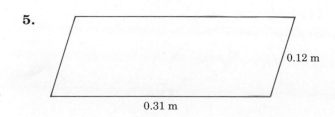

0.12 m

0.31 m

6.

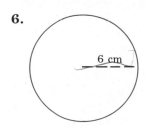

6 cm

7.

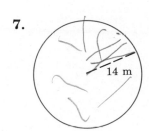

14 m

8.

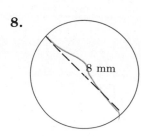

8 mm

9.

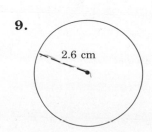

2.6 cm

10.

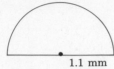

1.1 mm

11.

0.03 cm

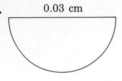

12.

5 in.

13.

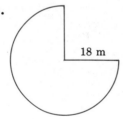

18 m

14.

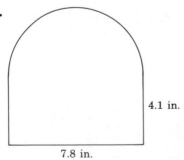

4.1 in.

7.8 in.

15.

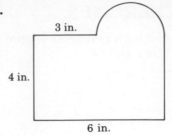

3 in.

4 in.

6 in.

16.

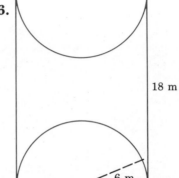

18 m

6 m

17.

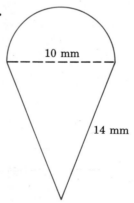

10 mm

14 mm

18.

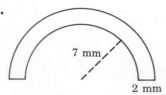

7 mm

2 mm

19.

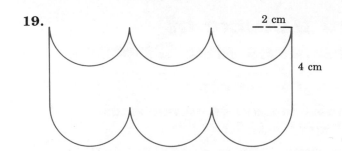

20. Find the outside perimeter.

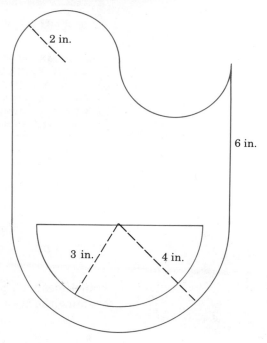

EXERCISES FOR REVIEW

(4.4) **21.** Find the value of $2\dfrac{8}{13} \cdot \sqrt{10\dfrac{9}{16}}$.

(5.2) **22.** Find the value of $\dfrac{8}{15} + \dfrac{7}{10} + \dfrac{21}{60}$.

(6.8) **23.** Convert $\dfrac{7}{8}$ to a decimal.

(9.1) **24.** What is the name given to a quantity that is used as a comparison to determine the measure of another quantity?

(9.3) **25.** Add 42 min 26 sec to 53 min 40 sec and simplify the result.

★ **Answers to Practice Sets (9.4)**

A. **1.** 20 ft   **2.** 26.8 m   **3.** 49.89 mi

B. **1.** 9.1π in.   **2.** 5.652 mm   **3.** 126.228 m   **4.** 41.634 mm

9.5 Area and Volume of Geometric Figures and Objects

Section Overview

☐ **THE MEANING AND NOTATION FOR AREA**
☐ **AREA FORMULAS**
☐ **FINDING AREAS OF SOME COMMON GEOMETRIC FIGURES**
☐ **THE MEANING AND NOTATION FOR VOLUME**
☐ **VOLUME FORMULAS**
☐ **FINDING VOLUMES OF SOME COMMON GEOMETRIC OBJECTS**

Quite often it is necessary to multiply one denominate number by another. To do so, we multiply the number parts together and the unit parts together. For example,

$$8 \text{ in.} \cdot 8 \text{ in.} = 8 \cdot 8 \cdot \text{in.} \cdot \text{in.}$$
$$= 64 \text{ in.}^2$$

$$4 \text{ mm} \cdot 4 \text{ mm} \cdot 4 \text{ mm} = 4 \cdot 4 \cdot 4 \cdot \text{mm} \cdot \text{mm} \cdot \text{mm}$$
$$= 64 \text{ mm}^3$$

Sometimes the product of units has a physical meaning. In this section, we will examine the meaning of the products (length unit)^2 and (length unit)^3.

☐ THE MEANING AND NOTATION FOR AREA

The product (length unit) $\cdot$ (length unit) = (length unit)^2, or, square length unit (sq length unit), can be interpreted physically as the *area* of a surface.

Area

> The **area** of a surface is the amount of square length units contained in the surface.

For example, 3 sq in. means that 3 squares, 1 inch on each side, can be placed precisely on some surface. (The squares may have to be cut and rearranged so they match the shape of the surface.)

We will examine the area of the following geometric figures.

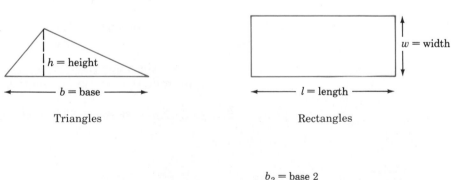

Triangles

Rectangles

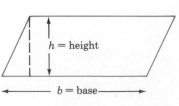

Parallelograms

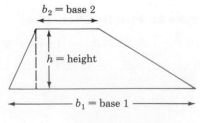

Trapezoids

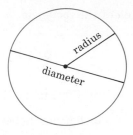

Circles

☑ AREA FORMULAS

We can determine the areas of these geometric figures using the following formulas.

Figure	Area Formula	Statement
Triangle	$A_T = \dfrac{1}{2} \cdot b \cdot h$	Area of a triangle is one half the base times the height.
Rectangle	$A_R = 1 \cdot w$	Area of a rectangle is the length times the width.
Parallelogram	$A_P = b \cdot h$	Area of a parallelogram is the base times the height.
Trapezoid	$A_{Trap} = \dfrac{1}{2} \cdot (b_1 + b_2) \cdot h$	Area of a trapezoid is one half the sum of the two bases times the height.
Circle	$A_c = \pi r^2$	Area of a circle is π times the square of the radius.

☑ FINDING AREAS OF SOME COMMON GEOMETRIC FIGURES

☆ **SAMPLE SET A**

1. Find the area of the triangle.

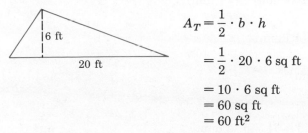

$$A_T = \frac{1}{2} \cdot b \cdot h$$

$$= \frac{1}{2} \cdot 20 \cdot 6 \text{ sq ft}$$

$$= 10 \cdot 6 \text{ sq ft}$$
$$= 60 \text{ sq ft}$$
$$= 60 \text{ ft}^2$$

The area of this triangle is 60 sq ft, which is often written as 60 ft^2.

Continued

2. Find the area of the rectangle.

8 in.

4 ft 2 in.

Let's first convert 4 ft 2 in. to inches. Since we wish to convert to inches, we'll use the unit fraction $\dfrac{12 \text{ in.}}{1 \text{ ft}}$ since it has inches in the numerator. Then,

$$4 \text{ ft} = \frac{4 \text{ ft}}{1} \cdot \frac{12 \text{ in.}}{1 \text{ ft}}$$

$$= \frac{4 \cancel{\text{ft}}}{1} \cdot \frac{12 \text{ in.}}{1 \cancel{\text{ft}}}$$

$$= 48 \text{ in.}$$

Thus, 4 ft 2 in. = 48 in. + 2 in. = 50 in.

$$A_R = l \cdot w$$
$$= 50 \text{ in.} \cdot 8 \text{ in.}$$
$$= 400 \text{ sq in.}$$

The area of this rectangle is 400 sq in.

3. Find the area of the parallelogram.

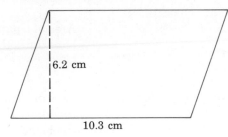

6.2 cm

10.3 cm

$$A_P = b \cdot h$$
$$= 10.3 \text{ cm} \cdot 6.2 \text{ cm}$$
$$= 63.86 \text{ sq cm}$$

The area of this parallelogram is 63.86 sq cm.

4. Find the area of the trapezoid.

14.5 mm

4.1 mm

20.4 mm

$$A_{Trap} = \frac{1}{2} \cdot (b_1 + b_2) \cdot h$$

$$= \frac{1}{2} \cdot (14.5 \text{ mm} + 20.4 \text{ mm}) \cdot (4.1 \text{ mm})$$

$$= \frac{1}{2} \cdot (34.9 \text{ mm}) \cdot (4.1 \text{ mm})$$

$$= \frac{1}{2} \cdot (143.09 \text{ sq mm})$$

$$= 71.545 \text{ sq mm}$$

The area of this trapezoid is 71.545 sq mm.

5. Find the approximate area of the circle.

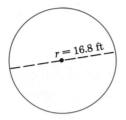

$r = 16.8$ ft

$A_c = \pi \cdot r^2$
$\approx (3.14) \cdot (16.8 \text{ ft})^2$
$\approx (3.14) \cdot (282.24 \text{ sq ft})$
≈ 886.23 sq ft

The area of this circle is approximately 886.23 sq ft.

★ **PRACTICE SET A**

Find the area of each of the following geometric figures.

1.

4 cm

18 cm

2.

4.05 mm

9.26 mm

3.

2.6 in.

5.1 in.

4.

17 mi

$h = 15$ mi

32 mi

5.

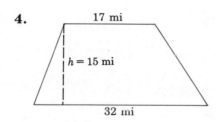

$r = 12$ ft

(approximate)

6.

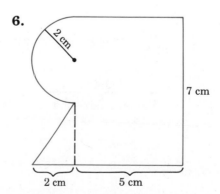

2 cm

7 cm

2 cm   5 cm

☐ THE MEANING AND NOTATION FOR VOLUME

The product (length unit) · (length unit) · (length unit) = (length unit)^3, or cubic length unit (cu length unit), can be interpreted physically as the *volume* of a three-dimensional object.

Volume

> The **volume** of an object is the amount of cubic length units contained in the object.

For example, 4 cu mm means that 4 cubes, 1 mm on each side, would precisely fill some three-dimensional object. (The cubes may have to be cut and rearranged so they match the shape of the object.)

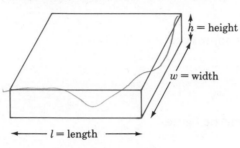

Rectangular solid

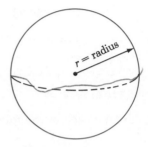

Sphere

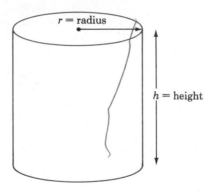

Cylinder

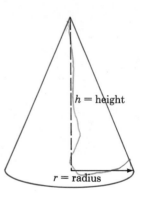

Cone

☐ VOLUME FORMULAS

Figure	Volume Formula	Statement
Rectangular solid	$V_R = l \cdot w \cdot h$ $= \text{(area of base)} \cdot \text{(height)}$	The volume of a rectangular solid is the length times the width times the height.
Sphere	$V_S = \dfrac{4}{3} \cdot \pi \cdot r^3$	The volume of a sphere is $\dfrac{4}{3}$ times π times the cube of the radius.

Cylinder

$$V_{Cyl} = \pi \cdot r^2 \cdot h$$
$$= (\text{area of base}) \cdot (\text{height})$$

The volume of a cylinder is π times the square of the radius times the height.

Cone

$$V_c = \frac{1}{3} \cdot \pi \cdot r^2 \cdot h$$
$$= (\text{area of base}) \cdot (\text{height})$$

The volume of a cone is $\frac{1}{3}$ times π times the square of the radius times the height.

☐ FINDING VOLUMES OF SOME COMMON GEOMETRIC OBJECTS

☆ **SAMPLE SET B**

1. Find the volume of the rectangular solid.

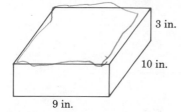

3 in.

10 in.

9 in.

$$V_R = l \cdot w \cdot h$$
$$= 9 \text{ in.} \cdot 10 \text{ in.} \cdot 3 \text{ in.}$$
$$= 270 \text{ cu in.}$$
$$= 270 \text{ in.}^3$$

The volume of this rectangular solid is 270 cu in.

2. Find the approximate volume of the sphere.

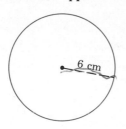

6 cm

$$V_S = \frac{4}{3} \cdot \pi \cdot r^3$$

$$\approx \left(\frac{4}{3}\right) \cdot (3.14) \cdot (6 \text{ cm})^3$$

$$\approx \left(\frac{4}{3}\right) \cdot (3.14) \cdot (216 \text{ cu cm})$$

$$\approx 904.32 \text{ cu cm}$$

The approximate volume of this sphere is 904.32 cu cm, which is often written as 904.32 cm³.

3. Find the approximate volume of the cylinder.

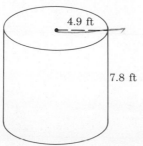

4.9 ft

7.8 ft

$$V_{Cyl} = \pi \cdot r^2 \cdot h$$
$$\approx (3.14) \cdot (4.9 \text{ ft})^2 \cdot (7.8 \text{ ft})$$
$$\approx (3.14) \cdot (24.01 \text{ sq ft}) \cdot (7.8 \text{ ft})$$
$$\approx (3.14) \cdot (187.278 \text{ cu ft})$$
$$\approx 588.05292 \text{ cu ft}$$

The volume of this cylinder is approximately 588.05292 cu ft. The volume is approximate because we approximated π with 3.14.

Continued

4. Find the approximate volume of the cone. Round to two decimal places.

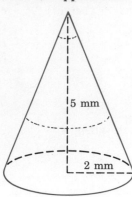

$$V_c = \frac{1}{3} \cdot \pi \cdot r^2 \cdot h$$

$$\approx \left(\frac{1}{3}\right) \cdot (3.14) \cdot (2 \text{ mm})^2 \cdot (5 \text{ mm})$$

$$\approx \left(\frac{1}{3}\right) \cdot (3.14) \cdot (4 \text{ sq mm}) \cdot (5 \text{ mm})$$

$$\approx \left(\frac{1}{3}\right) \cdot (3.14) \cdot (20 \text{ cu mm})$$

$$\approx 20.9\overline{3} \text{ cu mm}$$

$$\approx 20.93 \text{ cu mm}$$

The volume of this cone is approximately 20.93 cu mm. The volume is approximate because we approximated π with 3.14.

★ **PRACTICE SET B**

Find the volume of each geometric object. If π is required, approximate it with 3.14 and find the approximate volume.

1.

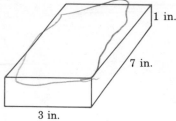

2. Sphere

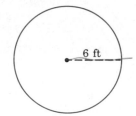

3.

4.

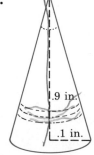

Section 9.5 EXERCISES

For problems 1–26, find each indicated measurement.

1. Area

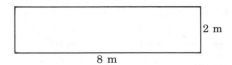

2 m
8 m

2. Area

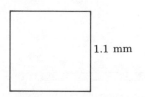

2.3 in.
4.1 in.

3. Area

1.1 mm

4. Area

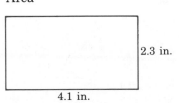

3 cm
8 cm

5. Area

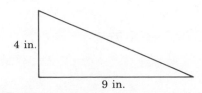

4 in.
9 in.

6. Area

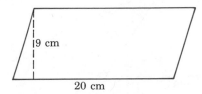

9 cm
20 cm

7. Exact area

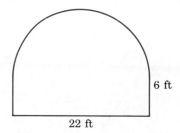

6 ft
22 ft

8. Approximate area

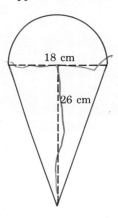

18 cm
26 cm

9. Area

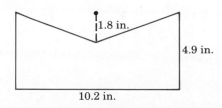

1.8 in.
4.9 in.
10.2 in.

10. Area

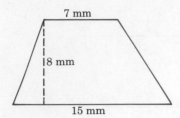

7 mm

8 mm

15 mm

14. Exact area

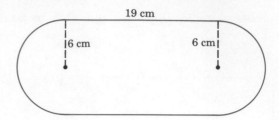

19 cm

6 cm   6 cm

11. Approximate area

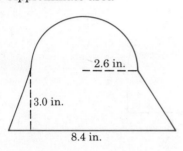

2.6 in.

3.0 in.

8.4 in.

15. Approximate area

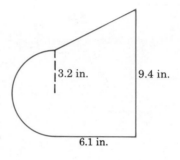

3.2 in.   9.4 in.

6.1 in.

12. Exact area

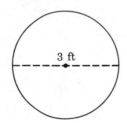

3 ft

16. Area

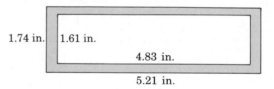

1.74 in.   1.61 in.

4.83 in.

5.21 in.

13. Approximate area

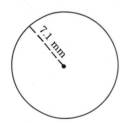

7.1 mm

17. Approximate area

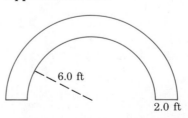

6.0 ft

2.0 ft

18. Volume

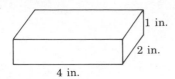

1 in.
2 in.
4 in.

19. Volume

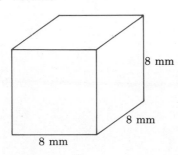

8 mm
8 mm
8 mm

20. Exact volume

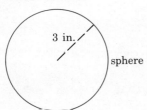

3 in.
sphere

21. Approximate volume

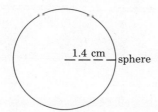

1.4 cm
sphere

22. Approximate volume

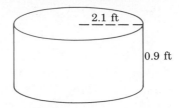

2.1 ft
0.9 ft

23. Exact volume

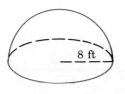

8 ft

24. Approximate volume

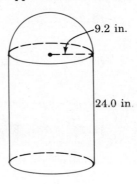

9.2 in.
24.0 in.

25. Approximate volume

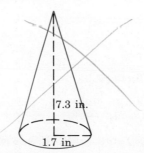

7.3 in.
1.7 in.

26. Approximate volume

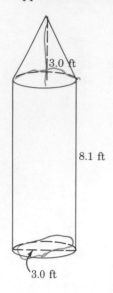

3.0 ft

8.1 ft

3.0 ft

EXERCISES FOR REVIEW

(1.1) **27.** In the number 23,426, how many hundreds are there?

(3.3) **28.** List all the factors of 32.

(5.3) **29.** Find the value of $4\frac{3}{4} - 3\frac{5}{6} + 1\frac{2}{3}$.

(5.5) **30.** Find the value of $\dfrac{5 + \dfrac{1}{3}}{2 + \dfrac{2}{15}}$.

(9.4) **31.** Find the perimeter.

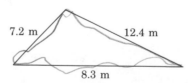

7.2 m   12.4 m

8.3 m

★ **Answers to Practice Sets (9.5)**

A. **1.** 36 sq cm   **2.** 37.503 sq mm   **3.** 13.26 sq in.   **4.** 367.5 sq mi   **5.** 452.16 sq ft
6. 44.28 sq cm

B. **1.** 21 cu in.   **2.** 904.32 cu ft   **3.** 157 cu m   **4.** 0.00942 cu in.

Chapter 9   SUMMARY OF KEY CONCEPTS

Measurement (9.1)

Measurement is comparison to some standard.

Standard Unit of Measure (9.1)

A quantity that is used for comparison is called a *standard unit of measure.*

Two Types of Measurement Systems (9.1)

There are two major types of measurement systems in use today. They are the *United States system* and the *metric system.*

Unit Fraction (9.1)

A *unit fraction* is a fraction that has a value of 1. Unit fractions can be used to convert from one unit of measure to another.

Meter
Liter
Gram
  kilo
  hecto
  deka
  deci
  centi
  milli
(9.2)

Common units of measure in the metric system are the *meter* (m), for length, the *liter* (L), for volume, and the *gram* (g), for mass. To each of these units, a prefix can be attached.

kilo — thousand	deci — tenth
hecto — hundred	centi — hundredth
deka — ten	milli — thousandth

Metric Conversions (9.2)

To *convert* from one metric unit to another:

1. Determine the location of the original number on the metric scale.
2. Move the decimal point of the original number in the same direction and the same number of places as is necessary to move to the metric unit you wish to convert to.

Denominate Numbers (9.3)

Numbers that have units of measure associated with them are *denominate numbers.* The number 25 mg is a denominate number since the mg unit is associated with the pure number 25. The number 82 is not a denominate number since it has no unit of measure associated with it.

Simplified Denominate Number (9.3)

A denominate number is *simplified* when the number of standard units of measure associated with it does not exceed the next higher type of unit.

55 min is simplified, whereas 65 min is not simplified

Addition and Subtraction of Denominate Numbers (9.3)

Denominate numbers can be *added* or *subtracted* by

1. writing the numbers vertically so that the like units appear in the same column.
2. adding or subtracting the number parts, carrying along the unit.
3. simplifying the sum or difference.

Multiplying a Denominate Number by a Whole Number (9.3)

To *multiply* a denominate number by a whole number, multiply the number part of each unit by the whole number and affix the unit to the product.

Dividing a Denominate Number by a Whole Number (9.3)

To *divide* a denominate number by a whole number, divide the number part of each unit by the whole number beginning with the largest unit. Affix the unit to this quotient. Carry the remainder to the next unit.

Polygon (9.4)

A *polygon* is a closed plane (flat) figure whose sides are line segments (portions of straight lines).

Perimeter (9.4)

The *perimeter* of a polygon is the distance around the polygon.

Circumference
Diameter
Radius
(9.4)

The *circumference* of a circle is the distance around the circle. The *diameter* of a circle is any line segment that passes through the center of the circle and has its endpoints on the circle. The *radius* of a circle is one half the diameter of the circle.

The number π (9.4)

The symbol π, read "pi," represents the nonterminating, nonrepeating decimal number 3.14159 . . . . For computational purposes, π is often approximated by the number 3.14.

421

Formula **(9.4)**	A *formula* is a rule for performing a task. In mathematics, a formula is a rule that directs us in computations.

Circumference Formulas **(9.4)**

$C = \pi \cdot d \qquad C \approx (3.14)d$

$C = 2 \cdot \pi \cdot r \qquad C \approx 2(3.14)r$

Area **(9.5)**	The *area* of a surface is the amount of square length units contained in the surface.
Volume **(9.5)**	The *volume* of an object is a measure of the amount of cubic length units contained in the object.

Area Formulas **(9.5)**

Triangle: $\quad A = \dfrac{1}{2} b \cdot h \qquad$ *Rectangle:* $\quad A = l \cdot w$

Parallelogram: $\quad A = b \cdot h \qquad$ *Trapezoid:* $\quad A = \dfrac{1}{2} \cdot (b_1 + b_2) \cdot h$

Circle: $\quad A = \pi \cdot r^2$

Volume Formulas **(9.5)**

Rectangular solid: $\quad V = l \cdot w \cdot h \qquad$ *Sphere:* $\quad V = \dfrac{4}{3} \cdot \pi \cdot r^3$

Cylinder: $\quad V = \pi \cdot r^2 \cdot h \qquad$ *Cone:* $\quad V = \dfrac{1}{3} \cdot \pi \cdot r^2 \cdot h$

EXERCISE SUPPLEMENT

Section 9.1

1. What is measurement?

For problems 2–6, make each conversion. Use the conversion table given in Section 9.1.

2. 9 ft = _____ yd

3. 32 oz = _____ lb

4. 1,500 mg = _____ g

5. 12,000 lb = _____ T

6. 5,280 ft = _____ mi

For problems 7–23, make each conversion.

7. 23 yd to ft

8. $2\frac{1}{2}$ mi to yd

9. 8 in. to ft

10. 51 in. to mi

11. 3 qt to pt

12. 8 lb to oz

13. 5 cups to tbsp

14. 9 da to hr

15. $3\frac{1}{2}$ min to sec

16. $\frac{3}{4}$ wk to min

Section 9.2

17. 250 mL to L

18. 18.57 cm to m

19. 0.01961 kg to mg

20. 52,211 mg to kg

21. 54.006 dag to g

22. 1.181 hg to mg

23. 3.5 kL to mL

Section 9.3

For problems 24–31, perform the indicated operations. Simplify, if possible.

24. Add 8 min 50 sec to 5 min 25 sec.

25. Add 3 wk 3 da to 2 wk 5 da

26. Subtract 4 gal 3 qt from 5 gal 2 qt.

27. Subtract 2 gal 3 qt 1 pt from 8 gal 2 qt.

28. Subtract 5 wk 4 da 21 hr from 12 wk 3 da 14 hr.

29. Subtract 2 T 1,850 lb from 10 T 1,700 lb.

30. Subtract the sum of 2 wk 3 da 15 hr and 5 wk 2 da 9 hr from 10 wk.

31. Subtract the sum of 20 hr 15 min and 18 hr 18 min from the sum of 8 da 1 hr 16 min 5 sec.

For problems 32–43, simplify, if necessary.

32. 18 in.   **33.** 4 ft

34. 23 da   **35.** 3,100 lb

36. 135 min   **37.** 4 tsp

38. 10 fl oz   **39.** 7 pt

40. 9 qt   **41.** 2,300 mm

42. 14,780 mL   **43.** 1,050 m

Sections 9.4 and 9.5

For problems 44–58, find the perimeter, circumference, area, or volume.

44. Perimeter, area

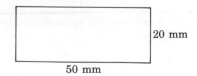
20 mm
50 mm

45. Approximate circumference

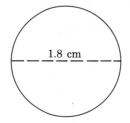
1.8 cm

46. Approximate volume

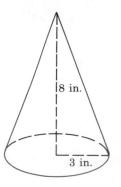
8 in.
3 in.

47. Approximate volume

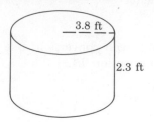
3.8 ft
2.3 ft

48. Exact area

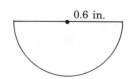
5 m

49. Exact area

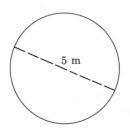
0.6 in.

50. Exact volume

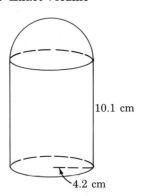
10.1 cm
4.2 cm

51. Approximate volume

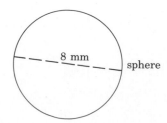
8 mm   sphere

52. Area

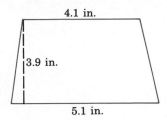

53. Volume

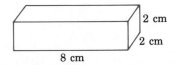

54. Exact area

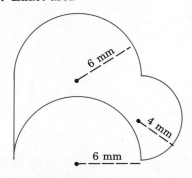

55. Approximate area

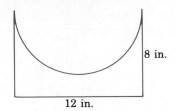

56. Exact area

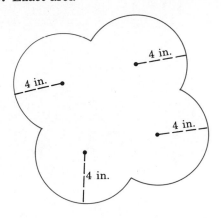

57. Approximate area

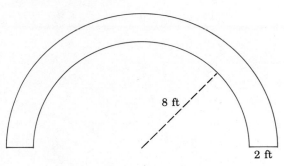

58. Approximate area

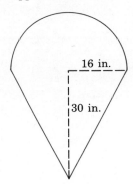

1. _____

1. (9.1) The process of determining, by comparison to some standard, the size of something is called _____ .

For problems 2–9, make each conversion.

2. _____

2. (9.1) 14 yards to feet

3. _____

3. (9.1) 51 feet to inches

4. _____

4. (9.1) $\frac{1}{3}$ yard to feet

5. _____

5. (9.1) $2\frac{1}{4}$ minutes to seconds

6. _____

6. (9.2) 8,500 mg to cg

7. _____

7. (9.2) 5.8623 L to kL

8. _____

8. (9.2) 213.1062 mm to m

9. _____

9. (9.2) 100,001 kL to mL

For problems 10–13, simplify each number.

10. _____

10. (9.3) 23 da

11. _____

11. **(9.3)** 88 ft

12. _____

12. **(9.3)** 4216 lb

13. **(9.3)** 7 qt

13. _____

For problems 14–18, perform the indicated operations. Simplify answers if possible.

14. _____

14. **(9.3)** Add 6 wk 3 da to 2 wk 2 da.

15. _____

15. **(9.3)** Add 9 gal 3 qt to 4 gal 3 qt.

16. _____

16. **(9.3)** Subtract 3 yd 2 ft 5 in. from 5 yd 8 ft 2 in.

17. **(9.3)** Subtract 2 hr 50 min 12 sec from 3 hr 20 min 8 sec.

17. _____

18. **(9.3)** Subtract the sum of 3 wk 6 da and 2 wk 3 da from 10 wk.

18. _____

For problems 19–30, find either the perimeter, circumference, area, or volume.

19. _____

19. **(9.4)** Perimeter

8.61 m

8.61 m

428

20. _____

20. (9.4) Perimeter

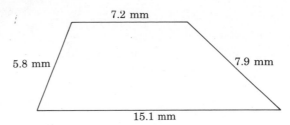

21. (9.4) Approximate circumference

21. _____

14 ft

22. (9.4) Approximate perimeter

22. _____

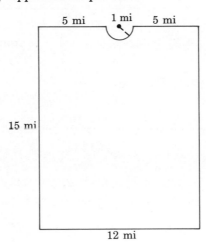

23. (9.5) Area

23. _____

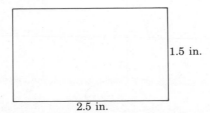

24. _____

24. (9.5) Approximate area

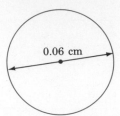

0.06 cm

25. _____

25. (9.5) Approximate area

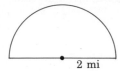

2 mi

26. _____

26. (9.5) Area

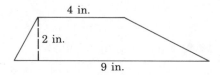

4 in.

2 in.

9 in.

27. (9.5) Exact area

27. _____

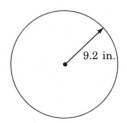

9.2 in.

28. _____

28. (9.5) Approximate volume

6 mm

2 mm

430

29. _____

29. (9.5) Exact volume

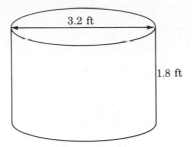

3.2 ft

1.8 ft

30. _____

30. (9.5) Approximate volume

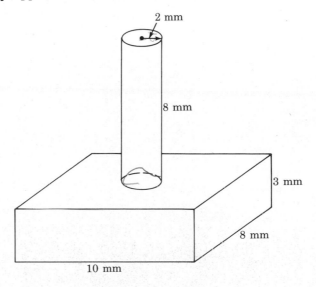

2 mm

8 mm

3 mm

8 mm

10 mm

431

10
Signed Numbers

After completing this chapter, you should

Section 10.1 Variables, Constants, and Real Numbers
- be able to distinguish between variables and constants
- be able to recognize a real number and particular subsets of the real numbers
- understand the ordering of the real numbers

Section 10.2 Signed Numbers
- be able to distinguish between positive and negative real numbers
- be able to read signed numbers
- understand the origin and use of the double-negative property

Section 10.3 Absolute Value
- understand the geometric and algebraic definitions of absolute value

Section 10.4 Addition of Signed Numbers
- be able to add numbers with like signs and with unlike signs
- be able to use the calculator for addition of signed numbers

Section 10.5 Subtraction of Signed Numbers
- understand the definition of subtraction
- be able to subtract signed numbers
- be able to use a calculator to subtract signed numbers

Section 10.6 Multiplication and Division of Signed Numbers
- be able to multiply and divide signed numbers
- be able to multiply and divide signed numbers using a calculator

10.1 Variables, Constants, and Real Numbers

Section Overview

- ☐ **VARIABLES AND CONSTANTS**
- ☐ **REAL NUMBERS**
- ☐ **SUBSETS OF REAL NUMBERS**
- ☐ **ORDERING REAL NUMBERS**

☐ VARIABLES AND CONSTANTS

A basic distinction between algebra and arithmetic is the use of symbols (usually letters) in algebra to represent numbers. So, algebra is a generalization of arithmetic. Let us look at two examples of situations in which letters are substituted for numbers:

1. Suppose that a student is taking four college classes, and each class can have at most 1 exam per week. In any 1-week period, the student may have 0, 1, 2, 3, or 4 exams. In algebra, we can let the letter x represent the number of exams this student may have in a 1-week period. The letter x may assume any of the *various* values 0, 1, 2, 3, 4.

2. Suppose that in writing a term paper for a biology class a student needs to specify the average lifetime, in days, of a male housefly. If she does not know this number off the top of her head, she might represent it (at least temporarily) on her paper with the letter t (which reminds her of *time*). Later, she could look up the average time in a reference book and find it to be 17 days. The letter t can assume only the one value, 17, and no other values. The value t is *constant*.

Variable

Constant

> 1. A letter or symbol that represents any member of a collection of two or more numbers is called a **variable.**
> 2. A letter or symbol that represents one specific number, known or unknown, is called a **constant.**

In example 1, the letter x is a variable since it can represent any of the numbers 0, 1, 2, 3, 4. The letter t in example 2 is a constant since it can only have the value 17.

☐ REAL NUMBERS

The study of mathematics requires the use of several collections of numbers. The **real number line** allows us to visually display (graph) the numbers in which we are interested.

Real Number Line

A line is composed of infinitely many points. To each point we can associate a unique number, and with each number, we can associate a particular point.

Coordinate

> The *number* associated with a point on the number line is called the **coordinate** of the point.
>
> The *point* on a number line that is associated with a particular number is called the **graph** of that number.

Graph

We construct a real number line as follows:

Constructing a Real Number Line

1. Draw a horizontal line.

   ⟵————————————————————⟶

2. Choose any point on the line and label it 0. This point is called the **origin.**

   ⟵————————————|————————————⟶
   0

Origin

3. Choose a convenient length. Starting at 0, mark this length off in both directions, being careful to have the lengths look like they are about the same.

We now define a real number.

Real Number

> A **real number** is any number that is the coordinate of a point on the real number line.

Positive Numbers
Negative Numbers

Real numbers whose graphs are to the right of 0 are called *positive real numbers,* or more simply, **positive numbers.** Real numbers whose graphs appear to the left of 0 are called *negative real numbers,* or more simply, **negative numbers.**

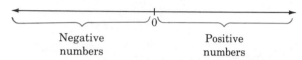

Negative   Positive
numbers   numbers

The number 0 is neither positive nor negative.

☐ SUBSETS OF REAL NUMBERS

The set of real numbers has many subsets. Some of the subsets that are of interest in the study of algebra are listed below along with their notations and graphs.

Natural Numbers, Counting Numbers

The **natural** or **counting numbers (N):** 1, 2, 3, 4, . . .

Read "and so on."

$$\overset{}{\underset{}{\longleftarrow}} \quad 1 \quad 2 \quad 3 \quad 4 \quad 5 \quad 6 \quad 7$$

Whole Numbers

The **whole numbers (W):** 0, 1, 2, 3, 4, . . .

$$\overset{}{\underset{}{\longleftarrow}} \quad 0 \quad 1 \quad 2 \quad 3 \quad 4 \quad 5 \quad 6 \quad 7$$

Notice that every natural number is a whole number.

Integers

The **integers (Z):** . . . $-3, -2, -1, 0, 1, 2, 3, \ldots$

$$\overset{}{\underset{}{\longleftarrow}} \quad -4 \quad -3 \quad -2 \quad -1 \quad 0 \quad 1 \quad 2 \quad 3 \quad 4$$

Notice that every whole number is an integer.

Rational Numbers (Fractions)

The **rational numbers (Q):** Rational numbers are sometimes called **fractions.** They are numbers that can be written as the *quotient* of two integers. They have decimal representations that either terminate or do not terminate but contain a repeating block of digits. Some examples are

$$\underbrace{\frac{-3}{4} = -0.75}_{\text{Terminating}} \qquad \underbrace{8\frac{11}{27} = 8.407407407 \ldots}_{\text{Nonterminating, but repeating}}$$

Some rational numbers are graphed below.

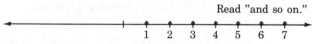

Notice that *every integer is a rational number.*

Notice that there are still a great many points on the number line that have not yet been assigned a type of number. We will not examine these other types of numbers in this text. They are examined in detail in algebra. An example of these numbers is the number π, whose decimal representation does not terminate nor contain a repeating block of digits. An approximation for π is 3.14.

1. Is every whole number a natural number?

   No. The number 0 is a whole number but it is not a natural number.

2. Is there an integer that is not a natural number?

   Yes. Some examples are $0, -1, -2, -3$, and -4.

3. Is there an integer that is a whole number?

   Yes. In fact, every whole number is an integer.

★ **PRACTICE SET A**

1. Is every natural number a whole number? _____
2. Is every whole number an integer? _____
3. Is every integer a real number? _____
4. Is there an integer that is a whole number? _____
5. Is there an integer that is not a natural number? _____

☐ ORDERING REAL NUMBERS

Ordering Real Numbers

A real number b is said to be *greater* than a real number a, denoted $b > a$, if b is to the right of a on the number line. Thus, as we would expect, $5 > 2$ since 5 is to the right of 2 on the number line. Also, $-2 > -5$ since -2 is to the right of -5 on the number line.

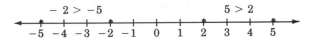

If we let a and b represent two numbers, then a and b are related in exactly one of three ways: Either

Equality Symbol

$$a = b \quad a \text{ and } b \text{ are equal} \qquad (8 = 8)$$
$$\begin{cases} a > b & a \text{ is greater than } b & (8 > 5) \\ a < b & a \text{ is less than } b & (5 < 8) \\ \text{Some variations of these symbols are} \end{cases}$$

Inequality Symbols

$$\begin{cases} a \neq b & a \text{ is not equal to } b & (8 \neq 5) \\ a \geq b & a \text{ is greater than or equal to } b & (a \geq 8) \\ a \leq b & a \text{ is less than or equal to } b & (a \leq 8) \end{cases}$$

1. What integers can replace x so that the following statement is true?

   $$-3 \leq x < 2$$

   The integers are $-3, -2, -1, 0, 1$.

2. Draw a number line that extends from -3 to 5. Place points at all whole numbers between and including -1 and 3.

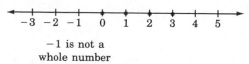

-1 is not a
whole number

★ **PRACTICE SET B**

1. What integers can replace x so that the following statement is true?

$-5 \leq x < 1$

2. Draw a number line that extends from -4 to 3. Place points at all natural numbers between, but not including, -2 to 2.

Answers to Practice Sets are on p. 439.

Section 10.1 EXERCISES

For problems 1–8, next to each real number, note all collections to which it belongs by writing N for natural number, W for whole number, or Z for integer. Some numbers may belong to more than one collection.

1. 6

2. 12

3. 0

4. 1

5. -3

6. -7

7. -805

8. -900

9. Is the number 0 a positive number, a negative number, neither, or both?

10. An integer is an even integer if it is evenly divisible by 2. Draw a number line that extends from -5 to 5 and place points at all negative even integers and all positive odd integers.

11. Draw a number line that extends from -5 to 5. Place points at all integers that satisfy $-3 \leq x < 4$.

12. Is there a largest two digit number? If so, what is it?

13. Is there a smallest two digit number? If so, what is it?

For the pairs of real numbers in problems 14–18, write the appropriate symbol ($<, >, =$) in place of the □.

14. $-7 \,\square\, -2$

15. $-5 \,\square\, 0$

16. $-1 \,\square\, 4$

17. $6 \,\square\, -1$

18. $10 \,\square\, 10$

For problems 19–23, what numbers can replace m so that the following statements are true?

19. $-1 \leq m \leq 5$, m an integer.

20. $-7 < m < -1$, m an integer.

21. $-3 \leq m < 2$, m a natural number.

22. $-15 < m \leq -1$, m a natural number.

23. $-5 \leq m < 5$, m a whole number.

For problems 24–28, on the number line, how many units are there between the given pair of numbers?

24. 0 and 3

25. -4 and 0

26. -1 and 6

27. -6 and 2

28. -3 and 3

29. Are all positive numbers greater than zero?

30. Are all positive numbers greater than all negative numbers?

31. Is 0 greater than all negative number?

32. Is there a largest natural number?

33. Is there a largest negative integer?

EXERCISES FOR REVIEW

(4.2)   **34.** Convert $6\frac{5}{8}$ to an improper fraction.

(4.4)   **35.** Find the value: $\frac{3}{11}$ of $\frac{33}{5}$.

(5.2)   **36.** Find the sum of $\frac{4}{5} + \frac{3}{8}$.

(9.2)   **37.** Convert 30.06 cm to m.

(9.5)   **38.** Find the area of the triangle.

3 mm

16 mm

A. The answer to each question is yes.

B. **1.** $-5, -4, -3, -2, -1, 0$ **2.**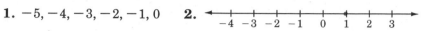

10.2 Signed Numbers

Section Overview	☐ **POSITIVE AND NEGATIVE NUMBERS**
	☐ **READING SIGNED NUMBERS**
	☐ **OPPOSITES**
	☐ **THE DOUBLE-NEGATIVE PROPERTY**

☐ POSITIVE AND NEGATIVE NUMBERS

Positive and Negative Numbers

Each real number other than zero has a **sign** associated with it. A real number is said to be a **positive number** if it is to the right of 0 on the number line and **negative** if it is to the left of 0 on the number line.

+ and − Notation

> **THE NOTATION OF SIGNED NUMBERS**
>
> A number is denoted as **positive** if it is directly preceded by a plus sign or no sign at all.
>
> A number is denoted as **negative** if it is directly preceded by a minus sign.

☐ READING SIGNED NUMBERS

The plus and minus signs now have *two meanings:*

The plus sign can denote the operation of addition *or* a positive number.
The minus sign can denote the operation of subtraction *or* a negative number.

To avoid any confusion between "sign" and "operation," it is preferable to read the sign of a number as "positive" or "negative." When "+" is used as an operation sign, it is read as "plus." When "−" is used as an operation sign, it is read as "minus."

☆ **SAMPLE SET A**

Read each expression so as to avoid confusion between "operation" and "sign."

1. -8 should be read as "negative eight" rather than "minus eight."

2. $4 + (-2)$ should be read as "four plus negative two" rather than "four plus minus two."

3. $-6 + (-3)$ should be read as "negative six plus negative three" rather than "minus six plus minus three."

4. $-15 - (-6)$ should be read as "negative fifteen minus negative six" rather than "minus fifteen minus minus six."

5. $-5 + 7$ should be read as "negative five plus seven" rather than "minus five plus seven."

6. $0 - 2$ should be read as "zero minus two."

★ **PRACTICE SET A**

Write each expression in words.

1. $6 + 1$

2. $2 + (-8)$

3. $-7 + 5$

4. $-10 - (+3)$

5. $-1 - (-8)$

6. $0 + (-11)$

▢ OPPOSITES

Opposites

On the number line, each real number, other than zero, has an image on the opposite side of 0. For this reason, we say that each real number has an opposite. **Opposites** are the same distance from zero but have opposite signs.

The opposite of a real number is denoted by placing a negative sign directly in front of the number. Thus, if a is any real number, then $-a$ is its opposite.

Note: The letter "a" is a variable. Thus, "a" need not be positive, and "$-a$" need not be negative.

If a is any real number, $-a$ is opposite a on the number line.

a positive a negative

$-a$ 0 a a 0 $-a$

▢ THE DOUBLE-NEGATIVE PROPERTY

The number a is opposite $-a$ on the number line. Therefore, $-(-a)$ is opposite $-a$ on the number line. This means that

$$-(-a) = a$$

From this property of opposites, we can suggest the **double-negative property** for real numbers.

Double-Negative Property:

$-(-a) = a$

> If a is a real number, then
>
> $-(-a) = a$

☆ **SAMPLE SET B**

Find the opposite of each number.

1. If $a = 2$, then $-a = -2$. Also, $-(-a) = -(-2) = 2$.

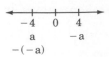

-2 0 2
$-a$ a
          $-(-a)$

2. If $a = -4$, then $-a = -(-4) = 4$. Also, $-(-a) = a = -4$.

-4 0 4
a $-a$
$-(-a)$

★ **PRACTICE SET B**

Find the opposite of each number.

1. 8   **2.** 17   **3.** -6 **4.** -15 **5.** $-(-1)$ **6.** $-[-(-7)]$

7. Suppose a is a positive number. Is $-a$ positive or negative?

8. Suppose a is a negative number. Is $-a$ positive or negative?

9. Suppose we do not know the sign of the number k. Is $-k$ positive, negative, or do we not know?

Answers to Practice Sets are on p. 442.

Section 10.2 EXERCISES

1. A number is denoted as positive if it is directly preceded by _____.

2. A number is denoted as negative if it is directly preceded by _____.

How should the number in problems 3–8 be read? (Write in words.)

3. -7

4. -5

5. 15

6. 11

7. $-(-1)$

8. $-(-5)$

For problems 9–14, write each expression in words.

9. $5 + 3$

10. $3 + 8$

11. $15 + (-3)$

12. $1 + (-9)$

13. $-7 - (-2)$

14. $0 - (-12)$

For problems 15–20, rewrite each number in simpler form.

15. $-(-2)$

16. $-(-16)$

17. $-[-(-8)]$

18. $-[-(-20)]$

19. $7 - (-3)$

20. $6 - (-4)$

EXERCISES FOR REVIEW

(6.6) **21.** Find the quotient: $8 \div 27$.

(7.2) **22.** Solve the proportion: $\dfrac{5}{9} = \dfrac{60}{x}$.

(8.1) **23.** Use the method of rounding to estimate the sum: $5829 + 8767$.

(9.1) **24.** Use a unit fraction to convert 4 yd to feet.

(9.2) **25.** Convert 25 cm to hm.

★ **Answers to Practice Sets (10.2)**

A. **1.** six plus one   **2.** two plus negative eight   **3.** negative seven plus five
    4. negative ten minus three   **5.** negative one minus negative eight   **6.** zero plus negative eleven

B. **1.** -8 **2.** -17 **3.** 6   **4.** 15   **5.** -1 **6.** 7   **7.** $-a$ is negative   **8.** $-a$ is positive
    9. We must say that we do not know.

10.3 Absolute Value

Section Overview	☐ GEOMETRIC DEFINITION OF ABSOLUTE VALUE ☐ ALGEBRAIC DEFINITION OF ABSOLUTE VALUE

☐ GEOMETRIC DEFINITION OF ABSOLUTE VALUE

Absolute Value — Geometric Approach

> Geometric definition of absolute value:
>
> The **absolute value** of a number a, denoted $|a|$, is the distance from a to 0 on the number line.

Absolute value answers the question of "how far," and not "which way." The phrase "how far" implies "length" and *length is always a nonnegative quantity.* Thus, the absolute value of a number is a nonnegative number.

☆ **SAMPLE SET A**

Determine each value.

1. $|4| = 4$

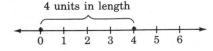
4 units in length

2. $|-4| = 4$

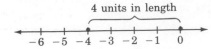
4 units in length

3. $|0| = 0$

4. $-|5| = -5$.   The quantity on the left side of the equal sign is read as "negative the absolute value of 5." The absolute value of 5 is 5. Hence, negative the absolute value of 5 is -5.

5. $-|-3| = -3$.   The quantity on the left side of the equal sign is read as "negative the absolute value of -3." The absolute value of -3 is 3. Hence, negative the absolute value of -3 is $-(3) = -3$.

★ **PRACTICE SET A**

By reasoning geometrically, determine each absolute value.

1. $|7|$ **2.** $|-3|$ **3.** $|12|$ **4.** $|0|$ **5.** $-|9|$ **6.** $-|-6|$

☐ ALGEBRAIC DEFINITION OF ABSOLUTE VALUE

From the problems in Sample Set A, we can suggest the following algebraic definition of absolute value. Note that the definition has two parts.

Absolute Value — Algebraic Approach

> Algebraic definition of absolute value
>
> The absolute value of a number a is
> $$|a| = \begin{cases} a, & \text{if } a \geq 0 \\ -a, & \text{if } a < 0 \end{cases}$$

The algebraic definition takes into account the fact that the number a could be either positive or zero ($a \geq 0$) or negative ($a < 0$).

1. If the number a is positive or zero ($a \geq 0$), the upper part of the definition applies. The upper part of the definition tells us that if the number enclosed in the absolute value bars is a nonnegative number, the absolute value of the number is the number itself.
2. The lower part of the definition tells us that if the number enclosed within the absolute value bars is a negative number, the absolute value of the number is the opposite of the number. The opposite of a negative number is a positive number.

Note: The definition says that the vertical absolute value lines may be eliminated only if we know whether the number inside is positive or negative.

☆ **SAMPLE SET B**

Use the algebraic definition of absolute value to find the following values.

1. $|8|$.   The number enclosed within the absolute value bars is a nonnegative number, so the upper part of the definition applies. This part says that the absolute value of 8 is 8 itself.

$|8| = 8$

2. $|-3|$.   The number enclosed within absolute value bars is a negative number, so the lower part of the definition applies. This part says that the absolute value of -3 is the opposite of -3, which is $-(-3)$. By the definition of absolute value and the double-negative property,

$|-3| = -(-3) = 3$

★ **PRACTICE SET B**

Use the algebraic definition of absolute value to find the following values.

1. $|7|$ **2.** $|9|$ **3.** $|-12|$ **4.** $|-5|$

5. $-|8|$ **6.** $-|1|$ **7.** $-|-52|$ **8.** $-|-31|$

Answers to Practice Sets are on p. 445.

Section 10.3 EXERCISES

For problems 1–28, determine each of the values.

1. $|5|$

2. $|3|$

3. $|6|$

4. $|-9|$

5. $|-1|$

6. $|-4|$

7. $-|3|$

8. $-|7|$

9. $-|-14|$

10. $|0|$

11. $|-26|$

12. $-|-26|$

13. $-(-|4|)$

14. $-(-|2|)$

15. $-(-|-6|)$

16. $-(-|-42|)$

17. $|5| - |-2|$

18. $|-2|^3$

19. $|-(2 \cdot 3)|$

20. $|-2| - |-9|$

21. $(|-6| + |4|)^2$

22. $(|-1| - |1|)^3$

23. $(|4| + |-6|)^2 - (|-2|)^3$

24. $-[|-10| - 6]^2$

25. $-\{-[-|-4| + |-3|]^3\}^2$

26. A Mission Control Officer at Cape Canaveral makes the statement "lift-off, T minus 50 seconds." How long is it before lift-off?

27. Due to a slowdown in the industry, a Silicon Valley computer company finds itself in debt $2,400,000. Use absolute value notation to describe this company's debt.

28. A particular machine is set correctly if upon action its meter reads 0. One particular machine has a meter reading of −1.6 upon action. How far is this machine off its correct setting?

EXERCISES FOR REVIEW

(5.2) **29.** Find the sum: $\dfrac{9}{70} + \dfrac{5}{21} + \dfrac{8}{15}$.

(5.5) **30.** Find the value of $\dfrac{\dfrac{3}{10} + \dfrac{4}{12}}{\dfrac{19}{20}}$.

(6.2) **31.** Convert $3.2\dfrac{3}{5}$ to a fraction.

(7.2) **32.** The ratio of acid to water in a solution is $\dfrac{3}{8}$. How many mL of acid are there in a solution that contain 112 mL of water?

(10.2) **33.** Find the value of $-6 - (-8)$.

★ **Answers to Practice Sets (10.3)**

A. **1.** 7   **2.** 3   **3.** 12   **4.** 0   **5.** −9   **6.** −6

B. **1.** 7   **2.** 9   **3.** 12   **4.** 5   **5.** −8   **6.** −1   **7.** −52   **8.** −31

10.4 Addition of Signed Numbers

Section Overview

- ☐ **ADDITION OF NUMBERS WITH LIKE SIGNS**
- ☐ **ADDITION WITH ZERO**
- ☐ **ADDITION OF NUMBERS WITH UNLIKE SIGNS**
- ☐ **CALCULATORS**

☐ ADDITION OF NUMBERS WITH LIKE SIGNS

The addition of the *two positive numbers* 2 and 3 is performed on the number line as follows.

Begin at 0, the origin.
Since 2 is positive, move 2 units to the right.
Since 3 is positive, move 3 more units to the right.
We are now located at 5.

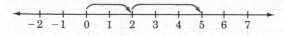

Thus, $2 + 3 = 5$.
Summarizing, we have

(2 positive units) + (3 positive units) = (5 positive units)

The addition of the *two negative numbers* -2 and -3 is performed on the number line as follows.

Begin at 0, the origin.
Since -2 is negative, move 2 units to the left.
Since -3 is negative, move 3 more units to the left.
We are now located at -5.

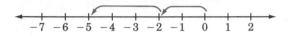

Thus, $(-2) + (-3) = -5$.
Summarizing, we have

(2 negative units) + (3 negative units) = (5 negative units)

Observing these two examples, we can suggest these relationships:

(positive number) + (positive number) = (positive number)

(negative number) + (negative number) = (negative number)

Adding Numbers with the Same Sign

Addition of numbers with like sign:

To add two real numbers that have the *same* sign, add the absolute values of the numbers and associate with the sum the common sign.

Find the sums.

1. $3 + 7$

$\left.\begin{array}{l} |3| = 3 \\ |7| = 7 \end{array}\right\}$ **Add these absolute values.**

$3 + 7 = 10$

The common sign is "+."

Thus, $3 + 7 = +10$, or $3 + 7 = 10$.

2. $(-4) + (-9)$

$\left.\begin{array}{l} |-4| = 4 \\ |-9| = 9 \end{array}\right\}$ **Add these absolute values.**

$4 + 9 = 13$

The common sign is "−."

Thus, $(-4) + (-9) = -13$.

★ PRACTICE SET A

Find the sums.

1. $8 + 6$

2. $41 + 11$

3. $(-4) + (-8)$

4. $(-36) + (-9)$

5. $-14 + (-20)$

6. $-\dfrac{2}{3} + \left(-\dfrac{5}{3}\right)$

7. $-2.8 + (-4.6)$

8. $0 + (-16)$

❑ ADDITION WITH ZERO

Notice that

Addition with Zero

$(0) +$ (a positive number) $=$ (that same positive number).
$(0) +$ (a negative number) $=$ (that same negative number).

Since adding zero to a real number leaves that number unchanged, zero is called the **additive identity.**

The Additive Identity Is Zero

❑ ADDITION OF NUMBERS WITH UNLIKE SIGNS

The addition $2 + (-6)$, *two numbers with unlike signs,* can also be illustrated using the number line.

Begin at 0, the origin.
Since 2 is positive, move 2 units to the right.
Since -6 is negative, move, from 2, 6 units to the left.
We are now located at -4.

We can suggest a rule for adding two numbers that have *unlike signs* by noting that if the signs are disregarded, 4 can be obtained by subtracting 2 from 6. But 2 and 6 are precisely the absolute values of 2 and -6. Also, notice that the sign of the number with the larger absolute value is negative and that the sign of the resulting sum is negative.

Adding Numbers with Unlike Signs

> Addition of numbers with unlike signs:
>
> To add two real numbers that have *unlike signs*, subtract the smaller absolute value from the larger absolute value and associate with this difference the sign of the number with the larger absolute value.

☆ SAMPLE SET B

Find the following sums.

1. $7 + (-2)$

$$\underbrace{|7| = 7}_{\substack{\text{Larger absolute} \\ \text{value. Sign is positive.}}} \qquad \underbrace{|-2| = 2}_{\substack{\text{Smaller absolute} \\ \text{value.}}}$$

Subtract absolute values: $7 - 2 = 5$.
Attach the proper sign: "$+$."

Thus, $7 + (-2) = +5$ or $7 + (-2) = 5$.

Continued

2. $3 + (-11)$

$$\underbrace{|3| = 3}_{\substack{\text{Smaller absolute} \\ \text{value.}}} \qquad \underbrace{|-11| = 11}_{\substack{\text{Larger absolute} \\ \text{value. Sign is negative.}}}$$

Subtract absolute values: $11 - 3 = 8$.
Attach the proper sign: "$-$."

Thus, $3 + (-11) = -8$.

3. The morning temperature on a winter's day in Lake Tahoe was -12 degrees. The afternoon temperature was 25 degrees warmer. What was the afternoon temperature?

We need to find $-12 + 25$.

$$\underbrace{|-12| = 12}_{\substack{\text{Smaller absolute} \\ \text{value.}}} \qquad \underbrace{|25| = 25}_{\substack{\text{Larger absolute} \\ \text{value. Sign is positive.}}}$$

Subtract absolute values: $25 - 12 = 13$.
Attach the proper sign: "$+$."

Thus, $-12 + 25 = 13$.

★ **PRACTICE SET B**

Find the sums.

1. $4 + (-3)$ **2.** $-3 + 5$ **3.** $15 + (-18)$

4. $0 + (-6)$ **5.** $-26 + 12$ **6.** $35 + (-78)$

7. $15 + (-10)$ **8.** $1.5 + (-2)$ **9.** $-8 + 0$

10. $0 + (0.57)$ **11.** $-879 + 454$

❑ **CALCULATORS**

Calculators having the $\boxed{+/-}$ key can be used for finding sums of signed numbers.

☆ **SAMPLE SET C**

Use a calculator to find the sum of -147 and 84.

		Display Reads	
Type	147	147	
Press	$\boxed{+/-}$	-147	**This key changes the sign of**
Press	$\boxed{+}$	-147	**a number. It is different than** $\boxed{-}$.
Type	84	84	
Press	$\boxed{=}$	-63	

★ **PRACTICE SET C**

Use a calculator to find each sum.

1. $673 + (-721)$ **2.** $-8,291 + 2,206$ **3.** $-1,345.6 + (-6,648.1)$

Answers to Practice Sets are on p. 450.

Section 10.4 EXERCISES

Find the sums in problems 1–27. If possible, use a calculator to check each result.

1. $4 + 12$

2. $8 + 6$

3. $(-3) + (-12)$

4. $(-6) + (-20)$

5. $10 + (-2)$

6. $8 + (-15)$

7. $-16 + (-9)$

8. $-22 + (-1)$

9. $0 + (-12)$

10. $0 + (-4)$

11. $0 + (24)$

12. $-6 + 1 + (-7)$

13. $-5 + (-12) + (-4)$

14. $-5 + 5$

15. $-7 + 7$

16. $-14 + 14$

17. $4 + (-4)$

18. $9 + (-9)$

19. $84 + (-61)$

20. $13 + (\ 56)$

21. $452 + (-124)$

22. $636 + (-989)$

23. $1,811 + (-935)$

24. $-373 + (-14)$

25. $-1,211 + (-44)$

26. $-47.03 + (-22.71)$

27. $-1.998 + (-4.086)$

28. In order for a small business to break even on a project, it must have sales of $21,000. If the amount of sales was $15,000, by how much money did this company fall short?

29. Suppose a person has $56 in his checking account. He deposits $100 into his checking account by using the automatic teller machine. He then writes a check for $84.50. If an error causes the deposit not bę listed into this person's account, what is this person's checking balance?

30. A person borrows $7 on Monday and then $12 on Tuesday. How much has this person borrowed?

31. A person borrows $11 on Monday and then pays back $8 on Tuesday. How much does this person owe?

EXERCISES FOR REVIEW

(4.5) **32.** Find the reciprocal of $8\frac{5}{6}$.

(5.2) **33.** Find the value of $\frac{5}{12} + \frac{7}{18} - \frac{1}{3}$.

(6.3) **34.** Round 0.01628 to the nearest tenth.

(7.4) **35.** Convert 62% to a fraction.

(10.3) **36.** Find the value of $|-12|$.

★ **Answers to Practice Sets (10.4)**

A. **1.** 14 **2.** 52 **3.** -12 **4.** -45 **5.** -34 **6.** $-\frac{7}{3}$ **7.** -7.4 **8.** -16

B. **1.** 1 **2.** 2 **3.** -3 **4.** -6 **5.** -14 **6.** -43 **7.** 5 **8.** -0.5 **9.** -8 **10.** 0.57
11. -425

C. **1.** -48 **2.** $-6,085$ **3.** $-7,993.7$

10.5 Subtraction of Signed Numbers

Section Overview
- ❑ **DEFINITION OF SUBTRACTION**
- ❑ **THE PROCESS OF SUBTRACTION**
- ❑ **CALCULATORS**

❑ DEFINITION OF SUBTRACTION

We know from experience with arithmetic that the subtraction $5 - 2$ produces 3, that is, $5 - 2 = 3$. We can suggest a rule for subtracting signed numbers by illustrating this process on the number line.

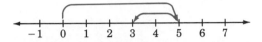

Begin at 0, the origin.
Since 5 is positive, move 5 units to the right.
Then, move *2 units to the left* to get to 3. (This reminds us of addition with a negative number.)

From this illustration we can see that $5 - 2$ is the same as $5 + (-2)$.
    This leads us directly to the definition of subtraction.

Definition of Subtraction

> If a and b are real numbers, $a - b$ is the same as $a + (-b)$, where $-b$ is the opposite of b.

❑ THE PROCESS OF SUBTRACTION

From this definition, we suggest the following rule for subtracting signed numbers.

Subtraction of Signed Numbers

> To perform the subtraction $a - b$, add the opposite of b to a, that is, change the sign of b and add.

☆ SAMPLE SET A

Perform the indicated subtractions.

1. $5 - 3 = 5 + (-3) = 2$

2. $4 - 9 = 4 + (-9) = -5$

3. $-4 - 6 = -4 + (-6) = -10$

4. $-3 - (-12) = -3 + 12 = 9$

5. $0 - (-15) = 0 + 15 = 15$

6. The high temperature today in Lake Tahoe was $26°F$. The low temperature tonight is expected to be $-7°F$. How many degrees is the temperature expected to drop?

We need to find the difference between 26 and -7.

$26 - (-7) = 26 + 7 = 33$

Thus, the expected temperature drop is $33°F$.

Continued

7. $-6 - (-5) - 10 = -6 + 5 + (-10)$
$$= (-6 + 5) + (-10)$$
$$= -1 + (-10)$$
$$= -11$$

★ **PRACTICE SET A**

Perform the indicated subtractions.

1. $9 - 6$ **2.** $6 - 9$ **3.** $0 - 7$

4. $1 - 14$ **5.** $-8 - 12$ **6.** $-21 - 6$

7. $-6 - (-4)$ **8.** $8 - (-10)$ **9.** $1 - (-12)$

10. $86 - (-32)$ **11.** $0 - 16$ **12.** $0 - (-16)$

13. $0 - (8)$ **14.** $5 - (-5)$ **15.** $24 - [-(-24)]$

☐ CALCULATORS

Calculators can be used for subtraction of signed numbers. The most efficient calculators are those with a $\boxed{+/-}$ key.

☆ SAMPLE SET B

Use a calculator to find each difference.

1. $3,187 - 8,719$

		Display Reads
Type	3187	3187
Press	$\boxed{-}$	3187
Type	8719	8719
Press	$\boxed{=}$	−5532

Thus, $3,187 - 8,719 = -5,532$.

2. $-156 - (-211)$

Method A:

		Display Reads
Type	156	156
Press	$\boxed{+/-}$	-156
Press	$\boxed{-}$	-156
Type	211	211
Press	$\boxed{+/-}$	-211
Press	$\boxed{=}$	55

Thus, $-156 - (-211) = 55$.

Method B:

We manually change the subtraction to an addition and change the sign of the number to be subtracted.

$-156 - (-211)$ becomes $-156 + 211$

		Display Reads
Type	156	156
Press	$\boxed{+/-}$	-156
Press	$\boxed{+}$	-156
Type	211	211
Press	$\boxed{=}$	55

★ **PRACTICE SET B**

Use a calculator to find each difference.

1. $44 - 315$

2. $12.756 - 15.003$

3. $-31.89 - 44.17$

4. $-0.797 - (-0.615)$

Answers to Practice Sets are on p. 454.

Section 10.5 EXERCISES

For problems 1–18, perform each subtraction. Use a calculator to check each result.

1. $8 - 3$

2. $12 - 7$

3. $5 - 6$

4. $14 - 30$

5. $-6 - 8$

6. $-1 - 12$

7. $-5 - (-3)$

8. $-11 - (-8)$

9. $0 - 6$

10. $0 - 15$

11. $0 - (-7)$

12. $0 - (-10)$

13. $67 - 38$

14. $142 - 85$

15. $816 - 1140$

16. $105 - 421$

17. $-550 - (-121)$

18. $-15.016 - (4.001)$

For problems 19–22, perform the indicated operations.

19. $-26 + 7 - 52$

20. $-15 - 21 - (-2)$

21. $-104 - (-216) - (-52)$

23. When a particular machine is operating properly, its meter will read 34. If a broken bearing in the machine causes the meter reading to drop by 45 units, what is the meter reading?

22. $-0.012 - (-0.111) - (0.035)$

24. The low temperature today in Denver was $-4°F$ and the high was $42°F$. What is the temperature difference?

EXERCISES FOR REVIEW

(6.2) **25.** Convert $16.02\frac{1}{5}$ to a decimal.

(6.5) **26.** Find 4.01 of 6.2.

(7.4) **27.** Convert $\frac{5}{16}$ to a percent.

(8.3) **28.** Use the distributive property to compute the product: $15 \cdot 82$.

(10.4) **29.** Find the sum: $16 + (-21)$.

★ **Answers to Practice Sets (10.5)**

A. **1.** 3   **2.** -3 **3.** -7 **4.** -13 **5.** -20 **6.** -27 **7.** -2 **8.** 18   **9.** 13   **10.** 118   **11.** -16 **12.** 16   **13.** -8 **14.** 10   **15.** 0

B. **1.** -271 **2.** -2.247 **3.** -76.06 **4.** -0.182

10.6 Multiplication and Division of Signed Numbers

Section Overview	☐ **MULTIPLICATION OF SIGNED NUMBERS** ☐ **DIVISION OF SIGNED NUMBERS** ☐ **CALCULATORS**

☐ MULTIPLICATION OF SIGNED NUMBERS

Let us consider first, the product of two positive numbers. Multiply: $3 \cdot 5$.

$3 \cdot 5$ means $5 + 5 + 5 = 15$

This suggests* that

(positive number) $\cdot$ (positive number) = (positive number)

---

* In later mathematics courses, the word "suggests" turns into the word "proof." One example does not prove a claim. Mathematical proofs are constructed to validate a claim for all possible cases.

More briefly,

$(+)(+) = (+)$

$$(+)(+) = (+)$$

Now consider the product of a positive number and a negative number. Multiply: $(3)(-5)$.

$(3)(-5)$ means $(-5) + (-5) + (-5) = -15$

This suggests that

(positive number) · (negative number) = (negative number)

More briefly,

$(+)(-) = (-)$

$$(+)(-) = (-)$$

By the commutative property of multiplication, we get

(negative number) · (positive number) = (negative number)

More briefly,

$(-)(+) = (-)$

$$(-)(+) = (-)$$

The sign of the product of two negative numbers can be suggested after observing the following illustration.

Multiply -2 by, respectively, $4, 3, 2, 1, 0, -1, -2, -3, -4$.

When this number   this product increases
    decreases by 1,   by 2.

$$
\begin{aligned}
4(-2) &= -8 \\
3(-2) &= -6 \\
2(-2) &= -4 \\
1(-2) &= -2 \\
0(-2) &= 0 \\
-1(-2) &= 2 \\
-2(-2) &= 4 \\
-3(-2) &= 6 \\
-4(-2) &= 8
\end{aligned}
$$

As we know,
$\rightarrow (+)(-) = (-)$

As we know,
$\longrightarrow (0) \cdot \text{(any number)} = 0$

The pattern suggested is
$\rightarrow$ $(-)(-) = (+)$

$(-)(-) = (+)$

We have the following rules for multiplying signed numbers.

Rules for Multiplying Signed Numbers

Multiplying signed numbers:

1. To multiply two real numbers that have the *same sign*, multiply their absolute values. The product is positive.

    $(+)(+) = (+)$
    $(-)(-) = (+)$

2. To multiply two real numbers that have *opposite signs*, multiply their absolute values. The product is negative.

    $(+)(-) = (-)$
    $(-)(+) = (-)$

Find the following products.

1. $8 \cdot 6$

$\left.\begin{array}{l} |8| = 8 \\ |6| = 6 \end{array}\right\}$ **Multiply these absolute values.**

$8 \cdot 6 = 48$

Since the numbers have the same sign, the product is positive.

Thus, $8 \cdot 6 = +48$, or $8 \cdot 6 = 48$.

2. $(-8)(-6)$

$\left.\begin{array}{l} |-8| = 8 \\ |-6| = 6 \end{array}\right\}$ **Multiply these absolute values.**

$8 \cdot 6 = 48$

Since the numbers have the same sign, the product is positive.

Thus, $(-8)(-6) = +48$, or $(-8)(-6) = 48$.

3. $(-4)(7)$

$\left.\begin{array}{l} |-4| = 4 \\ |7| = 7 \end{array}\right\}$ **Multiply these absolute values.**

$4 \cdot 7 = 28$

Since the numbers have opposite signs, the product is negative.

Thus, $(-4)(7) = -28$.

4. $6(-3)$

$\left.\begin{array}{l} |6| = 6 \\ |-3| = 3 \end{array}\right\}$ **Multiply these absolute values.**

$6 \cdot 3 = 18$

Since the numbers have opposite signs, the product is negative.

Thus, $6(-3) = -18$.

★ **PRACTICE SET A**

Find the following products.

1. $3(-8)$ **2.** $4(16)$ **3.** $(-6)(-5)$

4. $(-7)(-2)$ **5.** $(-1)(4)$ **6.** $(-7)7$

☐ DIVISION OF SIGNED NUMBERS

To determine the signs in a division problem, recall that

$$\frac{12}{3} = 4 \qquad \text{since} \qquad 12 = 3 \cdot 4$$

This suggests that

$\frac{(+)}{(+)} = (+)$

$$\frac{(+)}{(+)} = (+) \qquad \text{since} \qquad (+) = (+)(+)$$

What is $\frac{12}{-3}$?

$12 = (-3)(-4) \qquad$ suggests that $\qquad \frac{12}{-3} = -4.$ That is,

$\frac{(+)}{(-)} = (-)$

$$(+) = (-)(-) \qquad \text{suggests that} \qquad \frac{(+)}{(-)} = (-)$$

What is $\frac{-12}{3}$?

$-12 = (3)(-4) \qquad$ suggests that $\qquad \frac{-12}{3} = -4.$ That is,

$\frac{(-)}{(+)} = (-)$

$$(-) = (+)(-) \qquad \text{suggests that} \qquad \frac{(-)}{(+)} = (-)$$

What is $\frac{-12}{-3}$?

$-12 = (-3)(4) \qquad$ suggests that $\qquad \frac{-12}{-3} = 4.$ That is,

$\frac{(-)}{(-)} = (+)$

$$(-) = (-)(+) \qquad \text{suggests that} \qquad \frac{(-)}{(-)} = (+)$$

We have the following rules for dividing signed numbers.

Rules for Dividing Signed
Numbers

Dividing signed numbers:

1. To divide two real numbers that have the *same sign,* divide their absolute values. The quotient is positive.

$$\frac{(+)}{(+)} = (+) \qquad \frac{(-)}{(-)} = (+)$$

2. To divide two real numbers that have *opposite signs,* divide their absolute values. The quotient is negative.

$$\frac{(-)}{(+)} = (-) \qquad \frac{(+)}{(-)} = (-)$$

☆ SAMPLE SET B

Find the following quotients.

1. $\dfrac{-10}{2}$

$\left.\begin{array}{l} |-10| = 10 \\ |2| = 2 \end{array}\right\}$ **Divide these absolute values.**

$\dfrac{10}{2} = 5$

Since the numbers have opposite signs, the quotient is negative.

Thus, $\dfrac{-10}{2} = -5.$

2. $\dfrac{-35}{-7}$

$\left.\begin{array}{l} |-35| = 35 \\ |-7| = 7 \end{array}\right\}$ **Divide these absolute values.**

$\dfrac{35}{7} = 5$

Since the numbers have the same signs, the quotient is positive.

Thus, $\dfrac{-35}{-7} = 5.$

3. $\dfrac{18}{-9}$

$\left.\begin{array}{l} |18| = 18 \\ |-9| = 9 \end{array}\right\}$ **Divide these absolute values.**

$\dfrac{18}{9} = 2$

Since the numbers have opposite signs, the quotient is negative.

Thus, $\dfrac{18}{-9} = -2.$

★ PRACTICE SET B

Find the following quotients.

1. $\dfrac{-24}{-6}$ 2. $\dfrac{30}{-5}$ 3. $\dfrac{-54}{27}$ 4. $\dfrac{51}{17}$

☆ SAMPLE SET C

Find the value of $\dfrac{-6(4-7)-2(8-9)}{-(4+1)+1}$.

Using the order of operations and what we know about signed numbers, we get,

$$\frac{-6(4-7)-2(8-9)}{-(4+1)+1} = \frac{-6(-3)-2(-1)}{-(5)+1}$$

$$= \frac{18+2}{-5+1}$$

$$= \frac{20}{-4}$$

$$= -5$$

★ PRACTICE SET C

Find the value of $\dfrac{-5(2-6)-4(-8-1)}{2(3-10)-9(-2)}$.

☐ CALCULATORS

Calculators with the $\boxed{+/-}$ key can be used for multiplying and dividing signed numbers.

☆ SAMPLE SET D

Use a calculator to find each quotient or product.

1. $(-186) \cdot (-43)$

Since this product involves a (negative) · (negative), we know the result should be a positive number. We'll illustrate this on the calculator.

		Display Reads
Type	186	186
Press	$\boxed{+/-}$	-186
Press	$\boxed{\times}$	-186
Type	43	43
Press	$\boxed{+/-}$	-43
Press	$\boxed{=}$	7998

Thus, $(-186) \cdot (-43) = 7{,}998$.

Continued

2. $\dfrac{158.64}{-54.3}$. Round to one decimal place.

		Display Reads
Type	158.64	158.64
Press	$\div$	158.64
Type	54.3	54.3
Press	$+/-$	-54.3
Press	$=$	-2.921546961

Rounding to one decimal place we get -2.9.

★ PRACTICE SET D

Use a calculator to find each value.

1. $(-51.3) \cdot (-21.6)$ **2.** $-2.5746 \div -2.1$

3. $(0.006) \cdot (-0.241)$. Round to three decimal places.

Answers to Practice Sets are on p. 463.

Section 10.6 EXERCISES

For problems 1–43, find the value of each of the following. Use a calculator to check each result.

1. $(-2)(-8)$

2. $(-3)(-9)$

3. $(-4)(-8)$

4. $(-5)(-2)$

5. $(3)(-12)$

6. $(4)(-18)$

7. $(10)(-6)$

8. $(-6)(4)$

9. $(-2)(6)$

10. $(-8)(7)$

11. $\dfrac{21}{7}$

20. $14 - (-20)$

21. $20 - (-8)$

12. $\dfrac{42}{6}$

22. $-4 - (-1)$

13. $\dfrac{-39}{3}$

23. $0 - 4$

14. $\dfrac{-20}{10}$

24. $0 - (-1)$

15. $\dfrac{-45}{-5}$

25. $-6 + 1 - 7$

16. $\dfrac{16}{-8}$

26. $15 - 12 - 20$

17. $\dfrac{25}{-5}$

27. $1 - 6 - 7 + 8$

18. $\dfrac{36}{-4}$

28. $2 + 7 - 10 + 2$

19. $8 - (-3)$

29. $3(4 - 6)$

30. $8(5 - 12)$

31. $-3(1 - 6)$

32. $-8(4 - 12) + 2$

33. $-4(1 - 8) + 3(10 - 3)$

34. $-9(0 - 2) + 4(8 - 9) + 0(-3)$

35. $6(-2 - 9) - 6(2 + 9) + 4(-1 - 1)$

36. $\dfrac{3(4 + 1) - 2(5)}{-2}$

37. $\dfrac{4(8 + 1) - 3(-2)}{-4 - 2}$

38. $\dfrac{-1(3 + 2) + 5}{-1}$

39. $\dfrac{-3(4 - 2) + (-3)(-6)}{-4}$

40. $-1(4 + 2)$

41. $-1(6 - 1)$

42. $-(8 + 21)$

43. $-(8 - 21)$

EXERCISES FOR REVIEW

(3.2) 44. Use the order of operations to simplify $(5^2 + 3^2 + 2) \div 2^2$.

(4.6) 45. Find $\frac{3}{8}$ of $\frac{32}{9}$.

(6.1) 46. Write this number in decimal form using digits: "fifty-two three-thousandths."

(7.3) 47. The ratio of chlorine to water in a solution is 2 to 7. How many mL of water are in a solution that contains 15 mL of chlorine?

(10.5) 48. Perform the subtraction: $-8 - (-20)$.

★ **Answers to Practice Sets (10.6)**

A. 1. -24 2. 64   3. 30   4. 14   5. -4 6. -49

B. 1. 4   2. -6 3. -2 4. 3

C. 14

D. 1. 1,108.08   2. 1.226   3. -0.001

Chapter 10 SUMMARY OF KEY CONCEPTS

Variables and Constants (10.1)

A *variable* is a letter or symbol that represents any member of a set of two or more numbers. A *constant* is a letter or symbol that represents a specific number. For example, the Greek letter π (pi) represents the constant 3.14159 . . . .

The Real Number Line (10.1)

The *real number line* allows us to visually display some of the numbers in which we are interested.

$$\longleftarrow \overset{\mid}{\underset{-3}{\mid}} \overset{\mid}{\underset{-2}{\mid}} \overset{\mid}{\underset{-1}{\mid}} \overset{\mid}{\underset{0}{\mid}} \overset{\mid}{\underset{1}{\mid}} \overset{\mid}{\underset{2}{\mid}} \overset{\mid}{\underset{3}{\mid}} \longrightarrow$$

Coordinate and Graph (10.1)

The number associated with a point on the number line is called the *coordinate* of the point. The point associated with a number is called the *graph* of the number.

Real Number (10.1)

A *real number* is any number that is the coordinate of a point on the real number line.

Types of Real Numbers (10.1)

The set *of real numbers* has many subsets. The ones of most interest to us are:

The *natural numbers:* $\{1, 2, 3, 4, \ldots\}$
The *whole numbers:* $\{0, 1, 2, 3, 4, \ldots\}$
The *integers:* $\{\ldots, -3, -2, -1, 0, 1, 2, 3, \ldots\}$
The *rational numbers:* {All numbers that can be expressed as the quotient of two integers.}

Positive and Negative Numbers (10.2)

A number is denoted as *positive* if it is directly preceded by a plus sign (+) or no sign at all. A number is denoted as *negative* if it is directly preceded by a minus sign (−).

Opposites (10.2)

Opposites are numbers that are the same distance from zero on the number line but have opposite signs. The numbers a and $-a$ are opposites.

Double-Negative Property (10.2)

$-(-a) = a$

Absolute Value (Geometric) (10.3)

The *absolute value* of a number a, denoted $|a|$, is the distance from a to 0 on the number line.

Absolute Value (Algebraic) (10.3)

$$|a| = \begin{cases} a, & \text{if } a \geq 0 \\ -a, & \text{if } a < 0 \end{cases}$$

Addition of Signed Numbers (10.4)

To *add two numbers* with

1. *like signs,* add the absolute values of the numbers and associate with the sum the common sign.
2. *unlike signs,* subtract the smaller absolute value from the larger absolute value and associate with the difference the sign of the larger absolute value.

Addition with Zero (10.4)

$0 +$ (any number) = that particular number.

Additive Identity (10.4)

Since adding 0 to any real number leaves that number unchanged, 0 is called the *additive identity*.

Definition of Subtraction (10.5)

$a - b = a + (-b)$

Subtraction of Signed Numbers (10.5)

To perform the *subtraction* $a - b$, add the opposite of b to a, that is, change the sign of b and follow the addition rules (Section 10.4).

Multiplication and Division of Signed Numbers (10.6)

$(+)(+) = (+)$ $\qquad \dfrac{(+)}{(+)} = (+)$ $\qquad \dfrac{(+)}{(-)} = (-)$

$(-)(-) = (+)$

$(+)(-) = (-)$ $\qquad \dfrac{(-)}{(-)} = (+)$ $\qquad \dfrac{(-)}{(+)} = (-)$

$(-)(+) = (-)$

EXERCISE SUPPLEMENT

Section 10.1

For problems 1–5, next to each real number, note all subsets of the real numbers to which it belongs by writing *N* for natural numbers, *W* for whole numbers, or *Z* for integers. Some numbers may belong to more than one subset.

1. 61

2. −14

3. 0

4. 1

5. Write all the integers that are strictly between −4 and 3.

6. Write all the integers that are between and including −6 and −1.

For each pair of numbers in problems 7–10, write the appropriate symbol (<, >, =) in place of the □.

7. −5 □ −1   **8.** 0 □ 2

9. −7 □ 0   **10.** −1 □ 0

For problems 11–15, what numbers can replace *x* so that each statement is true?

11. $-5 \leq x \leq -1$, *x* is an integer.

12. $-10 < x \leq 0$, *x* is a whole number.

13. $0 \leq x < 5$, *x* is a natural number.

14. $-3 < x < 3$, *x* is a natural number.

15. $-8 < x \leq -2$, *x* is a whole number.

For problems 16–20, how many units are there between the given pair of numbers?

16. 0 and 4   **17.** −1 and 3

18. −7 and −4   **19.** −6 and 0

20. −1 and 1

21. A number is positive if it is directly preceded by a _____ sign or no sign at all.

22. A number is negative if it is directly preceded by a _____ sign.

Section 10.2

For problems 23–26, how should each number be read?

23. −8   **24.** −(−4)

25. −(−1)   **26.** −2

For problems 27–31, write each expression in words.

27. $1 + (-7)$

28. $-2 - (-6)$

29. $-1 - (+4)$

30. $-(-(-3))$

31. $0 - (-11)$

For problems 32–36, rewrite each expression in simpler form.

32. $-(-4)$

33. $-(-15)$

34. $-[-(-7)]$

35. $1 - (-18)$

36. $0 - (-1)$

Section 10.3

For problems 37–52, determine each value.

37. $|9|$

38. $|16|$

39. $|-5|$

40. $|-8|$

41. $-|-2|$

42. $-|-1|$

43. $-(-|12|)$

44. $-(-|90|)$

45. $-(-|-16|)$

46. $-(-|0|)$

47. $|-4|^2$

48. $|-5|^2$

49. $|-2|^3$

50. $|-(3 \cdot 4)|$

51. $|-5| + |-2|$

52. $|-7| - |-10|$

Sections 10.4, 10.5, and 10.6

For problems 53–71, perform each operation.

53. $-6 + 4$

54. $-10 + 8$

55. $-1 - 6$

56. $8 - 12$

57. $0 - 14$

58. $5 \cdot (-2)$

59. $-8 \cdot (-6)$

60. $(-3) \cdot (-9)$

61. $14 \cdot (-3)$

62. $5 \cdot (-70)$

63. $-18 \div -6$

64. $72 \div -12$

65. $-16 \div -16$

66. $0 \div -8$

67. $-5 \div 0$

68. $\dfrac{-15}{-3}$

69. $\dfrac{-28}{7}$

70. $\dfrac{-120}{-|2|}$

71. $\dfrac{|-66|}{-|-3|}$

1. _____

1. (10.1) Write all integers that are strictly between -8 and -3.

2. _____

2. (10.1) Write all integers that are between and including -2 and 1.

3. _____

For problems 3–5, write the appropriate symbol ($<$, $>$, $=$) in place of the $\square$ for each pair of numbers.

3. (10.1) $-1 \square -1$

4. _____

4. (10.1) $0 \square 3$

5. _____

5. (10.1) $-1 \square -2$

6. _____

For problems 6 and 7, what numbers can replace x so that the statement is true?

6. (10.1) $-3 \le x < 0$, x is an integer.

7. _____

7. (10.1) $-4 \le x \le 2$, x is a natural number.

8. _____

8. (10.1) How many units are there between -3 and 2?

9. _____

For problems 9–20, find each value.

9. (10.3) $|-16|$

10. _____

10. (10.3) $-|-2|$

11. _____ 11. **(10.3)** $-(-|-4|^2)$

12. _____ 12. **(10.3)** $|-5| + |-10|$

13. _____ 13. **(10.4)** $-8 + 6$

14. _____ 14. **(10.4)** $-3 + (-8)$

15. _____ 15. **(10.5)** $0 - 16$

16. _____ 16. **(10.6)** $(-14) \cdot (-3)$

17. _____ 17. **(10.6)** $(-5 - 6)^2$

18. _____ 18. **(10.6)** $(-51) \div (-7)$

19. _____ 19. **(10.6)** $\dfrac{-42}{-7}$

20. _____ 20. **(10.6)** $\left| \dfrac{-32}{8} - \dfrac{-15 - 5}{5} \right|$

11

Algebraic Expressions and Equations

After completing this chapter, you should

Section 11.1 Algebraic Expressions
- be able to recognize an algebraic expression
- be able to distinguish between terms and factors
- understand the meaning and function of coefficients
- be able to perform numerical evaluation

Section 11.2 Combining Like Terms Using Addition and Subtraction
- be able to combine like terms in an algebraic expression

Section 11.3 Solving Equations of the Form $x + a = b$ and $x - a = b$
- understand the meaning and function of an equation
- understand what is meant by the solution to an equation
- be able to solve equations of the form $x + a = b$ and $x - a = b$

Section 11.4 Solving Equations of the Form $ax = b$ and $\dfrac{x}{a} = b$
- be familiar with the multiplication/division property of equality
- be able to solve equations of the form $ax = b$ and $\dfrac{x}{a} = b$
- be able to use combined techniques to solve equations

Section 11.5 Applications I: Translating Words to Mathematical Symbols
- be able to translate phrases and statements to mathematical expressions and equations

Section 11.6 Applications II: Solving Problems
- be more familiar with the five-step method for solving applied problems
- be able to use the five-step method to solve number problems and geometry problems

11.1 Algebraic Expressions

Section
Overview

- ❑ **ALGEBRAIC EXPRESSIONS**
- ❑ **TERMS AND FACTORS**
- ❑ **COEFFICIENTS**
- ❑ **NUMERICAL EVALUATION**

❑ ALGEBRAIC EXPRESSIONS

Numerical Expression

In arithmetic, a **numerical expression** results when numbers are connected by arithmetic operation signs $(+, -, \cdot, \div)$. For example, $8 + 5$, $4 - 9$, $3 \cdot 8$, and $9 \div 7$ are numerical expressions.

Algebraic Expression

In algebra, letters are used to represent numbers, and an **algebraic expression** results when an arithmetic operation sign associates a letter with a number or a letter with a letter. For example, $x + 8$, $4 - y$, $3 \cdot x$, $x \div 7$, and $x \cdot y$ are algebraic expressions.

Expressions

Numerical expressions and algebraic expressions are often referred to simply as **expressions.**

❑ TERMS AND FACTORS

In algebra, it is extremely important to be able to distinguish between terms and factors.

Distinction Between Terms and Factors

Terms are parts of *sums* and are therefore connected by $+$ signs.

Factors are parts of *products* and are therefore separated by $\cdot$ signs.

Note: While making the distinction between sums and products, we must remember that subtraction and division are functions of these operations.

1. In some expressions it will appear that terms are separated by minus signs. We must keep in mind that subtraction is addition of the opposite, that is,

$x - y = x + (-y)$

$$x - y = x + (-y)$$

2. In some expressions it will appear that factors are separated by division signs. We must keep in mind that

$\dfrac{x}{y} = x \cdot \dfrac{1}{y}$

$$\frac{x}{y} = \frac{x}{1} \cdot \frac{1}{y} = x \cdot \frac{1}{y}$$

☆ SAMPLE SET A

State the number of terms in each expression and name them.

1. $x + 4$.   In this expression, x and 4 are connected by a "$+$" sign. Therefore, they are terms.

   This expression consists of two terms.

2. $y - 8$.   The expression $y - 8$ can be expressed as $y + (-8)$. We can now see that this expression consists of the two terms y and -8.

   Rather than rewriting the expression when a subtraction occurs, we can identify terms more quickly by associating the $+$ or $-$ sign with the individual quantity.

3. $a + 7 - b - m$.   Associating the sign with the individual quantities, we see that this expression consists of the four terms a, 7, $-b$, $-m$.

4. $5m - 8n$.   This expression consists of the two terms, $5m$ and $-8n$. Notice that the term $5m$ is composed of the two *factors* 5 and m. The term $-8n$ is composed of the two factors -8 and n.

5. $3x$.   This expression consists of one term. Notice that $3x$ can be expressed as $3x + 0$ or $3x \cdot 1$ (indicating the connecting signs of arithmetic). Note that no operation sign is necessary for multiplication.

★ **PRACTICE SET A**

Specify the terms in each expression.

1. $x + 7$ **2.** $3m - 6n$ **3.** $5y$ **4.** $a + 2b - c$ **5.** $-3x - 5$

❏ COEFFICIENTS

We know that multiplication is a description of repeated addition. For example,

$5 \cdot 7$ describes $7 + 7 + 7 + 7 + 7$

Suppose some quantity is represented by the letter x. The multiplication $5x$ describes $x + x + x + x + x$. It is now easy to see that $5x$ specifies 5 of the quantities represented by x. In the expression $5x$, 5 is called the **numerical coefficient,** or more simply, the **coefficient** of x.

Coefficient

> The **coefficient** of a quantity records how many of that quantity there are.

Since constants alone do not record the number of some quantity, they are not usually considered as numerical coefficients. For example, in the expression $7x + 2y - 8z + 12$, the coefficient of

$7x$ is   7. (There are   7 x's.)
$2y$ is   2. (There are   2 y's.)
$-8z$ is -8. (There are -8 z's.)

The constant 12 is not considered a numerical coefficient.

$1x = x$

> When the numerical coefficient of a variable is 1, we write only the variable and not the coefficient. For example, we write x rather than $1x$. It is clear just by looking at x that there is only one.

❏ NUMERICAL EVALUATION

We know that a variable represents an unknown quantity. Therefore, any expression that contains a variable represents an unknown quantity. For example, if the value of x is unknown, then the value of $3x + 5$ is unknown. The value of $3x + 5$ depends on the value of x.

Numerical Evaluation

> **Numerical evaluation** is the process of determining the numerical value of an algebraic expression by replacing the variables in the expression with specified numbers.

Find the value of each expression.

1. $2x + 7y$, if $x = -4$ and $y = 2$.

Replace x with -4 and y with 2.

$$2x + 7y = 2(-4) + 7(2)$$
$$= -8 + 14$$
$$= 6$$

Thus, when $x = -4$ and $y = 2$, $2x + 7y = 6$.

2. $\dfrac{5a}{b} + \dfrac{8b}{12}$, if $a = 6$ and $b = -3$.

Replace a with 6 and b with -3.

$$\frac{5a}{b} + \frac{8b}{12} = \frac{5(6)}{-3} + \frac{8(-3)}{12}$$
$$= \frac{30}{-3} + \frac{-24}{12}$$
$$= -10 + (-2)$$
$$= -12$$

Thus, when $a = 6$ and $b = -3$, $\dfrac{5a}{b} + \dfrac{8b}{12} = -12$.

3. $6(2a - 15b)$, if $a = -5$ and $b = -1$.

Replace a with -5 and b with -1.

$$6(2a - 15b) = 6(2(-5) - 15(-1))$$
$$= 6(-10 + 15)$$
$$= 6(5)$$
$$= 30$$

Thus, when $a = -5$ and $b = -1$, $6(2a - 15b) = 30$.

4. $3x^2 - 2x + 1$, if $x = 4$.

Replace x with 4.

$$3x^2 - 2x + 1 = 3(4)^2 - 2(4) + 1$$
$$= 3 \cdot 16 - 2(4) + 1$$
$$= 48 - 8 + 1$$
$$= 41$$

Thus, when $x = 4$, $3x^2 - 2x + 1 = 41$.

5. $-x^2 - 4$, if $x = 3$.

Replace x with 3.

$$-x^2 - 4 = -3^2 - 4$$

Be careful to square only the 3. The exponent 2 is connected *only* to 3, not -3.

$$= -9 - 4$$
$$= -13$$

6. $(-x)^2 - 4$, if $x = 3$.

Replace x with 3.

$(-x)^2 - 4 = (-3)^2 - 4$

The exponent is connected to -3, not 3 as in problem 5 above.

$$= 9 - 4$$
$$= 5$$

★ PRACTICE SET B

Find the value of each expression.

1. $9m - 2n$, if $m = -2$ and $n = 5$.

2. $-3x - 5y + 2z$, if $x = -4$, $y = 3$, $z = 0$.

3. $\dfrac{10a}{3b} + \dfrac{4b}{2}$, if $a = -6$ and $b = 2$.

4. $8(3m - 5n)$, if $m = -4$ and $n = -5$.

5. $3[-40 - 2(4a - 3b)]$, if $a = -6$ and $b = 0$.

6. $5y^2 + 6y - 11$, if $y = -1$.

7. $-x^2 + 2x + 7$, if $x = 4$.

8. $(-x)^2 + 2x + 7$, if $x = 4$.

Answers to Practice Sets are on p. 476.

Section 11.1 EXERCISES

1. In an algebraic expression, terms are separated by _____ signs and factors are separated by _____ signs.

For problems 2–9, specify each term.

2. $3m + 7n$

3. $5x + 18y$

4. $4a - 6b + c$

5. $8s + 2r - 7t$

6. $m - 3n - 4a + 7b$

7. $7a - 2b - 3c - 4d$

8. $-6a - 5b$

9. $-x - y$

10. What is the function of a numerical coefficient?

11. Write $1m$ in a simpler way.

12. Write $1s$ in a simpler way.

13. In the expression $5a$, how many a's are indicated?

14. In the expression $-7c$, how many c's are indicated?

For problems 15–38, find the value of each expression.

15. $2m - 6n$, if $m = -3$ and $n = 4$

16. $5a + 6b$, if $a = -6$ and $b = 5$

17. $2x - 3y + 4z$, if $x = 1, y = -1$, and $z = -2$

18. $9a + 6b - 8x + 4y$, if $a = -2, b = -1, x = -2$, and $y = 0$

19. $\dfrac{8x}{3y} + \dfrac{18y}{2x}$, if $x = 9$ and $y = -2$

27. $3[16 - 3(a + 3b)]$, if $a = 3$ and $b = -2$

20. $\dfrac{-3m}{2n} - \dfrac{-6n}{m}$, if $m = -6$ and $n = 3$

28. $-2[5a + 2b(b - 6)]$, if $a = -2$ and $b = 3$

21. $4(3r + 2s)$, if $r = 4$ and $s = 1$

29. $-\{6x + 3y[-2(x + 4y)]\}$, if $x = 0$ and $y = 1$

22. $3(9a - 6b)$, if $a = -1$ and $b = -2$

23. $-8(5m + 8n)$, if $m = 0$ and $n = -1$

30. $-2\{19 - 6[4 - 2(a - b - 7)]\}$, if $a = 10$ and $b = 3$

24. $-2(-6x + y - 2z)$, if $x = 1$, $y = 1$, and $z = 2$

31. $x^2 + 3x - 1$, if $x = 5$

25. $-(10x - 2y + 5z)$, if $x = 2$, $y = 8$, and $z = -1$

26. $-(a - 3b + 2c - d)$, if $a = -5$, $b = 2$, $c = 0$, and $d = -1$

32. $m^2 - 2m + 6$, if $m = 3$

33. $6a^2 + 2a - 15$, if $a = -2$

36. $-8y^2 + 6y + 11$, if $y = 0$

34. $5s^2 + 6s + 10$, if $x = -1$

37. $(y - 6)^2 + 3(y - 5) + 4$, if $y = 5$

35. $16x^2 + 8x - 7$, if $x = 0$

38. $(x + 8)^2 + 4(x + 9) + 1$, if $x = -6$

EXERCISES FOR REVIEW

(5.2)   **39.** Perform the addition: $5\dfrac{3}{8} + 2\dfrac{1}{6}$.

(5.4)   **40.** Arrange the numbers in order from smallest to largest:

$$\frac{11}{32}, \frac{15}{48}, \text{ and } \frac{7}{16}$$

(5.6)   **41.** Find the value of $\left(\dfrac{2}{3}\right)^2 + \dfrac{8}{27}$.

(7.2)   **42.** Write the proportion in fractional form: "9 is to 8 as x is to 7."

(10.6)   **43.** Find the value of $-3(2 - 6) - 12$.

★ **Answers to Practice Sets (11.1)**

A.   **1.** $x, 7$ **2.** $3m, -6n$ **3.** $5y$ **4.** $a, 2b, -c$ **5.** $-3x, -5$

B.   **1.** -28 **2.** -3 **3.** -6 **4.** 104 **5.** 24 **6.** -12 **7.** -1 **8.** 31

11.2 Combining Like Terms Using Addition and Subtraction

Section Overview

☐ **COMBINING LIKE TERMS**

☐ **COMBINING LIKE TERMS**

From our examination of terms in Section 11.1, we know that **like terms** are terms in which the variable parts are identical. Like terms is an appropriate name since terms with identical variable parts and different numerical coefficients represent different amounts of the same quantity. When we are dealing with quantities of the same type, we may combine them using addition and subtraction.

Simplifying an Algebraic Expression

> An algebraic expression may be **simplified** by combining like terms.

This concept is illustrated in the following examples.

1. 8 records + 5 records = 13 records.

   Eight and 5 of the same type give 13 of that type. We have combined quantities of the same type.

2. 8 records + 5 records + 3 tapes = 13 records + 3 tapes.

   Eight and 5 of the same type give 13 of that type. Thus, we have 13 of one type and 3 of another type. We have combined only quantities of the same type.

3. Suppose we let the letter x represent "record." Then, $8x + 5x = 13x$. The terms $8x$ and $5x$ are like terms. So, 8 and 5 of the same type give 13 of that type. We have combined like terms.

4. Suppose we let the letter x represent "record" and y represent "tape." Then,

   $$8x + 5x + 3y = 13x + 5y$$

   We have combined only the like terms.

After observing the problems in these examples, we can suggest a method for simplifying an algebraic expression by combining like terms.

Combining Like Terms

> Like terms may be combined by adding or subtracting their coefficients and affixing the result to the common variable.

☆ **SAMPLE SET A**

Simplify each expression by combining like terms.

1. $2m + 6m - 4m$.   All three terms are alike. Combine their coefficients and affix this result to m: $2 + 6 - 4 = 4$.

Thus, $2m + 6m - 4m = 4m$.

2. $5x + 2y - 9y$.   The terms $2y$ and $-9y$ are like terms. Combine their coefficients: $2 - 9 = -7$.

Thus, $5x + 2y - 9y = 5x - 7y$.

Continued

3. $-3a + 2b - 5a + a + 6b.$ The like terms are

$$\underbrace{-3a, -5a, a}_{\substack{-3 - 5 + 1 = -7 \\ -7a}} \qquad \underbrace{2b, 6b}_{\substack{2 + 6 = 8 \\ 8b}}$$

Thus, $-3a + 2b - 5a + a + 6b = -7a + 8b.$

4. $r - 2s + 7s + 3r - 4r - 5s.$ The like terms are

$$\underbrace{\underbrace{r, 3r, -4r}_{\substack{1 + 3 - 4 = 0 \\ 0r}} \qquad \underbrace{-2s, 7s, -5s}_{\substack{-2 + 7 - 5 = 0 \\ 0s}}}_{0r + 0s = 0}$$

Thus, $r - 2s + 7s + 3r - 4r - 5s = 0.$

★ **PRACTICE SET A**

Simplify each expression by combining like terms.

1. $4x + 3x + 6x$ 　　　　**2.** $5a + 8b + 6a - 2b$ 　　　　**3.** $10m - 6n - 2n - m + n$

4. $16a + 6m + 2r - 3r - 18a + m - 7m$ 　　　　**5.** $5h - 8k + 2h - 7h + 3k + 5k$

Answers to the Practice Set are on p. 480.

Section 11.2 EXERCISES

For problems 1–20, simplify each expression by combining like terms.

1. $4a + 7a$

2. $3m + 5m$

3. $6h - 2h$

4. $11k - 8k$

5. $5m + 3n - 2m$

6. $7x - 6x + 3y$

7. $14s + 3s - 8r + 7r$

8. $-5m - 3n + 2m + 6n$

9. $7h + 3a - 10k + 6a - 2h - 5k - 3k$

10. $4x - 8y - 3z + x - y - z - 3y - 2z$

11. $11w + 3x - 6w - 5w + 8x - 11x$

12. $15r - 6s + 2r + 8s - 6r - 7s - s - 2r$

13. $|-7|m + |6|m + |-3|m$

14. $|-2|x + |-8|x + |10|x$

15. $(-4 + 1)k + (6 - 3)k + (12 - 4)h + (5 + 2)k$

16. $(-5 + 3)a - (2 + 5)b - (3 + 8)b$

17. $5\star + 2\triangle + 3\triangle - 8\star$

18. $9\boxtimes + 10\boxplus - 11\boxtimes - 12\boxplus$

19. $16x - 12y + 5x + 7 - 5x - 16 - 3y$

20. $-3y + 4z - 11 - 3z - 2y + 5 - 4(8 - 3)$

EXERCISES FOR REVIEW

(4.2) 21. Convert $\frac{24}{11}$ to a mixed number.

(4.3 22. Determine the missing numerator: $\frac{3}{8} = \frac{?}{64}$.

(5.5) 23. Simplify $\dfrac{\frac{5}{6} - \frac{1}{4}}{\frac{1}{12}}$.

(7.4) 24. Convert $\frac{5}{16}$ to a percent.

(11.1) 25. In the expression $6k$, how many k's are there?

★ **Answers to Practice Set (11.2)**

A. 1. $13x$ 2. $11a + 6b$ 3. $9m - 7n$ 4. $-2a - r$ 5. 0

11.3 Solving Equations of the Form $x + a = b$ and $x - a = b$

Section Overview	☐ **EQUATIONS** ☐ **SOLUTIONS AND EQUIVALENT EQUATIONS** ☐ **SOLVING EQUATIONS**

☐ EQUATIONS

Equation

> An **equation** is a statement that two algebraic expressions are equal.

The following are examples of equations:

$$\underbrace{x + 6}_{\substack{\text{This}\\\text{expression}}} = \underbrace{10}_{\substack{\text{This}\\\text{expression}}} \qquad \underbrace{x - 4}_{\substack{\text{This}\\\text{expression}}} = \underbrace{-11}_{\substack{\text{This}\\\text{expression}}} \qquad \underbrace{3y - 5}_{\substack{\text{This}\\\text{expression}}} = \underbrace{2 + 2y}_{\substack{\text{This}\\\text{expression}}}$$

Notice that $x + 6$, $x - 4$, and $3y - 5$ are *not* equations. They are expressions. They are not equations because there is no statement that each of these expressions is equal to another expression.

☐ SOLUTIONS AND EQUIVALENT EQUATIONS

The truth of some equations is conditional upon the value chosen for the variable.

Conditional Equations

Such equations are called **conditional equations**. There are two additional types of equations. They are examined in courses in algebra, so we will not consider them now.

Solutions and Solving an Equation

> The set of values that, when substituted for the variables, make the equation true, are called the **solutions** of the equation.
>
> An equation has been **solved** when all its solutions have been found.

☆ **SAMPLE SET A**

1. Verify that 3 is a solution to $x + 7 = 10$.

   When $\qquad x = 3,$
   $$x + 7 = 10$$
   becomes $3 + 7 = 10$
   $\qquad\qquad 10 = 10$, which is a *true* statement, verifying that 3 is a solution to $x + 7 = 10$.

2. Verify that -6 is a solution to $5y + 8 = -22$.

   When $\qquad y = -6,$
   $$5y + 8 = -22$$
   becomes $5(-6) + 8 = -22$
   $\qquad\qquad -30 + 8 = -22$
   $\qquad\qquad\quad -22 = -22$, which is a *true* statement, verifying that -6 is a solution to $5y + 8 = -22$.

3. Verify that 5 is not a solution to $a - 1 = 2a + 3$.

   When $\qquad a = 5,$
   $$a - 1 = 2a + 3$$
   becomes $5 - 1 = 2 \cdot 5 + 3$
   $\qquad\quad 5 - 1 = 10 + 3$
   $\qquad\qquad\quad 4 = 13$, a *false* statement, verifying that 5 is not a solution to $a - 1 = 2a + 3$.

4. Verify that -2 is a solution to $3m - 2 = -4m - 16$.

   When $\qquad m = -2,$
   $$3m - 2 = -4m - 16$$
   becomes $3(-2) - 2 = -4(-2) - 16$
   $\qquad\quad -6 - 2 = 8 - 16$
   $\qquad\qquad\quad -8 = -8$, which is a *true* statement, verifying that -2 is a solution to $3m - 2 = -4m - 16$.

★ **PRACTICE SET A**

1. Verify that 5 is a solution to $m + 6 = 11$.

2. Verify that -5 is a solution to $2m - 4 = -14$.

3. Verify that 0 is a solution to $5x + 1 = 1$.

4. Verify that 3 is not a solution to $-3y + 1 = 4y + 5$.

5. Verify that -1 is a solution to $6m - 5 + 2m = 7m - 6$.

Equivalent Equations

Some equations have precisely the same collection of solutions. Such equations are called **equivalent equations.** For example, $x - 5 = -1$, $x + 7 = 11$, and $x = 4$ are all equivalent equations since the only solution to each is $x = 4$. (Can you verify this?)

☐ SOLVING EQUATIONS

We know that the equal sign of an equation indicates that the number represented by the expression on the left side is the same as the number represented by the expression on the right side.

This number	is the same as	this number
↓	↓	↓
x	$=$	4
$x + 7$	$=$	11
$x - 5$	$=$	-1

Addition/Subtraction Property of Equality

From this, we can suggest the **addition/subtraction property of equality.**

> Given any equation,
>
> 1. We can obtain an equivalent equation by *adding* the *same* number to *both* sides of the equation.
> 2. We can obtain an equivalent equation by *subtracting* the *same* number from *both* sides of the equation.

The Idea Behind Equation Solving

The idea behind **equation solving** is to isolate the variable on one side of the equation. Signs of operation $(+, -, \cdot, \div)$ are used to associate two numbers. For example, in the expression $5 + 3$, the numbers 5 and 3 are associated by addition. An association can be *undone* by performing the opposite operation. The addition/subtraction property of equality can be used to undo an association that is made by addition or subtraction.

Subtraction is used to undo an addition.
Addition is used to undo a subtraction.

The procedure is illustrated in the problems of Sample Set B.

☆ SAMPLE SET B

Use the addition/subtraction property of equality to solve each equation.

1. $x + 4 = 6$.   **4 is associated with x by addition. Undo the association by *subtracting* 4 from *both* sides.**

$x + 4 - 4 = 6 - 4$
$x + 0 = 2$
$x = 2$

Check: When $x = 2$, $x + 4$ becomes
$2 + 4 \overset{?}{=} 6$
$6 \overset{?}{=} 6$.

The solution to $x + 4 = 6$ is $x = 2$.

2. $m - 8 = 5$.   **8 is associated with m by subtraction. Undo the association by *adding* 8 to *both* sides.**

$$m - 8 + 8 = 5 + 8$$
$$m + 0 = 13$$
$$m = 13$$

Check: When $m = 13$,
$$m - 8 = 5$$
$$\text{becomes } 13 - 8 \stackrel{?}{=} 5$$
$$5 \stackrel{\checkmark}{=} 5, \text{ a true statement.}$$

The solution to $m - 8 = 5$ is $m = 13$.

3. $-3 - 5 = y - 2 + 8$.   **Before we use the addition/subtraction property, we should simplify as much as possible.**

$$-3 - 5 = y - 2 + 8$$

$$-8 = y + 6$$ **6 is associated with y by addition. Undo the association by *subtracting* 6 from *both* sides.**

$$-8 - 6 = y + 6 - 6$$
$$-14 = y + 0$$
$$-14 = y$$ **This is equivalent to $y = -14$.**

Check: When $y = -14$,
$$-3 - 5 = y - 2 + 8$$
$$\text{becomes } -3 - 5 \stackrel{?}{=} -14 - 2 + 8$$
$$-8 \stackrel{?}{=} -16 + 8$$
$$-8 \stackrel{\checkmark}{=} -8, \text{ a true statement.}$$

The solution to $-3 - 5 = y - 2 + 8$ is $y = -14$.

4. $-5a + 1 + 6a = -2$.   **Begin by simplifying the left side of the equation.**

$$\underbrace{-5a + 1 + 6a}_{-5 + 6 = 1} = -2$$

$$a + 1 = -2$$ **1 is associated with a by addition. Undo the association by *subtracting* 1 from *both* sides.**

$$a + 1 - 1 = -2 - 1$$
$$a + 0 = -3$$
$$a = -3$$

Check: When $a = -3$,
$$-5a + 1 + 6a = -2$$
$$\text{becomes } -5(-3) + 1 + 6(-3) \stackrel{?}{=} -2$$
$$15 + 1 - 18 \stackrel{?}{=} -2$$
$$-2 \stackrel{\checkmark}{=} -2, \text{ a true statement.}$$

The solution to $-5a + 1 + 6a = -2$ is $a = -3$.

5. $7k - 4 = 6k + 1$.   **In this equation, the variable appears on both sides. We need to isolate it on one side. Although we can choose either side, it will be more convenient to choose the side with the larger coefficient. Since 7 is greater than 6, we'll isolate k on the left side.**

$$7k - 4 = 6k + 1$$ **Since $6k$ represents $+6k$, subtract $6k$ from each side.**

$$\underbrace{7k - 4 - 6k}_{7 - 6 = 1} = \underbrace{6k + 1 - 6k}_{6 - 6 = 0}$$

$$k - 4 = 1$$ **4 is associated with k by subtraction. Undo the association by *adding* 4 to *both* sides.**

$$k - 4 + 4 = 1 + 4$$
$$k = 5$$

Continued

Check: When $k = 5$,

$$7k - 4 = 6k + 1$$

becomes $7 \cdot 5 - 4 \overset{?}{=} 6 \cdot 5 + 1$

$$35 - 4 \overset{?}{=} 30 + 1$$

$$31 \overset{\checkmark}{=} 31, \text{ a true statement.}$$

The solution to $7k - 4 = 6k + 1$ is $k = 5$.

6. $-8 + x = 5$. -8 is associated with x by addition. Undo the by *subtracting* -8 from *both* sides. Subtracting -8 we get $-(-8) = +8$. We actually *add* 8 to both sides.

$$-8 + x + 8 = 5 + 8$$
$$x = 13$$

Check: When $x = 13$,

$$-8 + x = 5$$

becomes $-8 + 13 \overset{?}{=} 5$

$$5 \overset{\checkmark}{=} 5, \text{ a true statement.}$$

The solution to $-8 + x = 5$ is $x = 13$.

★ **PRACTICE SET B**

Solve each equation. Be sure to check each solution.

1. $y + 9 = 4$ **2.** $a - 4 = 11$ **3.** $-1 + 7 = x + 3$

4. $8m + 4 - 7m = (-2)(-3)$ **5.** $12k - 4 = 9k - 6 + 2k$ **6.** $-3 + a = -4$

Answers to Practice Sets are on p. 486.

Section 11.3 EXERCISES

For problems 1–10, verify that each given value is a solution to the given equation.

1. $x - 11 = 5$, $x = 16$

2. $y - 4 = -6$, $y = -2$

3. $2m - 1 = 1$, $m = 1$

4. $5y + 6 = -14$, $y = -4$

5. $3x + 2 - 7x = -5x - 6$, $x = -8$

6. $-6a + 3 + 3a = 4a + 7 - 3a$, $a = -1$

7. $-8 + x = -8, \quad x = 0$

8. $8b + 6 = 6 - 5b, \quad b = 0$

9. $4x - 5 = 6x - 20, \quad x = \dfrac{15}{2}$

10. $-3y + 7 = 2y - 15, \quad y = \dfrac{22}{5}$

For problems 11–40, solve each equation. Be sure to check each result.

11. $y - 6 = 5$

12. $m + 8 = 4$

13. $k - 1 = 4$

14. $h - 9 = 1$

15. $a + 5 = -4$

16. $b - 7 = -1$

17. $x + 4 - 9 = 6$

18. $y - 8 + 10 = 2$

19. $z + 6 = 6$

20. $w - 4 = -4$

21. $x + 7 - 9 = 6$

22. $y - 2 + 5 = 4$

23. $m + 3 - 8 = -6 + 2$

24. $z + 10 - 8 = -8 + 10$

25. $2 + 9 = k - 8$

26. $-5 + 3 = h - 4$

27. $3m - 4 = 2m + 6$

28. $5a + 6 = 4a - 8$

29. $8b + 6 + 2b = 3b - 7 + 6b - 8$

30. $12h - 1 - 3 - 5h = 2h + 5h + 3(-4)$

31. $-4a + 5 - 2a = -3a - 11 - 2a$

32. $-9n - 2 - 6 + 5n = 3n - (2)(-5) - 6n$

⊞ Calculator Exercises

33. $y - 2.161 = 5.063$

34. $a - 44.0014 = -21.1625$

35. $-0.362 - 0.416 = 5.63m - 4.63m$

36. $8.078 - 9.112 = 2.106y - 1.106y$

37. $4.23k + 3.18 = 3.23k - 5.83$

38. $6.1185x - 4.0031 = 5.1185x - 0.0058$

39. $21.63y + 12.40 - 5.09y = 6.11y - 15.66 + 9.43y$

40. $0.029a - 0.013 - 0.034 - 0.057 = -0.038 + 0.56 + 1.01a$

EXERCISES FOR REVIEW

(7.1) **41.** Is $\dfrac{7 \text{ calculators}}{12 \text{ students}}$ an example of a ratio or a rate?

(7.4) **42.** Convert $\dfrac{3}{8}\%$ to a decimal.

(7.5) **43.** 0.4% of what number is 0.014?

(8.2) **44.** Use the clustering method to estimate the sum:
$$89 + 93 + 206 + 198 + 91$$

(11.2) **45.** Combine like terms: $4x + 8y + 12y + 9x - 2y$.

★ **Answers to Practice Sets (11.3)**

A. **1.** Substitute 5 into $m + 6 = 11$.

$$5 + 6 \overset{?}{=} 11$$
$$11 \overset{✓}{=} 11$$

Thus, 5 is a solution.

2. Substitute -5 into $2m - 4 = -14$.

$$2(-5) - 4 \overset{?}{=} -14$$
$$-10 - 4 \overset{?}{=} -14$$
$$-14 \overset{✓}{=} -14$$

Thus, -5 is a solution.

3. Substitute 0 into $5x + 1 = 1$.

$$5(0) + 1 \overset{?}{=} 1$$
$$0 + 1 \overset{?}{=} 1$$
$$1 \overset{\checkmark}{=} 1$$

Thus, 0 is a solution.

4. Substitute 3 into $-3y + 1 = 4y + 5$.

$$-3(3) + 1 \overset{?}{=} 4(3) + 5$$
$$-9 + 1 \overset{?}{=} 12 + 5$$
$$-8 \neq 17$$

Thus, 3 is not a solution.

5. Substitute -1 into $6m - 5 + 2m = 7m - 6$.

$$6(-1) - 5 + 2(-1) \overset{?}{=} 7(-1) - 6$$
$$-6 - 5 - 2 \overset{?}{=} -7 - 6$$
$$-13 \overset{\checkmark}{=} -13$$

Thus, -1 is a solution.

B. **1.** $y = -5$ **2.** $a = 15$ **3.** $x = 3$ **4.** $m = 2$ **5.** $k = -2$ **6.** $a = -1$

11.4 Solving Equations of the Form $ax = b$ and $\dfrac{x}{a} = b$

Section Overview

☐ **MULTIPLICATION/DIVISION PROPERTY OF EQUALITY**
☐ **COMBINING TECHNIQUES IN EQUATION SOLVING**

☐ MULTIPLICATION/DIVISION PROPERTY OF EQUALITY

Recall that the equal sign of an equation indicates that the number represented by the expression on the left side is the same as the number represented by the expression on the right side. From this, we can suggest the multiplication/division property of equality.

Multiplication/Division Property of Equality

Given any equation,

1. We can obtain an equivalent equation by *multiplying both sides* of the equation by the *same nonzero* number, that is, if $c \neq 0$, then $a = b$ is equivalent to

$$a \cdot c = b \cdot c$$

2. We can obtain an equivalent equation by *dividing both sides* of the equation by the *same nonzero* number, that is, if $c \neq 0$, then $a = b$ is equivalent to

$$\frac{a}{c} = \frac{b}{c}$$

The multiplication/division property of equality can be used to undo an association with a number that multiplies or divides the variable.

Use the multiplication/division property of equality to solve each equation.

1. $6y = 54$.

6 is associated with y by multiplication. Undo the association by *dividing both* sides by 6.

$$\frac{6y}{6} = \frac{54}{6}$$

$$\frac{\cancel{6}y}{\cancel{6}} = \frac{\cancel{54}^{9}}{\cancel{6}}$$

$$y = 9$$

Check: When $y = 9$,

$$6y = 54$$

becomes $6 \cdot 9 \stackrel{?}{=} 54$

$$54 \stackrel{\checkmark}{=} 54, \text{ a true statement.}$$

The solution to $6y = 54$ is $y = 9$.

2. $\dfrac{x}{-2} = 27$.

-2 is associated with x by division. Undo the association by *multiplying both* sides by -2.

$$(-2)\,\frac{x}{-2} = (-2)27$$

$$(\cancel{-2})\,\frac{x}{\cancel{-2}} = (-2)27$$

$$x = -54$$

Check: When $x = -54$,

$$\frac{x}{-2} = 27$$

becomes $\dfrac{-54}{-2} \stackrel{?}{=} 27$

$$27 \stackrel{\checkmark}{=} 27, \text{ a true statement.}$$

The solution to $\dfrac{x}{-2} = 27$ is $x = -54$.

3. $\dfrac{3a}{7} = 6$.

We will examine two methods for solving equations such as this one.

Method 1: Use of dividing out common factors.

$$\frac{3a}{7} = 6$$

7 is associated with a by division. Undo the association by *multiplying both* sides by 7.

$$7 \cdot \frac{3a}{7} = 7 \cdot 6$$

Divide out the 7's.

$$\cancel{7} \cdot \frac{3a}{\cancel{7}} = 42$$

$3a = 42$

3 is associated with a by multiplication. Undo the association by *dividing* *both* sides by 3.

$$\frac{3a}{3} = \frac{42}{3}$$

$$\frac{\cancel{3}a}{\cancel{3}} = 14$$

$a = 14$

Check: When $a = 14$,

$$\frac{3a}{7} = 6$$

becomes $\dfrac{3 \cdot 14}{7} \stackrel{?}{=} 6$

$$\frac{42}{7} = 6$$

$6 \stackrel{\checkmark}{=} 6$, a true statement.

The solution to $\dfrac{3a}{7} = 6$ is $a = 14$.

Method 2: Use of reciprocals.

Recall that if the product of two numbers is 1, the numbers are **reciprocals.** Thus $\dfrac{3}{7}$ and $\dfrac{7}{3}$ are reciprocals.

$$\frac{3a}{7} = 6$$

Multiply *both* sides of the equation by $\dfrac{7}{3}$, the reciprocal of $\dfrac{3}{7}$.

$$\frac{7}{3} \cdot \frac{3a}{7} = \frac{7}{3} \cdot 6$$

$$\overset{1}{\underset{1}{\cancel{7}}} \cdot \overset{1}{\underset{1}{\cancel{3}a}} = \frac{7}{\cancel{3}} \cdot \overset{2}{\underset{1}{\cancel{6}}}$$

$1 \cdot a = 14$

$a = 14$

Notice that we get the same solution using either method.

4. $-8x = 24$.

-8 is associated with x by multiplication. Undo the association by *dividing* *both* sides by -8.

$$\frac{-8x}{-8} = \frac{24}{-8}$$

$$\frac{-8x}{-8} = \frac{24}{-8}$$

$x = -3$

Check: When $x = -3$,

$$-8x = 24$$

becomes $-8(-3) \stackrel{?}{=} 24$

$24 \stackrel{\checkmark}{=} 24$, a true statement.

Continued

5. $-x = 7$.

Since $-x$ is actually $-1 \cdot x$ and $(-1)(-1) = 1$, we can isolate x by multiplying *both* sides of the equation by -1.

$(-1)(-x) = -1 \cdot 7$
$x = -7$

Check: When $x = 7$,
$$-x = 7$$
becomes $-(-7) \stackrel{?}{=} 7$
$$7 \stackrel{\checkmark}{=} 7$$

The solution to $-x = 7$ is $x = -7$.

★ **PRACTICE SET A**

Use the multiplication/division property of equality to solve each equation. Be sure to check each solution.

1. $7x = 21$ **2.** $-5x = 65$ **3.** $\dfrac{x}{4} = -8$ **4.** $\dfrac{3x}{8} = 6$

5. $-y = 3$ **6.** $-k = -2$

☐ COMBINING TECHNIQUES IN EQUATION SOLVING

Having examined solving equations using the addition/subtraction and the multiplication/division principles of equality, we can combine these techniques to solve more complicated equations.

When beginning to solve an equation such as $6x - 4 = -16$, it is helpful to know which property of equality to use first, addition/subtraction or multiplication/division. Recalling that in equation solving *we are trying to isolate the variable* (disassociate numbers from it), it is helpful to note the following.

> To *associate* numbers and letters, we use the order of operations.
>
> 1. Multiply/divide
> 2. Add/subtract
>
> To *undo an association* between numbers and letters, we use the order of operations in reverse.
>
> 1. Add/subtract
> 2. Multiply/divide

☆ **SAMPLE SET B**

Solve each equation. (In these example problems, we will not show the checks.)

1. $6x - 4 = -16$. —4 is associated with x by subtraction. Undo the association by *adding* 4 to *both* sides.

$6x - 4 + 4 = -16 + 4$

$6x = -12$ 6 is associated with x by multiplication. Undo the association by *dividing both* sides by 6.

$\dfrac{6x}{6} = \dfrac{-12}{6}$

$x = -2$

2. $-8k + 3 = -45$. 3 is associated with k by addition. Undo the association by *subtracting* 3 from *both* sides.

$-8k + 3 - 3 = -45 - 3$

$-8k = -48$ —8 is associated with k by multiplication. Undo the association by *dividing both* sides by —8.

$\dfrac{-8k}{-8} = \dfrac{-48}{-8}$

$k = 6$

3. $5m - 6 - 4m = 4m - 8 + 3m$. **Begin solving this equation by combining like terms.**

$m - 6 = 7m - 8$ **Choose a side on which to isolate *m*. Since 7 is greater than 1, we'll isolate *m* on the right side.**

Subtract *m* from *both* sides.

$m - 6 - m = 7m - 8 - m$

$-6 = 6m - 8$ **8 is associated with *m* by subtraction. Undo the association by *adding* 8 to *both* sides.**

$-6 + 8 = 6m - 8 + 8$

$2 = 6m$ **6 is associated with *m* by multiplication. Undo the association by *dividing both* sides by 6.**

$\dfrac{2}{6} = \dfrac{6m}{6}$ **Reduce.**

$\dfrac{1}{3} = m$

Notice that if we had chosen to isolate *m* on the left side of the equation rather than the right side, we would have proceeded as follows:

$m - 6 = 7m - 8$ **Subtract *7m* from *both* sides.**

$m - 6 - 7m = 7m - 8 - 7m$

$-6m - 6 = -8$ **Add 6 to *both* sides,**

$-6m - 6 + 6 = -8 + 6$

$-6m = -2$ **Divide *both* sides by —6.**

$\dfrac{-6m}{-6} = \dfrac{-2}{-6}$

$$m = \frac{1}{3}$$

This is the same result as with the previous approach.

4. $\dfrac{8x}{7} = -2$ 7 is associated with x by division. Undo the association by *multiplying both* sides by 7.

$$7 \cdot \frac{8x}{7} = 7(-2)$$

$$\cancel{7} \cdot \frac{8x}{\cancel{7}} = -14$$

$8x = -14$ 8 is associated with x by multiplication. Undo the association by *dividing both* sides by 8.

$$\frac{8x}{8} = \frac{-14}{8}$$

$$\frac{\cancel{8}x}{\cancel{8}} = \frac{-7}{4}$$

$$x = \frac{-7}{4}$$

★ **PRACTICE SET B**

Solve each equation. Be sure to check each solution.

1. $5m + 7 = -13$ **2.** $-3a - 6 = 9$ **3.** $2a + 10 - 3a = 9$

4. $11x - 4 - 13x = 4x + 14$ **5.** $-3m + 8 = -5m + 1$ **6.** $5y + 8y - 11 = -11$

Answers to Practice Sets are on p. 495.

Section 11.4 EXERCISES

For problems 1–36, solve each equation. Be sure to check each result.

1. $7x = 42$

2. $8x = 81$

3. $10x = 120$

4. $11x = 121$

5. $-6a = 48$

6. $-9y = 54$

7. $-3y = -42$

8. $-5a = -105$

9. $2m = -62$

10. $3m = -54$

11. $\dfrac{x}{4} = 7$

12. $\dfrac{y}{3} = 11$

13. $\dfrac{-z}{6} = -14$

14. $\dfrac{-w}{5} = 1$

15. $3m - 1 = -13$

16. $4x + 7 = -17$

17. $2 + 9x = -7$

18. $5 - 11x = 27$

19. $32 = 4y + 6$

20. $-5 + 4 = -8m + 1$

21. $3k + 6 = 5k + 10$

22. $4a + 16 = 6a + 8a + 6$

23. $6x + 5 + 2x - 1 = 9x - 3x + 15$

24. $-9y - 8 + 3y + 7 = -7y + 8y - 5y + 9$

31. $\dfrac{5a}{7} = 10$

25. $-3a = a + 5$

32. $\dfrac{2m}{9} = 4$

26. $5b = -2b + 8b + 1$

33. $\dfrac{3x}{4} = \dfrac{9}{2}$

27. $-3m + 2 - 8m - 4 = -14m + m - 4$

34. $\dfrac{8k}{3} = 32$

28. $5a + 3 = 3$

35. $\dfrac{3a}{8} - \dfrac{3}{2} = 0$

29. $7x + 3x = 0$

30. $7g + 4 - 11g = -4g + 1 + g$

36. $\dfrac{5m}{6} - \dfrac{25}{3} = 0$

EXERCISES FOR REVIEW

(8.3) **37.** Use the distributive property to compute $40 \cdot 28$.

(9.4) **38.** Approximating π by 3.14, find the approximate circumference of the circle.

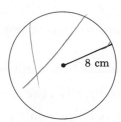

8 cm

(9.5) **39.** Find the area of the parallelogram.

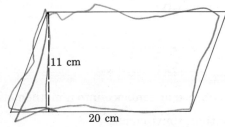

11 cm

20 cm

(10.6) **40.** Find the value of $\dfrac{-3(4-15)-2}{-5}$.

(11.3) **41.** Solve the equation $x - 14 + 8 = -2$.

★ **Answers to Practice Sets (11.4)**

A. **1.** $x = 3$ **2.** $x = -13$ **3.** $x = -32$ **4.** $x = 16$ **5.** $y = -3$ **6.** $k = 2$

B. **1.** $m = -4$ **2.** $a = -5$ **3.** $a = 1$ **4.** $x = -3$ **5.** $m = -\dfrac{7}{2}$ **6.** $y = 0$

11.5 Applications I: Translating Words to Mathematical Symbols

Section Overview	☐ **TRANSLATING WORDS TO SYMBOLS**

☐ TRANSLATING WORDS TO SYMBOLS

Practical problems seldom, if ever, come in equation form. The job of the problem solver is to translate the problem from phrases and statements into mathematical expressions and equations, and then to solve the equations.

As problem solvers, our job is made simpler if we are able to translate verbal phrases to mathematical expressions and if we follow the five-step method of solving applied problems. To help us translate from words to symbols, we can use the following Mathematics Dictionary.

MATHEMATICS DICTIONARY

Word or Phrase	Mathematical Operation
Sum, sum of, added to, increased by, more than, and, plus	$+$
Difference, minus, subtracted from, decreased by, less, less than	$-$
Product, the product of, of, multiplied by, times, per	$\cdot$
Quotient, divided by, ratio, per	$\div$
Equals, is equal to, is, the result is, becomes	$=$
A number, an unknown quantity, an unknown, a quantity	x (or any symbol)

☆ SAMPLE SET A

Translate each phrase or sentence into a mathematical expression or equation.

1. Nine more than some number.
   (9) (+) (x)

   Translation: $9 + x$.

2. Eighteen minus a number.
   (18) (−) (x)

   Translation: $18 - x$.

3. A quantity less five.
   (y) (−) (5)

   Translation: $y - 5$.

4. Four times a number is sixteen.
   (4) (·) (x) (=) (16)

   Translation: $4x = 16$

5. One fifth of a number is thirty.
   ($\frac{1}{5}$) (·) (n) (=) (30)

   Translation: $\frac{1}{5}n = 30$, or $\frac{n}{5} = 30$.

6. Five times a number is two more than twice the number.
   (5) (·) (x) (=) (2) (+) (2·) (x)

   Translation: $5x = 2 + 2x$.

★ PRACTICE SET A

Translate each phrase or sentence into a mathematical expression or equation.

1. Twelve more than a number.

2. Eight minus a number.

3. An unknown quantity less fourteen.

4. Six times a number is fifty-four.

5. Two ninths of a number is eleven.

6. Three more than seven times a number is nine more than five times the number.

7. Twice a number less eight is equal to one more than three times the number.

☆ **SAMPLE SET B**

1. Sometimes the structure of the sentence indicates the use of grouping symbols. We'll be alert for *commas*. They set off terms.

$$\underbrace{\text{A number}}_{(x}\ \underbrace{\text{divided by}}_{\div}\ \underbrace{\text{four,}}_{4)}\ \underbrace{\text{minus}}_{-}\ \underbrace{\text{six,}}_{6}\ \underbrace{\text{is}}_{=}\ \underbrace{\text{twelve.}}_{12}$$

Translation: $\dfrac{x}{4} - 6 = 12$.

2. Some phrases and sentences do not translate directly. We must be careful to read them properly. The word *from* often appears in such phrases and sentences. The word **from** means "a point of departure for motion." The following translation will illustrate this use.

$$\underbrace{\text{Twenty}}\ \underbrace{\text{is subtracted from}}\ \underbrace{\text{some number.}}$$
$$x \longleftarrow \hspace{1.5cm} \longrightarrow 20$$

Translation: $x - 20$.

The word *from* indicates the motion (subtraction) is to begin at the point of "some number."

3. Ten less than some number.   Notice that *less than* can be replaced by *from*.

Ten from some number.

Translation: $x - 10$.

★ **PRACTICE SET B**

Translate each phrase or sentence into a mathematical expression or equation.

1. A number divided by eight, plus seven, is fifty.

2. A number divided by three, minus the same number multiplied by six, is one more than the number.

3. Nine from some number is four.

4. Five less than some quantity is eight.

Answers to Practice Sets are on p. 500.

Section 11.5 EXERCISES

For problems 1–40, translate each phrase or sentence to a mathematical expression or equation.

1. A quantity less twelve.

2. Six more than an unknown number.

3. A number minus four.

4. A number plus seven.

5. A number increased by one.

6. A number decreased by ten.

7. Negative seven added to some number.

8. Negative nine added to a number.

9. A number plus the opposite of six.

10. A number minus the opposite of five.

11. A number minus the opposite of negative one.

12. A number minus the opposite of negative twelve.

13. Eleven added to three times a number.

14. Six plus five times an unknown number.

15. Twice a number minus seven equals four.

16. Ten times a quantity increased by two is nine.

17. When fourteen is added to two times a number the result is six.

18. Four times a number minus twenty-nine is eleven.

19. Three fifths of a number plus eight is fifty.

20. Two ninths of a number plus one fifth is forty-one.

21. When four thirds of a number is increased by twelve, the result is five.

22. When seven times a number is decreased by two times the number, the result is negative one.

23. When eight times a number is increased by five, the result is equal to the original number plus twenty-six.

24. Five more than some number is three more than four times the number.

25. When a number divided by six is increased by nine, the result is one.

26. A number is equal to itself minus three times itself.

27. A number divided by seven, plus two, is seventeen.

28. A number divided by nine, minus five times the number, is equal to one more than the number.

29. When two is subtracted from some number, the result is ten.

30. When four is subtracted from some number, the result is thirty one.

31. Three less than some number is equal to twice the number minus six.

32. Thirteen less than some number is equal to three times the number added to eight.

33. When twelve is subtracted from five times some number, the result is two less than the original number.

34. When one is subtracted from three times a number, the result is eight less than six times the original number.

35. When a number is subtracted from six, the result is four more than the original number.

36. When a number is subtracted from twenty-four, the result is six less than twice the number.

37. A number is subtracted from nine. This result is then increased by one. The result is eight more than three times the number.

38. Five times a number is increased by two. This result is then decreased by three times the number. The result is three more than three times the number.

39. Twice a number is decreased by seven. This result is decreased by four times the number. The result is negative the original number, minus six.

40. Fifteen times a number is decreased by fifteen. This result is then increased by two times the number. The result is negative five times the original number minus the opposite of ten.

EXERCISES FOR REVIEW

(4.6) 41. $\frac{8}{9}$ of what number is $\frac{2}{3}$?

(5.2) 42. Find the value of $\frac{21}{40} + \frac{17}{30}$.

(5.3) 43. Find the value of $3\frac{1}{12} + 4\frac{1}{3} + 1\frac{1}{4}$.

(6.2) 44. Convert $6.11\frac{1}{5}$ to a fraction.

(11.4) 45. Solve the equation $\frac{3x}{4} + 1 = -5$.

★ Answers to Practice Sets (11.5)

A. 1. $12 + x$ 2. $8 - x$ 3. $x - 14$ 4. $6x = 54$ 5. $\frac{2}{9}x = 11$ 6. $3 + 7x = 9 + 5x$

7. $2x - 8 = 3x + 1$ or $2x - 8 = 1 + 3x$

B. 1. $\frac{x}{8} + 7 = 50$ 2. $\frac{x}{3} - 6x = x + 1$ 3. $x - 9 = 4$ 4. $x - 5 = 8$

11.6 Applications II: Solving Problems

Section Overview

- ☐ **THE FIVE-STEP METHOD**
- ☐ **NUMBER PROBLEMS**
- ☐ **GEOMETRY PROBLEMS**

☐ THE FIVE-STEP METHOD

We are now in a position to solve some applied problems using algebraic methods. The problems we shall solve are intended as logic developers. Although they may not seem to reflect real situations, they do serve as a basis for solving more complex, real situation, applied problems. To solve problems algebraically, we will use the five-step method.

Strategy for Reading Word Problems

When solving mathematical word problems, you may wish to apply the following **"reading strategy."** Read the problem quickly to get a feel for the situation. Do not pay close attention to details. At the first reading, too much attention to details may be overwhelming and lead to confusion and discouragement. After the first, brief reading, read the problem carefully in *phrases*. Reading phrases introduces information more slowly and allows us to absorb and put together important information. We can look for the unknown quantity by reading one phrase at a time.

Five-Step Method for Solving Word Problems

1. Let x (or some other letter) represent the unknown quantity.
2. Translate the words to mathematical symbols and form an equation. Draw a picture if possible.
3. Solve the equation.

4. Check the solution by substituting the result into the original statement, not equation, of the problem.
5. Write a conclusion.

If it has been your experience that word problems are difficult, then follow the five-step method carefully. Most people have trouble with word problems for two reasons:

1. They are not able to translate the words to mathematical symbols. (See Section 11.4.)
2. They neglect step 1. After working through the problem phrase by phrase, to become familiar with the situation,

INTRODUCE A VARIABLE

❑ NUMBER PROBLEMS

☆ **SAMPLE SET A**

1. What number decreased by six is five?

Step 1:   Let n represent the unknown number.

Step 2:   Translate the words to mathematical symbols and construct an equation. Read phrases.

$$
\left.\begin{array}{ll}
\text{What number:} & n \\
\text{decreased by:} & - \\
\text{six:} & 6 \\
\text{is:} & = \\
\text{five:} & 5
\end{array}\right\} \quad n - 6 = 5
$$

Step 3:   Solve this equation.

$n - 6 = 5$ **Add 6 to *both* sides.**

$n - 6 + 6 = 5 + 6$
$n = 11$

Step 4:   Check the result.

When 11 is decreased by 6, the result is $11 - 6$, which is equal to 5. The solution checks.

Step 5:   The number is 11.

2. When three times a number is increased by four, the result is eight more than five times the number.

Step 1:   Let x = the unknown number.

Step 2:   Translate the phrases to mathematical symbols and construct an equation.

$$
\left.\begin{array}{ll}
\text{When three times a number:} & 3x \\
\text{is increased by:} & + \\
\text{four:} & 4 \\
\text{the result is:} & = \\
\text{eight:} & 8 \\
\text{more than:} & + \\
\text{five times the number:} & 5x
\end{array}\right\} \quad 3x + 4 = 5x + 8
$$

Continued

Step 3: $3x + 4 = 5x + 8.$ **Subtract $3x$ from *both* sides.**

$3x + 4 - 3x = 5x + 8 - 3x$

$4 = 2x + 8$ **Subtract 8 from *both* sides.**

$4 - 8 = 2x + 8 - 8$

$-4 = 2x$ **Divide *both* sides by 2.**

$-2 = x$

Step 4: Check this result.

Three times -2 is -6. Increasing -6 by 4 results in $-6 + 4 = -2$. Now, five times -2 is -10. Increasing -10 by 8 results in $-10 + 8 = -2$. The results agree, and the solution checks.

Step 5: The number is -2.

3. Consecutive integers have the property that if

$n =$ the smallest integer, then
$n + 1 =$ the next integer, and
$n + 2 =$ the next integer, and so on.

Consecutive odd or even integers have the property that if

$n =$ the smallest integer, then
$n + 2 =$ the next odd or even integer (since odd or even numbers differ by 2), and
$n + 4 =$ the next odd or even integer, and so on.

The sum of three consecutive odd integers is equal to one less than twice the first odd integer. Find the three integers.

Step 1: Let $n =$ the first odd integer. Then,
$n + 2 =$ the second odd integer, and
$n + 4 =$ the third odd integer.

Step 2: Translate the words to mathematical symbols and construct an equation. Read phrases.

The sum of:	add some numbers	
three consecutive odd integers:	$n, n + 2, n + 4$	
is equal to:	$=$	$n + (n + 2) + (n + 4) = 2n - 1$
one less than:	subtract 1 from	
twice the first odd integer:	$2n$	

Step 3: $n + n + 2 + n + 4 = 2n - 1$

$3n + 6 = 2n - 1$ **Subtract $2n$ from *both* sides.**

$3n + 6 - 2n = 2n - 1 - 2n$

$n + 6 = -1$ **Subtract 6 from *both* sides.**

$n + 6 - 6 = -1 - 6$

$n = -7$ **The first integer is -7.**

$n + 2 = -7 + 2 = -5$ **The second integer is -5.**

$n + 4 = -7 + 4 = -3$ **The third integer is -3.**

Step 4: Check this result.

The sum of the three integers is

$$-7 + (-5) + (-3) = -12 + (-3)$$
$$= -15$$

One less than twice the first integer is $2(-7) - 1 = -14 - 1 = -15$. Since these two results are equal, the solution checks.

Step 5: The three odd integers are $-7, -5, -3$.

★ PRACTICE SET A

1. When three times a number is decreased by 5, the result is -23. Find the number.

Step 1: Let $x =$
Step 2:

Step 3:

Step 4: Check:

Step 5: The number is _____ .

2. When five times a number is increased by 7, the result is five less than seven times the number. Find the number.

Step 1: Let $n =$
Step 2:

Step 3:

Step 4: Check:

Step 5: The number is _____ .

3. Two consecutive numbers add to 35. Find the numbers.

Step 1:
Step 2:

Step 3:

Step 4:   Check:

Step 5:   The numbers are _____ and _____ .

4. The sum of three consecutive even integers is six more than four times the middle integer. Find the integers.

Step 1:   Let x = smallest integer.
      _____ = next integer.
      _____ = largest integer.

Step 2:

Step 3:

Step 4:   Check:

Step 5:   The integers are _____ , _____ , and _____ .

❑ GEOMETRY PROBLEMS

☆ **SAMPLE SET B**

The perimeter (length around) of a rectangle is 20 meters. If the length is 4 meters longer than the width, find the length and width of the rectangle.

Step 1:   Let x = the width of the rectangle. Then,
      $x + 4$ = the length of the rectangle.

Step 2:   We can draw a picture.

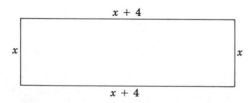

The length around the rectangle is

$$\underbrace{x}_{\text{width}} + \underbrace{(x + 4)}_{\text{length}} + \underbrace{x}_{\text{width}} + \underbrace{(x + 4)}_{\text{length}} = 20$$

Step 3: $x + x + 4 + x + x + 4 = 20$

$4x + 8 = 20$ **Subtract 8 from *both* sides.**

$4x = 12$ **Divide *both* sides by 4.**

$x = 3$ **Then,**

$x + 4 = 3 + 4 = 7$

Step 4:   Check:

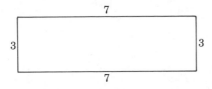

$$3 + 7 + 3 + 7 \overset{?}{=} 20$$
$$20 \overset{?}{=} 20$$

Step 5:   The length of the rectangle is 7 meters.
          The width of the rectangle is 3 meters.

★ **PRACTICE SET B**

The perimeter of a triangle is 16 inches. The second leg is 2 inches longer than the first leg, and the third leg is 5 inches longer than the first leg. Find the length of each leg.

Step 1:   Let x = length of the first leg.
          _____ = length of the second leg.
          _____ = length of the third leg.

Step 2:   We can draw a picture.

Step 3:

Step 4:   Check:

Step 5:   The lengths of the legs are _____, _____, and _____.

Answers to Practice Sets are on p. 515.

Section 11.6 EXERCISES

For problems 1–17, find each solution using the five-step method.

1. What number decreased by nine is fifteen?

    Step 1:   Let n = the number.
    Step 2:

    Step 3:

Step 4:   Check:

Step 5: The number is _____ .

2. What number increased by twelve is twenty?

Step 1:   Let n = the number.
Step 2:

Step 3:

Step 4:   Check:

Step 5:   The number is _____ .

3. If five more than three times a number is thirty-two, what is the number?

Step 1:   Let x = the number.
Step 2:

Step 3:

Step 4:   Check:

Step 5:   The number is _____ .

4. If four times a number is increased by fifteen, the result is five. What is the number?

Step 1:   Let x =
Step 2:

Step 3:

Step 4:   Check:

Step 5:   The number is _____ .

5. When three times a quantity is decreased by five times the quantity, the result is negative twenty. What is the quantity?

Step 1:   Let $x =$
Step 2:

Step 3:

Step 4:   Check:

Step 5:   The quantity is _____ .

6. If four times a quantity is decreased by nine times the quantity, the result is ten. What is the quantity?

Step 1:   Let $y =$
Step 2:

Step 3:

Step 4:   Check:

Step 5:   The quantity is _____ .

7. When five is added to three times some number, the result is equal to five times the number decreased by seven. What is the number?

Step 1:   Let $n =$
Step 2:

Step 3:

Step 4:   Check:

Step 5:   The number is _____ .

8. When six times a quantity is decreased by two, the result is six more than seven times the quantity. What is the quantity?

Step 1:   Let $x =$
Step 2:

Step 3:

Step 4:   Check:

Step 5:   The quantity is _____ .

9. When four is decreased by three times some number, the result is equal to one less than twice the number. What is the number?

Step 1:

Step 2:

Step 3:

Step 4:   Check:

Step 5:

10. When twice a number is subtracted from one, the result is equal to twenty-one more than the number. What is the number?

Step 1:

Step 2:

Step 3:

Step 4:

Step 5:

11. The perimeter of a rectangle is 36 inches. If the length of the rectangle is 6 inches more than the width, find the length and width of the rectangle.

Step 1:   Let w = the width.
_____ = the length.

Step 2:   We can draw a picture.

Step 3:

Step 4:   Check:

Step 5:   The length of the rectangle is _____ inches, and the width is _____ inches.

12. The perimeter of a rectangle is 48 feet. Find the length and the width of the rectangle if the length is 8 feet more than the width.

Step 1:   Let w = the width.
_____ = the length.

Step 2:   We can draw a picture.

Step 3:

Step 4: Check:

Step 5: The length of the rectangle is _____ feet, and the width is _____ feet.

13. The sum of three consecutive integers is 48. What are they?

Step 1: Let n = the smallest integer.
_____ = the next integer.
_____ = the next integer.

Step 2:

Step 3:

Step 4: Check:

Step 5: The three integers are _____, _____, and _____.

14. The sum of three consecutive integers is -27. What are they?

Step 1: Let n = the smallest integer.
_____ = the next integer.
_____ = the next integer.

Step 2:

Step 3:

Step 4: Check:

Step 5: The three integers are _____, _____, and _____.

15. The sum of five consecutive integers is zero. What are they?

Step 1: Let n =

Step 2:

Step 3:

Step 4:

Step 5:   The five integers are _____, _____, _____, _____, and _____.

16. The sum of five consecutive integers is −5. What are they?

Step 1:   Let $n =$

Step 2:

Step 3:

Step 4:

Step 5:   The five integers are _____, _____, _____, _____, and _____.

For problems 17–35, continue using the five-step procedure to find the solutions.

17. The perimeter of a rectangle is 18 meters. Find the length and width of the rectangle if the length is 1 meter more than three times the width.

18. The perimeter of a rectangle is 80 centimeters. Find the length and width of the rectangle if the length is 2 meters less than five times the width.

19. Find the length and width of a rectangle with perimeter 74 inches, if the width of the rectangle is 8 inches less than twice the length.

20. Find the length and width of a rectangle with perimeter 18 feet, if the width of the rectangle is 7 feet less than three times the length.

21. A person makes a mistake when copying information regarding a particular rectangle. The copied information is as follows: The length of a rectangle is 5 inches less than two times the width. The perimeter of the rectangle is 2 inches. What is the mistake?

22. A person makes a mistake when copying information regarding a particular triangle. The copied information is as follows: Two sides of a triangle are the same length. The third side is 10 feet less than three times the length of one of the other sides. The perimeter of the triangle is 5 feet. What is the mistake?

23. The perimeter of a triangle is 75 meters. If each of two legs is exactly twice the length of the shortest leg, how long is the shortest leg?

24. If five is subtracted from four times some number the result is negative twenty-nine. What is the number?

25. If two is subtracted from ten times some number, the result is negative two. What is the number?

26. If three less than six times a number is equal to five times the number minus three, what is the number?

27. If one is added to negative four times a number the result is equal to eight less than five times the number. What is the number?

28. Find three consecutive integers that add to -57.

29. Find four consecutive integers that add to negative two.

30. Find three consecutive even integers that add to -24.

31. Find three consecutive odd integers that add to -99.

32. Suppose someone wants to find three consecutive odd integers that add to 120. Why will that person not be able to do it?

33. Suppose someone wants to find two consecutive even integers that add to 139. Why will that person not be able to do it?

34. Three numbers add to 35. The second number is five less than twice the smallest. The third number is exactly twice the smallest. Find the numbers.

35. Three numbers add to 37. The second number is one less than eight times the smallest. The third number is two less than eleven times the smallest. Find the numbers.

EXERCISES FOR REVIEW

(6.6) **36.** Find the decimal representation of $0.34992 \div 4.32$.

(7.3) **37.** A 5-foot woman casts a 9-foot shadow at a particular time of the day. How tall is a person that casts a 10.8-foot shadow at the same time of the day?

(8.4) **38.** Use the method of rounding to estimate the sum: $4\frac{5}{12} + 15\frac{1}{25}$.

(9.2) **39.** Convert 463 mg to cg.

(11.5) **40.** Twice a number is added to 5. The result is 2 less than three times the number. What is the number?

★ **Answers to Practice Sets (11.6)**

A. **1.** -6 **2.** 6   **3.** 17 and 18   **4.** $-8, -6,$ and -4

B. 3 inches, 5 inches, and 8 inches

Chapter 11   SUMMARY OF KEY CONCEPTS

Numerical Expression (11.1)

A *numerical expression* results when numbers are associated by arithmetic operation signs. The expressions $3 + 5, 9 - 2, 5 \cdot 6,$ and $8 \div 5$ are numerical expressions.

Algebraic Expressions (11.1)

When an arithmetic operation sign connects a letter with a number or a letter with a letter, an *algebraic expression* results. The expressions $4x + 1, x - 5, 7x \cdot 6y,$ and $4x \div 3$ are algebraic expressions.

Terms and Factors (11.1)

Terms are parts of *sums* and are therefore separated by addition (or subtraction) signs. In the expression, $5x - 2y$, $5x$ and $-2y$ are the terms.

Factors are parts of *products* and are therefore separated by multiplication signs. In the expression $5a$, 5 and a are the factors.

Coefficients (11.1)

The *coefficient* of a quantity records how many of that quantity there are. In the expression $7x$, the coefficient 7 indicates that there are seven x's.

Numerical Evaluation (11.1)

Numerical evaluation is the process of determining the value of an algebraic expression by replacing the variables in the expression with specified values.

Combining Like Terms (11.2)

An algebraic expression may be simplified by combining like terms. *To combine like terms,* we simply add or subtract their coefficients then affix the variable. For example $4x + 9x = (4 + 9)x = 13x$.

Equation (11.3)

An *equation* is a statement that two expressions are equal. The statements $5x + 1 = 3$ and $\dfrac{4x}{5} + 4 = \dfrac{2}{5}$ are equations. The expressions represent the same quantities.

Conditional Equation (11.3)

A *conditional equation* is an equation whose truth depends on the value selected for the variable. The equation $3x = 9$ is a conditional equation since it is only true on the condition that 3 is selected for x.

Solutions and Solving an Equation (11.3)

The values that when substituted for the variables make the equation true are called the *solutions* of the equation.

An equation has been *solved* when all its solutions have been found.

Equivalent Equations (11.3)

Equations that have precisely the same solutions are called *equivalent equations*. The equations $6y = 18$ and $y = 3$ are equivalent equations.

Addition/Subtraction Property of Equality (11.3)

Given any equation, we can obtain an equivalent equation by

1. adding the same number to both sides, or
2. subtracting the same number from both sides.

Solving $x + a = b$ and $x - a = b$ (11.3)

To solve $x + a = b$, subtract a from both sides.

$x + a = b$
$x + a - a = b - a$
$x = b - a$

To solve $x - a = b$, add a to both sides.

$x - a = b$
$x - a + a = b + a$
$x = b + a$

Multiplication/Division Property of Equality (11.4)

Given any equation, we can obtain an *equivalent equation* by

1. multiplying both sides by the same nonzero number, that is, if $c \neq 0$, $a = b$ and $a \cdot c = b \cdot c$ are equivalent.

2. dividing both sides by the same nonzero number, that is, if $c \neq 0$, $a = b$ and $\dfrac{a}{c} = \dfrac{b}{c}$

   are equivalent.

516

Solving $ax = b$ and $\dfrac{x}{a} = b$

(11.4)

To solve $ax = b$, $a \neq 0$, *divide both sides by* a.

$$ax = b$$

$$\frac{ax}{a} = \frac{b}{a}$$

$$\frac{\cancel{a}x}{\cancel{a}} = \frac{b}{a}$$

$$x = \frac{b}{a}$$

To solve $\dfrac{x}{a} = b$, $a \neq 0$, *multiply* both sides by a.

$$\frac{x}{a} = b$$

$$a \cdot \frac{x}{a} = a \cdot b$$

$$\cancel{a} \cdot \frac{x}{\cancel{a}} = a \cdot b$$

$$x = a \cdot b$$

Translating Words to Mathematics **(11.5)**

In solving applied problems, it is important to be able to translate phrases and sentences to mathematical expressions and equations.

The Five-Step Method for Solving Applied Problems **(11.6)**

To solve problems algebraically, it is a good idea to use the following *five-step procedure.*

After working your way through the problem carefully, phrase by phrase:

1. Let x (or some other letter) represent the unknown quantity.
2. Translate the phrases and sentences to mathematical symbols and form an equation. Draw a picture if possible.
3. Solve this equation.
4. Check the solution by substituting the result into the original statement of the problem.
5. Write a conclusion.

EXERCISE SUPPLEMENT

Section 11.1

For problems 1–10, specify each term.

1. $6a - 2b + 5c$

2. $9x - 6y + 1$

3. $7m - 3n$

4. $-5h + 2k - 8 + 4m$

5. $x + 2n - z$

6. $y - 5$

7. $-y - 3z$

8. $-a - b - c - 1$

9. -4

10. -6

11. Write $1k$ in a simpler way.

12. Write $1x$ is a simpler way.

13. In the expression $7r$, how many r's are indicated?

14. In the expression $12m$, how many m's are indicated?

15. In the expression $-5n$, how many n's are indicated?

16. In the expression $-10y$, how many y's are indicated?

For problems 17–46, find the value of each expression.

17. $5a - 2s$, if $a = -5$ and $s = 1$

18. $7n - 3r$, if $n = -6$ and $r = 2$

19. $9x + 2y - 3s$, if $x = -2$, $y = 5$, and $s = -3$

20. $10a - 2b + 5c$, if $a = 0$, $b = -6$, and $c = 8$

21. $-5s - 2t + 1$, if $s = 2$ and $t = -2$

22. $-3m - 4n + 5$, if $m = -1$ and $n = -1$

23. $m - 4$, if $m = 4$

24. $n = 2$, if $n = 2$

25. $-x + 2y$, if $x = -7$ and $y = -1$

26. $-a + 3b - 6$, if $a = -3$ and $b = 0$

27. $5x - 4y - 7y + y - 7x$, if $x = 1$ and $y = -2$

28. $2a - 6b - 3a - a + 2b$, if $a = 4$ and $b = -2$

29. $a^2 - 6a + 4$, if $a = -2$

30. $m^2 - 8m - 6$, if $m = -5$

31. $4y^2 + 3y + 1$, if $y = -2$

32. $5a^2 - 6a + 11$, if $a = 0$

33. $-k^2 - k - 1$, if $k = -1$

34. $-h^2 - 2h - 3$, if $h = -4$

518

35. $\dfrac{m}{6} + 5m$, if $m = -18$

36. $\dfrac{a}{8} - 2a + 1$, if $a = 24$

37. $\dfrac{5x}{7} + 3x - 7$, if $x = 14$

38. $\dfrac{3k}{4} - 5k + 18$, if $k = 16$

39. $\dfrac{-6a}{5} + 3a + 10$, if $a = 25$

40. $\dfrac{-7h}{9} - 7h - 7$, if $h = -18$

41. $5(3a + 4b)$, if $a = -2$ and $b = 2$

42. $7(2y - x)$, if $x = -1$ and $y = 2$

43. $-(a - b)$, if $a = 0$ and $b = -6$

44. $-(x - x - y)$, if $x = 4$ and $y = -4$

45. $(y + 2)^2 - 6(y + 2) - 6$, if $y = 2$

46. $(a - 7)^2 - 2(a - 7) - 2$, if $a = 7$

Section 11.2

For problems 47–56, simplify each expression by combining like terms.

47. $4a + 5 - 2a + 1$

48. $7x + 3x - 14x$

49. $-7n + 4m - 3 + 3n$

50. $-9k - 8h - k + 6h$

51. $-x + 5y - 8x - 6x + 7y$

52. $6n - 2n + 6 - 2 - n$

53. $0m + 3k - 5s + 2m - s$

54. $|-8|a + |2|b - |-4|a$

55. $|6|h - |-7|k + |-12|h + |4| \cdot |-5|h$

56. $|0|a - 0a + 0$

Sections 11.4, 11.5, and 11.6

For problems 57–140, solve each equation.

57. $x + 1 = 5$ **58.** $y - 3 = -7$

59. $x + 12 = 10$ **60.** $x - 4 = -6$

61. $5x = 25$ **62.** $3x = 17$

63. $\dfrac{x}{2} = 6$ **64.** $\dfrac{x}{-8} = 3$

65. $\dfrac{x}{15} = -1$ **66.** $\dfrac{x}{-4} = -3$

67. $-3x = 9$ **68.** $-2x = 5$

69. $-5x = -5$ **70.** $-3x = -1$

71. $\dfrac{x}{-3} = 9$ **72.** $\dfrac{a}{-5} = 2$

73. $-7 = 3y$ **74.** $-7 = \dfrac{x}{3}$

75. $\dfrac{m}{4} = \dfrac{-2}{5}$

76. $4y = \dfrac{1}{2}$

77. $\dfrac{-1}{3} = -5x$

78. $\dfrac{-1}{9} = \dfrac{k}{3}$

79. $\dfrac{-1}{6} = \dfrac{s}{-6}$

80. $\dfrac{0}{4} = 4s$

81. $x + 2 = -1$

82. $x - 5 = -6$

83. $\dfrac{-3}{2}x = 6$

84. $3x + 2 = 7$

85. $-4x - 5 = -3$

86. $\dfrac{x}{6} + 1 = 4$

87. $\dfrac{a}{-5} - 3 = -2$

88. $\dfrac{4x}{3} = 7$

89. $\dfrac{2x}{5} + 2 = 8$

90. $\dfrac{3y}{2} - 4 = 6$

91. $m + 3 = 8$

92. $\dfrac{1x}{2} = 2$

93. $\dfrac{2a}{3} = 5$

94. $\dfrac{-3x}{7} - 4 = 4$

95. $\dfrac{5x}{-2} - 6 = -10$

96. $-4k - 6 = 7$

97. $\dfrac{-3x}{-2} + 1 = 4$

98. $\dfrac{-6x}{4} = 2$

99. $x + 9 = 14$

100. $y + 5 = 21$

101. $y + 5 = -7$

102. $4x = 24$

103. $4w = 37$

104. $6y - 11 = 13$

105. $-3x + 8 = -7$

106. $3z + 9 = -51$

107. $\dfrac{x}{-3} = 8$

108. $\dfrac{6y}{7} = 5$

109. $\dfrac{w}{2} - 15 = 4$

110. $\dfrac{x}{-2} - 23 = -10$

111. $\dfrac{2x}{3} - 5 = 8$

112. $\dfrac{3z}{4} = \dfrac{-7}{8}$

113. $-2 - \dfrac{2x}{7} = 3$

114. $3 - x = 4$

115. $-5 - y = -2$

116. $3 - z = -2$

117. $3x + 2x = 6$

118. $4x + 1 + 6x = 10$

119. $6y - 6 = -4 + 3y$

120. $3 = 4a - 2a + a$

121. $3m + 4 = 2m + 1$

122. $5w - 6 = 4 + 2w$

123. $8 - 3a = 32 - 2a$

124. $5x - 2x + 6x = 13$

125. $x + 2 = 3 - x$

126. $5y + 2y - 1 = 6y$

127. $x = 32$

128. $k = -4$

129. $\dfrac{3x}{2} + 4 = \dfrac{5x}{2} + 6$

135. $\dfrac{3x}{4} + 5 = \dfrac{-3x}{4} - 11$

130. $\dfrac{x}{3} + \dfrac{3x}{3} - 2 = 16$

136. $\dfrac{3x}{7} = \dfrac{-3x}{7} + 12$

131. $x - 2 = 6 - x$

137. $\dfrac{5y}{13} - 4 = \dfrac{7y}{26} + 1$

132. $\dfrac{-5x}{7} = \dfrac{2x}{7}$

138. $\dfrac{-3m}{5} = \dfrac{6m}{10} - 2$

133. $\dfrac{2x}{3} + 1 = 5$

139. $\dfrac{-3m}{2} + 1 = 5m$

134. $\dfrac{-3x}{5} + 3 = \dfrac{2x}{5} + 2$

140. $-3z = \dfrac{2z}{5}$

For problems 1 and 2, specify each term.

1. (11.1) $5x + 6y + 3z$

1. _____

2. (11.1) $8m - 2n - 4$

2. _____

3. (11.1) In the expression $-9a$, how many a's are indicated?

3. _____

For problems 4–9, find the value of each expression.

4. (11.1) $6a - 3b$, if $a = -2$ and $b = -1$.

4. _____

5. (11.1) $-5m + 2n - 6$, if $m = -1$ and $n = 4$.

5. _____

6. (11.1) $-x^2 + 3x - 5$, if $x = -2$.

6. _____

7. (11.1) $y^2 + 9y + 1$, if $y = 0$.

7. _____

8. (11.1) $-a^2 + 3a + 4$, if $a = 4$.

8. _____

9. (11.1) $-(5 - x)^2 + 7(m - x) + x - 2m$, if $x = 5$ and $m = 5$.

9. _____

For problems 10–12, simplify each expression by combining like terms.

10. **(11.2)** $6y + 5 - 2y + 1$

10. _____

11. **(11.2)** $14a - 3b + 5b - 6a - b$

11. _____

12. **(11.2)** $9x + 5y - 7 + 4x - 6y + 3(-2)$

12. _____

13. _____

For problems 13–22, solve each equation.

13. **(11.3)** $x + 7 = 15$

14. _____

14. **(11.3)** $y - 6 = 2$

15. _____

15. **(11.3)** $m + 8 = -1$

16. _____

16. **(11.3)** $-5 + a = -4$

17. _____

17. **(11.4)** $4x = 104$

18. _____

18. **(11.4)** $6y + 3 = -21$

19. _____

19. **(11.4)** $\dfrac{5m}{6} = \dfrac{10}{3}$

20. _____

20. **(11.4)** $\dfrac{7y}{8} + \dfrac{1}{4} = \dfrac{-13}{4}$

524

21. (11.4) $6x + 5 = 4x - 11$

22. (11.4) $4y - 8 - 6y = 3y + 1$

23. (11.5 and 11.6) Three consecutive even integers add to -36. What are they?

24. (11.5 and 11.6) The perimeter of a rectangle is 38 feet. Find the length and width of the rectangle if the length is 5 feet less than three times the width.

25. (11.5 and 11.6) Four numbers add to -2. The second number is three more than twice the negative of the first number. The third number is six less than the first number. The fourth number is eleven less than twice the first number. Find the numbers.

Answers to Selected Exercises

CHAPTER 1

Section 1.1 Exercises

1. concept

3. Yes, since it is a symbol that represents a number.

5. positional; 10   7. units, tens, hundreds   9. 4

11. 0   13. 0   15. ten thousand

17. ten million   19. 1,340 (answers may vary)

21. 900   23. yes; zero   25. graphing

27.
    0 1 2 3 4   29 30 31 32 33 34

29. 61, 99, 100, 102

Section 1.2 Exercises

1. nine hundred twelve

3. one thousand, four hundred ninety-one

5. thirty-five thousand, two hundred twenty-three

7. four hundred thirty-seven thousand, one hundred five

9. eight million, one thousand, one

11. seven hundred seventy million, three hundred eleven thousand, one hundred one

13. one hundred six billion, one hundred million, one thousand ten

15. eight hundred billion, eight hundred thousand

17. four; one thousand, four hundred sixty

19. twenty billion

21. four hundred twelve; fifty-two; twenty-one thousand, four hundred twenty-four

23. one thousand, nine hundred seventy-nine; eighty-five thousand; two million, nine hundred five thousand

25. one thousand, nine hundred eighty; two hundred seventeen

27. one thousand, nine hundred eighty-one; one million, nine hundred fifty-six thousand

29. one thousand, nine hundred eighty; thirteen thousand, one hundred

31. one thousand, nine hundred eighty-one; twelve million, six hundred thirty thousand

33. 681   35. 7,201   37. 512,003

39. 35,007,101   41. 16,000,059,004

43. 23,000,000,000   45. 100,000,000,000,001

47. 4   49. yes, zero

Section 1.3 Exercises

1. 1,600; 2,000; 0; 0   3. 91,800; 92,000; 90,000; 0

5. 200; 0; 0; 0   7. 900; 1,000; 0; 0

9. 900; 1,000; 0; 0   11. 1,000; 1,000; 0; 0

13. 551,061,300; 551,061,000; 551,060,000; 551,000,000

15. 106,999,413,200; 106,999,413,000; 106,999,410,000; 106,999,000,000

17. 8,006,000; 8,006,000; 8,010,000; 8,000,000

19. 33,500; 33,000; 30,000; 0

21. 388,600; 389,000; 390,000; 0

23. 8,200; 8,000; 10,000; 0   25. 19,310,000

27. 29,000,000   29. 70% or 75%

31. $5,500,000,000   33. 230,000   35. 5,400,000

37. graphing

39. forty-two thousand, one hundred nine

41. 4,000,000,008

Section 1.4 Exercises

1. 19   3. 48   5. 98   7. 978   9. 368

11. 12,777   13. 58,738   15. 45,169,739

17. 33   19. 81   21. 91   23. 809   25. 862

27. 2,321   29. 1,141,204   31. 562,364,111

33. 942,302,364,207,060   35. 234   37. 95,365

39. 1,972,128   41. 3,700   43. 3,101,500

45. 100   47. 0   49. 1,000   51. 5   53. 19

55. 88   57. 60,511,000

59. 5,682,651 square miles   61. 1,190,000

63. 271,564,000   65. 20   67. 25   69. 40

71. 50   73. 50   75. 50   77. 0   79. 6,800

Section 1.5 Exercises

1. 7   3. 6   5. 3   7. 41   9. 209

11. 21,001   13. 279,253   15. 77,472   17. 24

19. 188   21. 2,377   23. 26,686

25. 63,143,259   27. 8,034   29. 33

31. 32,611,636,567   33. 3,938   35. 8,273,955

37. 51   39. 3,405   41. 26   43. 72,069

45. 3,197   **47.** 29   **49.** 10,385   **51.** 15%

53. 11,247,000   **55.** 436   **57.** 57,187,000

59. 165,000   **61.** 74   **63.** 4,547   **65.** 11

67. 10   **69.** 12   **71.** 76   **73.** 51   **75.** 60

77. 165   **79.** 214   **81.** 330 (answers may vary)

83. 27,000,000

Section 1.6   Exercises

1. 37   **3.** 45   **5.** 568   **7.** 122,323   **9.** 45

11. 100   **13.** 556   **15.** 43,461

17. $132 + 6 = 80 + 58 = 138$

19. $987 + 171 = 731 + 427 = 1,158$ **21.** identity

23. $15 + 8 = 8 + 15 = 23$

25. . . . because its partner in addition remains identically the same after that addition.

27. two thousand, two hundred eighteen   **29.** 550

Chapter 1   Exercise Supplement

1. 937   **3.** 565   **5.** 559   **7.** 1,342   **9.** 2,001

11. 1,963   **13.** 79,456   **15.** 96,953

17. 791,824   **19.** 8,301   **21.** 140,381

23. 76,224   **25.** 4,955   **27.** 185,611   **29.** 2,238

31. 1,338   **33.** 878   **35.** 618,227   **37.** 1,621

39. 484,601   **41.** 19,853   **43.** 1,702   **45.** 1,114

47. 1,300   **49.** 2,718   **51.** 7,356   **53.** 3,415

55. 11,827   **57.** 407,262   **59.** 718,478

61. 9,941   **63.** $626 + 1,242 = 1,242 + 626 = 1,868$

Chapter 1   Proficiency Exam

1. 9   **2.** ones, tens, hundreds   **3.** 8   **4.** no

5.

6. sixty-three thousand, four hundred twenty-five

7. 18,359,072   **8.** 400   **9.** 19,000   **10.** 500

11. 675   **12.** 4,027   **13.** 188   **14.** 23,501

15. 90   **16.** 304   **17.** 70,123   **18.** 391   **19.** 182

20. yes, commutative property of addition

CHAPTER 2

Section 2.1   Exercises

1. 24   **3.** 48   **5.** 6   **7.** 225   **9.** 270

11. 582   **13.** 960   **15.** 7,695   **17.** 3,648

19. 46,488   **21.** 2,530   **23.** 6,888   **25.** 24,180

27. 73,914   **29.** 68,625   **31.** 511,173

33. 1,352,550   **35.** 5,441,712   **37.** 36,901,053

39. 24,957,200   **41.** 0   **43.** 41,384   **45.** 0

47. 73,530   **49.** 6,440,000,000   **51.** 4,440,000

53. 641,900,000   **55.** 80,000   **57.** 384 reports

59. 3,330 problems   **61.** 114 units

63. 3,600 seconds

65. 5,865,696,000,000 miles per year   **67.** $110,055

69. 448,100,000   **71.** 712

Section 2.2   Exercises

1. 8   **3.** 3   **5.** 9   **7.** 7   **9.** 4   **11.** 5

13. 7   **15.** not defined   **17.** 3   **19.** 0   **21.** 5

23. 8   **25.** 9   **27.** $27 \div 9 = 3; \ 9\overline{)27} = 3; \ \dfrac{27}{9} = 3$

29. 7 is quotient; 8 is divisor; 56 is dividend

31. 12,124   **33.** $(2 + 3) + 7 = 2 + (3 + 7) = 12$
$5 + 7 = 2 + 10 = 12$

Section 2.3   Exercises

1. 13   **3.** 67   **5.** 52   **7.** 3   **9.** 70   **11.** 61

13. 59   **15.** 67   **17.** 87   **19.** 54   **21.** 52

23. 38   **25.** 45   **27.** 777   **29.** 342   **31.** 644

33. 533   **35.** 10,440   **37.** 8,147 remainder 847

39. 4 remainder 2   **41.** 66 remainder 1

43. 823 remainder 2   **45.** 40 remainder 28

47. 665 remainder 15   **49.** 957 remainder 34

51. 665 remainder 4   **53.** 458 remainder 13

55. 996 remainder 23   **57.** 42   **59.** 8,216

61. $1,975 per month   **63.** $485 each person invested

65. 14 cubes per hour   **67.** 8 bits in each byte

69. 38   **71.** 600,000   **73.** 625,600

Section 2.4   Exercises

1. 2, 3, 4, 6, 8   **3.** 2, 3, 5, 6, 10   **5.** 2   **7.** 2, 4

9. 3   **11.** none   **13.** 5   **15.** 5   **17.** 2, 3, 4, 6

19. 2, 3, 5, 6, 10   **21.** none   **23.** 2, 4   **25.** 2

27. none   **29.** none   **31.** 1

33. $(35 + 16) + 7 = 51 + 7 = 58$ **35.** 87
$35 + (16 + 7) = 35 + 23 = 58$

Section 2.5   Exercises

1. 234   **3.** 4,032   **5.** 326,000   **7.** 252

9. 21,340   **11.** 8,316   **13.** $32 \cdot 2 = 64 = 4 \cdot 16$

15. $23 \cdot 1{,}166 = 26{,}818 = 253 \cdot 106$ **17.** associative

19. $7 \cdot 9 = 63 = 9 \cdot 7$ **21.** 6   **23.** $4 + 15 = 19$
$15 + 4 = 19$

25. 2, 3, 4, 6

Chapter 2   Exercise Supplement

1. factors; product   **3.** divisor; quotient

5. an even digit (0, 2, 4, 6, or 8)   **7.** divisible by 4

9. 112   **11.** 7   **13.** 3,045   **15.** 4

17. 15,075   **19.** 14   **21.** 42,112   **23.** 63

25. 1,046,960   **27.** 101,010,000   **29.** 0

31. 428   **33.** 5 remainder 9   **35.** not defined

37. 32 remainder 3   **39.** 3,072,202   **41.** 0

43. 0   **45.** not defined   **47.** 1   **49.** $226 \cdot 114$

51. $(16 \cdot 14) \cdot 0$ **53.** 4,278,000   **55.** \$33

Chapter 2   Proficiency Exam

1. 8 and 7 are factors; 56 is the product

2. addition   **3.** 3 is the divisor; 4 is the quotient

4. 0, 2, 4, 6, or 8   **5.** commutative   **6.** 1   **7.** 84

8. 0   **9.** 352,000   **10.** 419,020   **11.** 252

12. not defined   **13.** 0   **14.** 17   **15.** 18

16. 142 remainder 32   **17.** 211   **18.** 216; 1,005

19. 216; 640   **20.** 1,005; 640

CHAPTER 3

Section 3.1   Exercises

1. 4^2 **3.** 9^4 **5.** 826^3 **7.** 6^{85} **9.** 1^{3008}

11. $7 \cdot 7 \cdot 7 \cdot 7$ **13.** $117 \cdot 117 \cdot 117 \cdot 117 \cdot 117$

15. $30 \cdot 30$ **17.** $4 \cdot 4 = 16$ **19.** $10 \cdot 10 = 100$

21. $12 \cdot 12 = 144$ **23.** $15 \cdot 15 = 225$

25. $3 \cdot 3 \cdot 3 \cdot 3 = 81$ **27.** $10 \cdot 10 \cdot 10 = 1{,}000$

29. $8 \cdot 8 \cdot 8 = 512$ **31.** $9 \cdot 9 \cdot 9 = 729$ **33.** $7^1 = 7$

35. $2 \cdot 2 \cdot 2 \cdot 2 \cdot 2 \cdot 2 \cdot 2 = 128$

37. $8 \cdot 8 \cdot 8 \cdot 8 = 4{,}096$

39. $6 \cdot 6 \cdot 6 \cdot 6 \cdot 6 \cdot 6 \cdot 6 \cdot 6 \cdot 6 = 10{,}077{,}696$

41. $42 \cdot 42 = 1{,}764$

43. $15 \cdot 15 \cdot 15 \cdot 15 \cdot 15 = 759{,}375$

45. $816 \cdot 816 = 665{,}856$ **47.** 4   **49.** 8   **51.** 12

53. 15   **55.** 2   **57.** 6   **59.** 20   **61.** 100

63. 60   **65.** 34   **67.** 4,158   **69.** 24   **71.** 4

73. 5   **75.** 81

77. 8 is the multiplier; 4 is the multiplicand   **79.** yes; 0

Section 3.2   Exercises

1. 26   **3.** 46   **5.** 1   **7.** 0   **9.** 3   **11.** 26

13. 97   **15.** 29   **17.** 1   **19.** 0   **21.** 90

23. 508   **25.** 19   **27.** 144   **29.** 1   **31.** 52

33. 25,001   **35.** $\dfrac{1}{25}$ **37.** 14   **39.** 0   **41.** 152

43. $\dfrac{4}{5}$ **45.** 2,690,730   **47.** 1

Section 3.3   Exercises

1. 2   **3.** 4   **5.** 11   **7.** $3 \cdot 2$ **9.** $2 \cdot 3 \cdot 5$

11. 1, 2, 4, 8, 16   **13.** 1, 2, 4, 7, 8, 14, 28, 56

15. 1, 2, 4, 5, 10, 11, 20, 22, 44, 55, 110, 220

17. 1, 2, 4, 8, 16, 32   **19.** 1, 2, 71, 142   **21.** prime

23. composite   **25.** prime   **27.** prime

29. prime   **31.** composite $(5 \cdot 11)$ **33.** composite

35. composite   **37.** composite $(11 \cdot 19)$

39. composite   **41.** $2 \cdot 13$ **43.** $2 \cdot 3^3$ **45.** $2^3 \cdot 7$

47. $2^5 \cdot 3 \cdot 5$ **49.** $3^4 \cdot 5^2$ **51.** 26,580

53. true   **55.** 14

Section 3.4   Exercises

1. 2   **3.** 4   **5.** 4   **7.** 5   **9.** 33   **11.** 9

13. 11   **15.** 3   **17.** 7   **19.** 25   **21.** 2

23. 11   **25.** 1   **27.** 1   **29.** 53   **31.** $8^6 = 262{,}144$

Section 3.5   Exercises

1. 24   **3.** 40   **5.** 12   **7.** 18   **9.** 30

11. 12   **13.** 63   **15.** 72   **17.** 720   **19.** 120

21. 216   **23.** 144   **25.** 105   **27.** 231

29. 126   **31.** 4,410   **33.** 240   **35.** 432

37. 144   **39.** 224   **41.** 193,050   **43.** 8

45. 36   **47.** 6,552   **49.** $84 \cdot 84 \cdot 84$

Chapter 3   Exercise Supplement

1. 27   **3.** 0   **5.** 144   **7.** 64   **9.** 32

11. 225   **13.** 625   **15.** 15   **17.** 2   **19.** 1

21. 12   **23.** 1   **25.** 2   **27.** 0   **29.** 325

31. 64   **33.** -3 **35.** $-\dfrac{9}{57}$ **37.** 146   **39.** 5

41. The sum of square roots is not necessarily equal to the square root of the sum.

43. 6   **45.** 8   **47.** 1, 2, 3, 6, 9, 18   **49.** 1, 11

51. 1, 3, 17, 51   **53.** 1, 2   **55.** $5 \cdot 11$ **57.** $2^4 \cdot 5$

59. $2^2 \cdot 5^2 \cdot 7$ **61.** $2 \cdot 3 \cdot 269$

63. 29 is a prime number **65.** 5 **67.** 5 **69.** 6

71. 20 **73.** 21 **75.** 11 **77.** 42 **79.** 180

81. 120 **83.** 79, 380 **85.** 450

87. 1, 2, 3, 4, 6, 8, 12, 24

89. 1, 2, 4, 5, 7, 8, 10, 14, 20, 25, 35, 40, 50, 56, 70, 100, 140, 175, 200, 280, 700, 1,400

91. yes

Chapter 3 Proficiency Exam

1. base; exponent **2.** 12^7

3. $9^4 = 9 \cdot 9 \cdot 9 \cdot 9 = 6,561$ **4.** 64 **5.** 1 **6.** 0

7. 64 **8.** 7 **9.** 3 **10.** 1 **11.** 20

12. 127 **13.** 24 **14.** 8 **15.** 5 **16.** $3^2 \cdot 2$

17. $2^2 \cdot 17$ **18.** $2 \cdot 71$ **19.** prime

20. $2^2 \cdot 3^2 \cdot 13$ **21.** 4 **22.** 45

23. 1, 2, 3, 4, 6, 9, 12, 18, 36 **24.** 1, 2, 3, 6, 9, 18

25. Yes, because one of the (prime) factors of the number is 7.

26. Yes, because it is one of the factors of the number.

27. No, because the prime 13 is not a factor of any of the listed factors of the number.

28. 2,160 **29.** 5,320

CHAPTER 4

Section 4.1 Exercises

1. numerator, 3; denominator, 4

3. numerator, 1; denominator, 5

5. numerator, 7; denominator, 7

7. numerator, 0; denominator, 12

9. numerator, 18; denominator, 1

11. $\frac{4}{5}$ **13.** $\frac{15}{20}$ **15.** $\frac{91}{107}$ **17.** $\frac{658}{134}$

19. $\frac{92}{1,000,000}$ **21.** five ninths **23.** eight fifteenths

25. seventy-five one hundredths

27. nine hundred sixteen one thousand fourteenths

29. eighteen thirty-one thousand six hundred eighths

31. $\frac{1}{2}$ **33.** $\frac{4}{7}$

35.

37.

39. numerator, 1; denominator, 4; one fourth

41. numerator, 4; denominator, 3; four thirds

43. numerator, 2; denominator, 7; two sevenths

45. numerator, 1; denominator, 56; one fifty-sixth

47. numerator, 125; denominator, 4; one hundred twenty-five fourths

49. $3 + 11 = 11 + 3 = 14$ **51.** 7^5 **53.** 144

Section 4.2 Exercises

1. improper fraction **3.** proper fraction

5. mixed number **7.** improper fraction

9. mixed number **11.** mixed number

13. proper fraction **15.** mixed number **17.** $4\frac{2}{3}$

19. $8\frac{3}{4}$ **21.** 9 **23.** $13\frac{9}{12}$ or $13\frac{3}{4}$ **25.** $555\frac{5}{9}$

27. $36\frac{1}{2}$ **29.** $7\frac{29}{41}$ **31.** $\frac{33}{8}$ **33.** $\frac{61}{9}$ **35.** $\frac{115}{11}$

37. $\frac{26}{3}$ **39.** $\frac{107}{5}$ **41.** $\frac{209}{21}$ **43.** $\frac{9001}{100}$ **45.** $\frac{159}{8}$

47. . . . because it may be written as $5\frac{0}{n}$, where n is any positive whole number.

49. $\frac{1,652}{61}$ **51.** $\frac{2,436}{23}$ **53.** $\frac{20,419}{25}$

55. $\frac{48,803,620}{8,117}$ **57.** 1,009,020 **59.** 252

Section 4.3 Exercises

1. equivalent **3.** equivalent **5.** not equivalent

7. not equivalent **9.** not equivalent

11. equivalent **13.** not equivalent

15. not equivalent **17.** 6 **19.** 12 **21.** 20

23. 75 **25.** 48 **27.** 80 **29.** 18 **31.** 154

33. 1,472 **35.** 1,850 **37.** $\frac{4}{5}$ **39.** $\frac{3}{7}$ **41.** $\frac{2}{7}$

43. $\frac{2}{3}$ **45.** $\frac{5}{2}$ **47.** $\frac{5}{3}$ **49.** $\frac{7}{3}$ **51.** $\frac{8}{35}$

53. $\frac{5}{3}$ **55.** $\frac{18}{5}$ **57.** $\frac{2}{3}$ **59.** $\frac{3}{4}$ **61.** $\frac{1}{2}$

63. $\frac{1}{3}$ **65.** 3   **67.** 2   **69.** already reduced

71. $\frac{9}{25}$ **73.** $\frac{2}{5}$ **75.** $\frac{9}{8}$ **77.** $\frac{23}{30}$ **79.** $\frac{20}{9}$

81. $\frac{1}{3}$ **83.** $\frac{17}{4}$ **85.** $\frac{17}{18}$ **87.** $\frac{7}{12}$ **89.** 16

91. Should be $\frac{1}{8}$; the cancellation is division, so the numerator should be 1.

93. Cancel factors only, not addends; $\frac{7}{15}$ is already reduced.

95. Same as 91; answer is $\frac{1}{1}$ or 1.   **97.** 0   **99.** 6

Section 4.4   Exercises

1. $\frac{1}{4}$

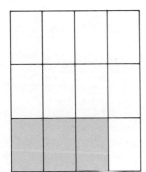

3. $\frac{1}{4}$

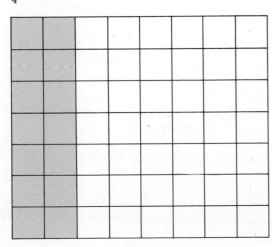

5. $\frac{1}{64}$

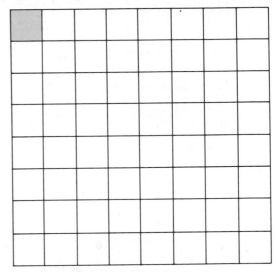

7. $\frac{2}{5}$ **9.** $\frac{2}{9}$ **11.** $\frac{4}{15}$ **13.** $\frac{3}{5}$ **15.** 2

17. $\frac{10}{3}$ or $3\frac{1}{3}$ **19.** 52   **21.** $\frac{4}{9}$ **23.** $\frac{1}{24}$

25. 126   **27.** $\frac{1}{4}$ **29.** $\frac{1}{3}$ **31.** $\frac{7}{9}$ **33.** 1

35. $\frac{28}{33}$ **37.** $\frac{4}{3}$ **39.** 12   **41.** $7\frac{13}{21}$ or $\frac{160}{21}$

43. 12   **45.** 18   **47.** $\frac{25}{3}$ or $8\frac{1}{3}$ **49.** $\frac{15}{8}=1\frac{7}{8}$

51. 6   **53.** $\frac{10}{9}=1\frac{1}{9}$ **55.** $\frac{9}{2}=4\frac{1}{2}$

57. $\frac{85}{6}=14\frac{1}{6}$ **59.** 72   **61.** $\frac{4}{9}$ **63.** $\frac{4}{121}$

65. $\frac{1}{4}$ **67.** $\frac{1}{15}$ **69.** $\frac{1}{25}$ **71.** $\frac{2}{3}$ **73.** $\frac{9}{11}$

75. $\frac{12}{5}=2\frac{2}{5}$ **77.** $\frac{1}{3}$ **79.** $\frac{7}{8}$ **81.** 2   **83.** yes

85. $\frac{6}{25}$

Section 4.5   Exercises

1. $\frac{5}{4}$ or $1\frac{1}{4}$ **3.** $\frac{9}{2}$ or $4\frac{1}{2}$ **5.** $\frac{4}{13}$ **7.** $\frac{7}{23}$ **9.** 1

11. $\frac{5}{8}$ **13.** $\frac{3}{10}$ **15.** $\frac{225}{196}$ or $1\frac{29}{196}$ **17.** $\frac{3}{5}$

19. 1   **21.** 1   **23.** $\frac{49}{100}$ **25.** $\frac{3}{5}$ **27.** $\frac{6}{7}$

29. $\frac{85}{6}$ or $14\frac{1}{6}$ **31.** $\frac{28}{18} = \frac{14}{9}$ or $1\frac{5}{9}$ **33.** 10

35. $\frac{10}{3}$ or $3\frac{1}{3}$ **37.** $\frac{4}{21}$ **39.** $\frac{3}{2}$ or $1\frac{1}{2}$ **41.** $\frac{4}{5}$

43. 3 **45.** 1 **47.** 321,600 **49.** 144

Section 4.6 Exercises

1. $\frac{1}{2}$ **3.** $\frac{1}{3}$ **5.** $\frac{9}{8}$ or $1\frac{1}{8}$ **7.** $\frac{1}{16}$ **9.** $\frac{3}{4}$

11. $\frac{1}{1,000}$ **13.** $\frac{10}{27}$ **15.** $\frac{1}{2}$ **17.** 2 **19.** $\frac{1}{6}$

21. $\frac{2}{3}$ **23.** 1 **25.** 1 **27.** $\frac{3}{2}$ or $1\frac{1}{2}$ **29.** 3

31. $\frac{5}{3}$ or $1\frac{2}{3}$ **33.** $\frac{27}{40}$ **35.** $\frac{1}{15}$ **37.** $\frac{3}{8}$ **39.** $\frac{9}{10}$

41. $\frac{9}{2} = 4\frac{1}{2}$ **43.** $\frac{16}{11}$ or $1\frac{5}{11}$ **45.** $\frac{30}{77}$

47. no **49.** $\frac{41}{12}$

Chapter 4 Exercise Supplement

1. $\frac{2}{6}$ or $\frac{1}{3}$ **3.** numerator, 4; denominator, 5

5. numerator, 1; denominator, 3 **7.** $\frac{8}{11}$

9. $\frac{200}{6,000}$ **11.** ten seventeenths

13. six hundred six, one thousand four hundred thirty-firsts

15. one sixteenth

17. numerator, 56; denominator, 14,190

19.

21. $2\frac{3}{4}$ **23.** $6\frac{3}{8}$ **25.** $118\frac{2}{3}$ **27.** $1\frac{1}{4}$

29. 3 **31.** $\frac{129}{8}$ **33.** $\frac{16}{5}$ **35.** $\frac{377}{21}$ **37.** $\frac{3}{2}$

39. $\frac{62}{7}$ **41.** because the whole number part is zero

43. equivalent **45.** not equivalent

47. not equivalent **49.** $\frac{8}{11}$ **51.** $\frac{5}{11}$ **53.** $\frac{3}{5}$

55. $\frac{9}{17}$ **57.** $\frac{35}{68}$ **59.** $\frac{65}{162}$ **61.** 15 **63.** 6

65. 27 **67.** 42 **69.** 168 **71.** 192 **73.** $\frac{3}{4}$

75. $\frac{1}{24}$ **77.** $\frac{5}{36}$ **79.** 1 **81.** $\frac{1}{48}$ **83.** $\frac{4}{35}$

85. $\frac{50}{7} = 7\frac{1}{7}$ **87.** 2 **89.** 3 **91.** $\frac{1}{2}$ **93.** 90

95. 1 **97.** $\frac{7}{2}$ or $3\frac{1}{2}$ **99.** 4 **101.** $\frac{12}{13}$

103. 4 **105.** $\frac{11}{12}$

Chapter 4 Proficiency Exam

1.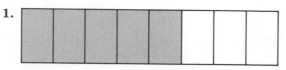

2. numerator, 5; denominator, 9 **3.** $\frac{5}{11}$

4. four fifths **5.** $\frac{5}{12}$ **6.** $\frac{25}{7}$ **7.** $3\frac{1}{5}$ **8.** yes

9. $\frac{3}{5}$ **10.** $\frac{5}{17}$ **11.** $\frac{13}{60}$ **12.** 20 **13.** 24

14. $\frac{3}{20}$ **15.** 55 **16.** $\frac{5}{6}$ **17.** $\frac{3}{4}$ **18.** $\frac{1}{2}$

19. $\frac{1}{30}$ **20.** $\frac{1}{11}$ **21.** $\frac{36}{25} = 1\frac{11}{25}$ **22.** $\frac{2}{5}$

23. $\frac{1}{4}$ **24.** $\frac{9}{5} = 1\frac{4}{5}$ **25.** $\frac{4}{3}$ or $1\frac{1}{3}$

CHAPTER 5

Section 5.1 Exercises

1. $\frac{5}{8}$ **3.** 1 **5.** $\frac{13}{15}$ **7.** 1 **9.** 0 **11.** 0

13. $\frac{9}{11}$ **15.** $\frac{15}{8}$ **17.** $\frac{1}{2}$ **19.** $\frac{3}{5}$ **21.** $\frac{13}{5}$

23. 10 **25.** $\frac{9}{11}$

27. $\frac{16}{30} = \frac{8}{5}$ (using the preposterous rule)

29. $\frac{13}{24}$ **31.** 2700 **33.** $3^2 \cdot 5 \cdot 11$ **35.** $\frac{2}{3}$

Section 5.2 Exercises

1. the same denominator **3.** $\frac{5}{8}$ **5.** $\frac{31}{24}$ **7.** $\frac{17}{28}$

9. $\dfrac{19}{36}$ **11.** $\dfrac{19}{39}$ **13.** $\dfrac{29}{60}$ **15.** $\dfrac{8}{81}$ **17.** $\dfrac{17}{65}$

19. $\dfrac{2}{63}$ **21.** $\dfrac{7}{16}$ **23.** $\dfrac{47}{18}$ **25.** $\dfrac{103}{30}$ **27.** $\dfrac{217}{264}$

29. $\dfrac{511}{720}$ **31.** $\dfrac{37}{72}$ **33.** $\dfrac{221}{150}$ **35.** $\dfrac{1,465}{2,016}$

37. $\dfrac{65}{204}$ **39.** $\dfrac{1}{5}$ **41.** $\dfrac{607}{180}$ **43.** $\dfrac{109}{520}$

45. $\$\dfrac{137}{8}$ or $\$17\dfrac{1}{8}$

47. No; 3 boxes add up to $26\dfrac{1''}{4}$, which is larger than $25\dfrac{1''}{5}$.

49. no pipe at all; inside diameter is greater than outside diameter

51. 449 **53.** 660 **55.** $\dfrac{7}{15}$

Section 5.3 Exercises

1. $7\dfrac{1}{2}$ **3.** $12\dfrac{1}{2}$ **5.** $21\dfrac{6}{11}$ **7.** $3\dfrac{7}{8}$ **9.** $1\dfrac{2}{3}$

11. $8\dfrac{11}{14}$ **13.** $4\dfrac{20}{21}$ **15.** $13\dfrac{17}{24}$ **17.** $2\dfrac{7}{12}$

19. $7\dfrac{17}{30}$ **21.** $74\dfrac{25}{42}$ **23.** $21\dfrac{1}{3}$ **25.** $5\dfrac{21}{32}$

27. $20\dfrac{1}{144}$ **29.** $7\dfrac{7}{12}$ **31.** $5\dfrac{13}{132}$ **33.** $1\dfrac{47}{112}$

35. $3\dfrac{1}{4}$ **37.** $2\dfrac{5}{8}$ gallons **39.** $8\dfrac{15}{16}$ pounds

41. $194\dfrac{3}{16}$ pounds **43.** 2 **45.** $\dfrac{7}{9}$

Section 5.4 Exercises

1. $\dfrac{3}{5}<\dfrac{5}{8}$ **3.** $\dfrac{3}{4}<\dfrac{5}{6}$ **5.** $\dfrac{3}{8}<\dfrac{2}{5}$ **7.** $\dfrac{1}{2}<\dfrac{4}{7}<\dfrac{3}{5}$

9. $\dfrac{3}{4}<\dfrac{7}{9}<\dfrac{5}{4}$ **11.** $\dfrac{3}{14}<\dfrac{2}{7}<\dfrac{3}{4}$ **13.** $5\dfrac{4}{7}<5\dfrac{3}{5}$

15. $9\dfrac{2}{3}<9\dfrac{4}{5}$ **17.** $1\dfrac{9}{16}<2\dfrac{1}{20}$ **19.** $2\dfrac{2}{9}<2\dfrac{3}{7}$

21. 270,000,000 **23.** $\dfrac{23}{7}$ **25.** $13\dfrac{5}{8}$ or $\dfrac{109}{8}$

Section 5.5 Exercises

1. 1 **3.** $\dfrac{3}{5}$ **5.** $\dfrac{5}{9}$ **7.** $\dfrac{5}{2}$ **9.** $\dfrac{31}{2}$ **11.** 7

13. 1 **15.** $\dfrac{1}{6}$ **17.** $\dfrac{52}{81}$ **19.** $\dfrac{16}{21}$ **21.** $\dfrac{686}{101}$

23. $\dfrac{1}{5}$ **25.** $8\dfrac{6}{7}$ **27.** $1\dfrac{13}{24}$ or $\dfrac{37}{24}$

Section 5.6 Exercises

1. $\dfrac{5}{4}$ **3.** $4\dfrac{2}{7}$ **5.** $\dfrac{2}{15}$ **7.** $\dfrac{1}{10}$ **9.** $6\dfrac{3}{14}$

11. $\dfrac{3}{8}$ **13.** $\dfrac{20}{27}$ **15.** 0 **17.** $\dfrac{2}{5}$ **19.** 1

21. $\dfrac{125}{72}$ **23.** $\dfrac{252}{19}$ **25.** $\dfrac{165}{256}$

27. multiplicative identity **29.** $\dfrac{241}{120}$ or $2\dfrac{1}{120}$

Chapter 5 Exercise Supplement

1. $\dfrac{11}{8}$ or $1\dfrac{3}{8}$ **3.** $\dfrac{1}{2}$ **5.** $\dfrac{59}{42}=1\dfrac{17}{42}$ **7.** $\dfrac{5}{8}$

9. $\dfrac{13}{21}$ **11.** $4\dfrac{7}{16}$ **13.** $8\dfrac{3}{20}$ **15.** $1\dfrac{3}{8}$ **17.** $24\dfrac{2}{3}$

19. $\dfrac{71}{12}=5\dfrac{11}{12}$ **21.** $\dfrac{13}{22}$ **23.** $11\dfrac{37}{60}$ **25.** $\dfrac{139}{144}$

27. $5\dfrac{1}{10}$ **29.** $\dfrac{1}{8}$ **31.** $\dfrac{11}{16}$ **33.** $8\dfrac{2}{11}$ **35.** 1

37. $7\dfrac{3}{10}$ **39.** $\dfrac{5}{9}$ **41.** $\dfrac{6}{7}$ **43.** $\dfrac{3}{2}$ or $1\dfrac{1}{2}$

45. $\dfrac{15}{28}$ **47.** $\dfrac{163}{108}$ or $1\dfrac{55}{108}$ **49.** $\dfrac{66}{13}$ or $5\dfrac{1}{13}$

51. $\dfrac{7}{40}$ **53.** $\dfrac{255}{184}$ or $1\dfrac{71}{184}$ **55.** $\dfrac{3}{32},\dfrac{1}{8}$

57. $\dfrac{3}{10},\dfrac{5}{6}$ **59.** $\dfrac{3}{8},\dfrac{8}{3},\dfrac{19}{6}$ **61.** $\dfrac{5}{9},\dfrac{4}{7}$

63. $\dfrac{5}{12},\dfrac{4}{9},\dfrac{7}{15}$ **65.** $\dfrac{5}{8},\dfrac{3}{4},\dfrac{13}{16}$

Chapter 5 Proficiency Exam

1. $\dfrac{5}{16}$ **2.** $7\dfrac{5}{6}$ **3.** 1 **4.** $\dfrac{8}{11}$ **5.** 8 **6.** $2\dfrac{13}{40}$

7. $\dfrac{49}{32}$ or $1\dfrac{17}{32}$ **8.** $\dfrac{7}{15}$ **9.** $3\dfrac{7}{16}$ **10.** $\dfrac{5}{32}$

11. $6\dfrac{1}{3}$ or $\dfrac{19}{3}$ **12.** $3\dfrac{3}{7}$ **13.** equivalent

14. not equivalent **15.** equivalent **16.** $\dfrac{6}{7},\dfrac{8}{9}$

17. $\dfrac{5}{8},\dfrac{7}{9}$ **18.** $11\dfrac{5}{16},11\dfrac{5}{12}$ **19.** $\dfrac{2}{15},\dfrac{1}{6},\dfrac{3}{10}$

20. $\dfrac{9}{16},\dfrac{19}{32},\dfrac{5}{8}$

CHAPTER 6

Section 6.1 Exercises

1. tenths; hundredths; thousandths

3. hundred thousandths; ten millionths

5. eight and one tenth

7. fifty-five and six hundredths

9. one and nine hundred four thousandths

11. 3.20 13. 1.8 15. 511.004 17. 0.947

19. 0.00071 21. seventy-five hundredths

23. four tenths 25. sixteen hundredths

27. one thousand eight hundred seventy-five ten thousandths

29. fifty-five hundredths 31. 2610 33. 12

35. $\dfrac{10}{9}$ or $1\dfrac{1}{9}$

Section 6.2 Exercises

1. $\dfrac{7}{10}$ 3. $\dfrac{53}{100}$ 5. $\dfrac{219}{1,000}$ 7. $4\dfrac{4}{5}$ 9. $16\dfrac{3}{25}$

11. $6\dfrac{1}{2,000}$ 13. $16\dfrac{1}{8}$ 15. $3\dfrac{1}{25}$ 17. $8\dfrac{9}{40}$

19. $9\dfrac{19,999}{20,000}$ 21. $\dfrac{3}{4}$ 23. $2\dfrac{13}{80}$ 25. $14\dfrac{337}{3,000}$

27. $1\dfrac{129}{320}$ 29. $2\dfrac{1}{25}$ 31. 14 33. $\dfrac{9}{10}$

35. thousandths

Section 6.3 Exercises

1. 20.0, 20.01, 20.011, 20.0107

3. 531.2, 531.22, 531.219, 531.2188

5. 2.0, 2.00, 2.000, 2.0000 7. 0.0, 0.00, 0.000, 0.0000

9. 9.2, 9.19, 9.192, 9.1919 11. 18.417

13. 18.41681 15. 18.42 17. 0.1 19. 0.83607

21. 5.333 23. ten million 25. 256 27. $3\dfrac{4}{25}$

Section 6.4 Exercises

1. 8.95 3. 39.846 5. 11.74931 7. 57.2115

9. 2.607 11. 0.41085 13. 9.135586

15. 27.351 17. 39.6660 19. 11.09 21. $23.23

23. $5.71 25. 869.37 27. 8,200,000

29. $\dfrac{20}{9} = \dfrac{5}{3}$ or $2\dfrac{2}{9}$

Section 6.5 Exercises

1. 31.28 3. 47.20 5. 0.152 7. 4.6324

9. 3.182 11. 0.0000273 13. 2.56 15. 0.81

17. 29.3045 19. 0.00000105486 21. 49.6

23. 4,218.842 25. 19.621 27. 3,596.168

29. 25,010 31. 28.382, 28.4, 28.38, 28.382

33. 134.216048, 134.2, 134.22, 134.216

35. 185.626, 185.6, 185.63, 185.626 37. 121.503

39. 1.2 41. 0.999702 43. 0.2528 45. 2.9544

47. $16.24 49. 0.24 51. 0.006099

53. 23.295102 55. 0.000144 57. 0.0000018

59. 0.0000000471 61. 0 63. $\dfrac{10}{77}$ 65. 1.78

Section 6.6 Exercises

1. 1.6 3. 3.7 5. 6.04 7. 9.38 9. 0.46

11. 8.6 13. 2.4 15. 6.21 17. 2.18

19. 1.001 21. 4 23. 14 25. 111

27. 643.51006 29. 0.064351006

31. 1.81, 1.8, 1.81, 1.810 33. 4.821, 4.8, 4.82, 4.821

35. 0.00351, 0.0, 0.00, 0.004 37. 0.01

39. 12.11 months 41. 27.2 miles per gallon

43. 3.7 45. 1.94 47. 29.120 49. 0.173

51. 1.111 53. $\dfrac{39}{8}$ 55. $\dfrac{47}{30}$ or $1\dfrac{17}{30}$ 57. 2.8634

Section 6.7 Exercises

1. $0.\overline{4}$ 3. 0.16 5. $0.\overline{142857}$ 7. 10.526

9. $0.1\overline{12}$ 11. $6.2\overline{1951}$ 13. $0.00\overline{81}$ 15. 0.835

17. $0.\overline{3}$ 19. $2.0\overline{5}$ 21. $0.\overline{7}$ 23. $0.\overline{518}$

25. $0.0\overline{45}$ 27. $0.\overline{72}$ 29. 0.7 31. 1

33. 4410 35. 8.6

Section 6.8 Exercises

1. 0.5 3. 0.875 5. 0.6 7. 0.04 9. 0.05

11. 0.02 13. $0.\overline{3}$ 15. 0.1875 17. $0.0\overline{37}$

19. $0.\overline{538461}$ 21. $7.\overline{6}$ 23. $1.1\overline{3}$ 25. 101.24

27. 0.24125 29. 810.31063125 31. 0.11111

33. 0.33333 35. 0.55556 37. 0.77778

39. 0.09091 41. 0.27273 43. 0.45455

45. 0.63636 47. 0.81818 49. 0.128

51. 0.9746 53. 0.0002 55. 38.7347 57. $\dfrac{15}{16}$

59. $3\frac{29}{40}$ or 3.725

Section 6.9   Exercises

1. 1   **3.** 0.112   **5.** 0.055   **7.** 0.1276   **9.** 0.7

11. 9.4625   **13.** 1.4075   **15.** 0.1875   **17.** 0.75

19. 0.615   **21.** 1.35   **23.** 0.125   **25.** $0.\overline{15}$

27. $2.\overline{6}$ **29.** yes   **31.** $2^3 \cdot 5^2 \cdot 7 \cdot 11$ **33.** $0.\overline{592}$

Chapter 6   Exercise Supplement

1. hundredths   **3.** seven and two tenths

5. sixteen and fifty-two hundredths

7. five thousandths   **9.** 9.12   **11.** 56.0035

13. 0.004081   **15.** $85\frac{63}{100}$ **17.** $\frac{11}{75}$ **19.** $4\frac{7}{675}$

21. 4.09   **23.** 17   **25.** 1   **27.** 8.946

29. 7.942   **31.** 0.03   **33.** 181,600.1   **35.** 1.4

37. 1.5079   **39.** $2.37\overline{2}$ **41.** 0.000   **43.** $0.1\overline{2}$

45. $2.\overline{702}$ **47.** 0.43   **49.** $9.\overline{571428}$ **51.** $1.\overline{3}$

53. 125.125125 (not repeating)   **55.** $0.08\overline{3}$

57. 0.255   **59.** 5.425   **61.** 0.09343

Chapter 6   Proficiency Exam

1. thousandth

2. fifteen and thirty-six thousandths

3. 81.12   **4.** 0.003017   **5.** $\frac{39}{50}$ **6.** $\frac{7}{8}$ **7.** 4.8

8. 200   **9.** 0.07   **10.** 20.997   **11.** 45.565

12. 77.76   **13.** 16   **14.** 2.4   **15.** 1.1256

16. 0   **17.** $6.1\overline{8}$ **18.** 0.055625   **19.** 5.45

20. $\frac{27}{128}$ or 0.2109375

CHAPTER 7

Section 7.1   Exercises

1. They are pure numbers or like denominate numbers.

3. rate   **5.** ratio   **7.** rate   **9.** ratio

11. two to five

13. 29 miles per 2 gallons or $14\frac{1}{2}$ miles per 1 gallon

15. 5 to 2   **17.** $\frac{12}{5}$ **19.** $\frac{42\ \text{plants}}{5\ \text{homes}}$ **21.** $\frac{16\ \text{pints}}{1\ \text{quart}}$

23. $\frac{2.54\ \text{cm}}{1\ \text{inch}}$ **25.** $\frac{5}{2}$ **27.** $\frac{1\ \text{hit}}{3\ \text{at bats}}$

29. $\frac{1,042\ \text{characters}}{1\ \text{page}}$ **31.** $5\frac{1}{3}$ **33.** $\frac{299}{1260}$

35. 3.3875

Section 7.2   Exercises

1. rates, proportion   **3.** $\frac{1}{11} = \frac{3}{33}$ **5.** $\frac{6}{90} = \frac{3}{45}$

7. $\frac{18\ \text{gr cobalt}}{10\ \text{gr silver}} = \frac{36\ \text{gr cobalt}}{20\ \text{gr silver}}$

9. $\frac{3\ \text{people absent}}{31\ \text{people present}} = \frac{15\ \text{people absent}}{155\ \text{people present}}$

11. 3 is to 4 as 15 is to 20

13. 3 joggers are to 100 feet as 6 joggers are to 200 feet

15. 40 miles are to 80 miles as 2 gallons are to 4 gallons

17. 1 person is to 1 job as 8 people are to 8 jobs

19. 2,000 pounds are to 1 ton as 60,000 pounds are to 30 tons

21. $x = 2$ **23.** $x = 8$ **25.** $x = 8$ **27.** $x = 27$

29. $x = 40$ **31.** $x = 30$ **33.** $x = 35$

35. $x = 30$ **37.** true   **39.** false   **41.** true

43. true   **45.** $5 + 7 = 12$ $7 + 5 = 12$ **47.** $\frac{17}{77}$

49. $\frac{8\ \text{sentences}}{1\ \text{paragraph}}$

Section 7.3   Exercises

1. 24   **3.** 45 inches   **5.** 33 parts   **7.** 2328.75

9. $\frac{21}{22}$ feet   **11.** 12,250   **13.** 12   **15.** $1\frac{5}{6}$

17. $1\frac{7}{8}$ **19.** $31\frac{1}{4}$ **21.** 0   **23.** $\frac{2}{3}$ **25.** 5

Section 7.4   Exercises

1. 25%   **3.** 48%   **5.** 77.1%   **7.** 258%

9. $1,618\frac{1}{4}$%   **11.** 200%   **13.** 0.15   **15.** 0.162

17. 0.0505   **19.** 0.0078   **21.** 0.0009   **23.** 20%

25. 62.5%   **27.** 28%   **29.** $49.\overline{09}$%   **31.** 164%

33. 945%   **35.** $54.\overline{54}$%   **37.** $\frac{4}{5}$ **39.** $\frac{1}{4}$

41. $\frac{13}{20}$ **43.** $\frac{1}{8}$ **45.** $\frac{41}{8}$ or $5\frac{1}{8}$ **47.** $\frac{1}{10}$

49. $\frac{2}{9}$ **51.** $\frac{4}{45}$ **53.** $\frac{129}{60}$ or $2\frac{9}{60} = 2\frac{3}{20}$

55. 36 inches

Section 7.5   Exercises

1. $\dfrac{3}{400}$ **3.** $\dfrac{1}{900}$ **5.** $\dfrac{5}{400}$ or $\dfrac{1}{80}$ **7.** $\dfrac{13}{700}$

9. $\dfrac{101}{400}$ **11.** $\dfrac{363}{500}$ **13.** $\dfrac{41}{30}$ **15.** $\dfrac{51}{500} = 0.102$

17. $\dfrac{31}{800} = 0.03875$ **19.** 0.004   **21.** 0.0627

23. 0.242   **25.** 0.1194   **27.** 8^5 **29.** $\dfrac{197}{210}$

31. 806.2%

Section 7.6   Exercises

1. 21.84   **3.** 534.1   **5.** 2.1   **7.** 51   **9.** 120

11. 20   **13.** 59   **15.** 91   **17.** 9.15   **19.** 568

21. 1.19351   **23.** 250   **25.** 0   **27.** 390.55

29. 84   **31.** 14   **33.** 160   **35.** $1,022.40

37. 36.28%   **39.** 61   **41.** 19,500   **43.** 14.85

45. marked correctly   **47.** $2\dfrac{4}{5}$ **49.** $4\dfrac{3}{500}$

Chapter 7   Exercise Supplement

1. 250 watts are 150 watts more than 100 watts

3. 98 radishes are 2.39 times as many radishes as 41 radishes

5. 100 tents are 20 times as many tents as 5 tents

7. ratio   **9.** $\dfrac{3}{1}$ **11.** $\dfrac{8\text{ ml}}{5\text{ ml}}$ **13.** 9 to 16

15. 1 diskette to 8 diskettes

17. 9 is to 16 as 18 is to 32

19. 8 items are to 4 dollars as 2 items are to 1 dollar

21. $\dfrac{20}{4} = \dfrac{5}{1}$

20 people are to 4 seats as 5 people are to 1 seat

23. 28   **25.** 10   **27.** 406   **29.** 80

31. $\dfrac{320}{3}$ or $106\dfrac{2}{3}$ **33.** 10   **35.** 16%

37. 535.36%   **39.** 300%   **41.** 0.0158

43. 0.0006   **45.** 0.053   **47.** 0.8286   **49.** 60%

51. 31.25%   **53.** 656.25%   **55.** $3688.\overline{8}\%$

57. $\dfrac{3}{25}$ **59.** $\dfrac{61}{160}$ **61.** $\dfrac{1}{160}$ **63.** $\dfrac{2977}{19800}$

65. 6.4   **67.** 0.625 or $\dfrac{5}{8}$ **69.** 9.588   **71.** 85

Chapter 7   Proficiency Exam

1. $\dfrac{4\text{ cassette tapes}}{7\text{ dollars}}$ **2.** rate   **3.** $\dfrac{11}{9}$

4. $\dfrac{5\text{ televisions}}{2\text{ radios}}$ **5.** 8 maps to 3 people

6. two psychologists to seventy-five people

7. 15   **8.** 1   **9.** 64   **10.** $37\dfrac{1}{2}$ **11.** 1.8

12. 21   **13.** 82%   **14.** $377\dfrac{7}{9}\%$ **15.** 0.02813

16. 0.00006   **17.** 840%   **18.** 12.5%

19. 1,000%   **20.** $\dfrac{3}{20}$ **21.** $\dfrac{4}{2,700}$ or $\dfrac{1}{675}$

22. 4.68   **23.** 206   **24.** $9.\overline{09}$ **25.** $1,660

CHAPTER 8

Section 8.1   Exercises

1. about 3,600; in fact 3,600

3. about 1,700; in fact 1,717

5. about 14,000; in fact 14,006

7. about 3,500; in fact 3,539

9. about 5,700; in fact 5,694

11. about 1,500; in fact 1,696

13. about 540,000; in fact 559,548

15. about 583,200,000; in fact 583,876,992

17. about 15; in fact 15.11

19. about 20; in fact 22

21. about 33; in fact 33.86

23. about 93.2; in fact 93.22

25. about 70; in fact 69.62

27. about 348.6; in fact 348.57

29. about 1,568.0; in fact 1,564.244

31. about 49.5; in fact 49.60956

33. about 43,776; in fact 43,833.258

35. about 6.21; in fact 6.0896

37. about 0.0519; in fact 0.05193

39. about 6.3; in fact 6.5193

41. about 4.5; in fact 4.896

43. about 0.8; in fact 0.74124

45. about 0.00008; in fact 0.000078

47. $\dfrac{23}{25}$ **49.** 60 feet tall

Section 8.2   Exercises

1. $2(30) + 2(50) = 160 \ (157)$

3. $2(90) + 2(60) = 300 \ (299)$

5. $3(20) + 90 = 150 \ (150)$ **7.** $3(90) + 30 = 300 \ (303)$

9. $40 + 3(60) = 220 \ (221)$

11. $3(60) + 2(30) = 240 \ (242)$

13. $3(100) + 2(70) = 440 \ (437)$

15. $3(20) + 3(90) = 330 \ (337)$

17. $2(500) + 2(700) = 2{,}400 \ (2{,}421)$

19. $2(20) + 4(80) = 360 \ (360)$ **21.** $7, 8, 9$

23. $\dfrac{3}{50}$ **25.** $4{,}900 + 2{,}700 = 7{,}600 \ (7{,}586)$

Section 8.3   Exercises

1. $15(10 + 3) = 150 + 45 = 195$

3. $25(10 + 1) = 250 + 25 = 275$

5. $15(20 - 4) = 300 - 60 = 240$

7. $45(80 + 3) = 3600 + 135 = 3735$

9. $25(40 - 2) = 1{,}000 - 50 = 950$

11. $75(10 + 4) = 750 + 300 = 1{,}050$

13. $65(20 + 6) = 1{,}300 + 390 = 1{,}690$ or $65(30 - 4) =$ $1{,}950 - 260 = 1{,}690$

15. $15(100 + 7) = 1{,}500 + 105 = 1{,}605$

17. $35(400 + 2) = 14{,}000 + 70 = 14{,}070$

19. $95(10 + 2) = 950 + 190 = 1{,}140$

21. $80(30 + 2) = 2{,}400 + 160 = 2{,}560$

23. $50(60 + 3) = 3{,}000 + 150 = 3{,}150$

25. $40(90 - 1) = 3{,}600 - 40 = 3{,}560$

27. $\dfrac{3}{26}$ **29.** $x = 42$

Section 8.4   Exercises

1. $1 + 1 = 2 \left(1\dfrac{17}{24}\right)$ **3.** $1 + \dfrac{1}{2} = 1\dfrac{1}{2} \left(1\dfrac{1}{2}\right)$

5. $\dfrac{1}{4} + \dfrac{1}{4} = \dfrac{1}{2} \left(\dfrac{39}{100}\right)$ **7.** $1 + 0 = 1 \left(1\dfrac{1}{48}\right)$

9. $\dfrac{1}{2} + 6\dfrac{1}{2} = 7 \left(6\dfrac{103}{132}\right)$ **11.** $1 + 2\dfrac{1}{2} = 3\dfrac{1}{2} \left(3\dfrac{11}{40}\right)$

13. $8\dfrac{1}{2} + 4 = 12\dfrac{1}{2} \left(12\dfrac{13}{20}\right)$ **15.** $9 + 7 = 16 \left(15\dfrac{13}{15}\right)$

17. $3\dfrac{1}{2} + 2\dfrac{1}{2} + 2 = 8 \left(7\dfrac{189}{200}\right)$ **19.** $1 - 1 = 0 \left(\dfrac{1}{16}\right)$

21. associative   **23.** $\dfrac{8}{9}$

25. $25(40 - 3) = 1000 - 75 = 925$

Chapter 8   Exercise Supplement

1. $600 \ (598)$ **3.** $(1{,}200)$ **5.** $4{,}900 \ (4{,}944)$

7. $3{,}800 \ (3{,}760)$ **9.** $6{,}050 \ (6{,}017)$

11. $6{,}700 \ (6{,}718)$ **13.** $61{,}400 \ (61{,}337)$

15. $89{,}700 \ (89{,}718)$ **17.** $213{,}000 \ (213{,}022)$

19. $3{,}900 \ (3{,}935)$ **21.** $2{,}400 \ (2{,}583)$

23. $600 \ (504)$ **25.** $150{,}123 \ \ 147{,}000 \ (150{,}123)$

27. $47{,}278 \ \ 48{,}800 \ (47{,}278)$

29. $12{,}771 \ \ 14{,}100 \ (12{,}771)$ **31.** $8{,}100 \ (9{,}768)$

33. $90{,}000 \ (86{,}784)$ **35.** $396{,}000 \ (386{,}496)$

37. $1{,}609{,}048 \ \ 1{,}560{,}000 \ (1{,}609{,}048)$

39. $22{,}050{,}000 \ (22{,}164{,}232)$ **41.** $18 \ (15)$

43. $14\dfrac{1}{2} \ (13)$ **45.** $50 \ (53)$ **47.** $54 \ (51)$

49. $130 \ (105)$ **51.** $41.25 \ (44)$ **53.** $42 \ (43)$

55. $14.4 \ (14)$ **57.** $60 \ (61)$ **59.** $1{,}000 \ (992)$

61. $71.1 \ (71.17)$ **63.** $119.4 \ (119.31)$

65. $272 \ (271.8)$ **67.** $2{,}001 \ (2{,}020.952)$

69. $4.16 \ (3.9596)$ **71.** $2(40) + 2(50) = 180 \ (178)$

73. $2(30) + 2(60) = 180 \ (179)$

75. $2(80) + 2(50) = 260 \ (263)$

77. $3(20) + 1(10) = 70 \ (79)$

79. $2(15) + 2(80) = 190 \ (201)$

81. $3(20) + 2(30) + 40 = 160 \ (160)$

83. $2(60) + 2(20) + 2(30) = 220 \ (219)$

85. $2(90) + 20 = 200 \ (200)$

87. $3(20) + 2(40) = 140 \ (140)$

89. $3(90) + 2(30) = 330 \ (321)$

91. $15(30 + 3) = 450 + 45 = 495$

93. $35(40 - 4) = 1400 - 140 = 1{,}260$

95. $85(20 + 3) = 1{,}700 + 255 = 1{,}955$

97. $30(10 + 4) = 300 + 120 = 420$

99. $75(20 + 3) = 1{,}500 + 225 = 1{,}725$

101. $15(20 - 3) = 300 - 45 = 255$

103. $65(10 + 4) = 650 + 260 = 910$

105. $60(40 + 2) = 2{,}400 + 120 = 2{,}520$

107. $15(100 + 5) = 1{,}500 + 75 = 1{,}575$

109. $45(300 + 6) = 13{,}500 + 270 = 13{,}770$

111. $\frac{1}{2} + 1 = 1\frac{1}{2}\left(1\frac{5}{24}\right)$

113. $\frac{1}{2} + \frac{1}{2} = 1\left(\frac{27}{30} \text{ or } \frac{9}{10}\right)$ **115.** $\frac{1}{2} + \frac{1}{4} = \frac{3}{4}\left(\frac{113}{150}\right)$

117. $\frac{1}{4} + 1 = 1\frac{1}{4}\left(\frac{39}{32} \text{ or } 1\frac{7}{32}\right)$

119. $2\frac{3}{4} + 6\frac{1}{2} = 9\frac{1}{4}\left(9\frac{7}{20}\right)$

121. $11\frac{1}{4} + 7\frac{1}{2} = 18\frac{3}{4}\left(18\frac{23}{30}\right)$

123. $6 + 2 + 8\frac{1}{4} = 16\frac{1}{4}\left(16\frac{11}{30}\right)$

125. $10\frac{1}{2} + 7 + 8\frac{1}{4} = 25\frac{3}{4}\left(25\frac{27}{40}\right)$

Chapter 8   Proficiency Exam

1. 10,500 (10,505)   **2.** 18,000 (18,038)

3. 5,000 (5,011)   **4.** 28,000 (27,406)

5. 3,200 (3,116)   **6.** 4,500 (4,692)

7. 300,000 (288,576)   **8.** 12 (14)   **9.** 90 (95)

10. 81 (81.00)   **11.** 505 (505.0065)

12. 4,371 (4,380.16)   **13.** $2(60) + 2(90) = 300$ (298)

14. $2(40) + 2(90) = 260$ (263)

15. $30 + 3(80) = 270$ (265)

16. $3(800) + 2(600) = 3,600$ (3,618)

17. $25(10 + 4) = 250 + 100 = 350$

18. $15(80 + 3) = 1,200 + 45 = 1,245$

19. $65(100 - 2) = 6,500 - 130 = 6,370$

20. $80(100 + 7) = 8,000 + 560 = 8,560$

21. $400(200 + 15) = 80,000 + 6,000 = 86,000$

22. $1 + \frac{1}{2} = 1\frac{1}{2}\left(1\frac{9}{16}\right)$ **23.** $0 + \frac{1}{2} + \frac{1}{2} = 1\left(1\frac{47}{300}\right)$

24. $8\frac{1}{2} + 14 = 22\frac{1}{2}\left(22\frac{31}{48}\right)$

25. $5\frac{1}{2} + 1\frac{1}{2} + 6\frac{1}{2} = 13\frac{1}{2}\left(13\frac{1}{3}\right)$

CHAPTER 9

Section 9.1   Exercises

1. 42 feet   **3.** 506,880 inches   **5.** 1.5 feet

7. 0.14 yard   **9.** 0.00 miles (to two decimal places)

11. 6 pints   **13.** 192,000 ounces   **15.** 937.5 pounds

17. 27 teaspoons   **19.** 80 fluid ounces

21. 0.16 quart   **23.** 480 teaspoons

25. 1,080 seconds   **27.** $\frac{1}{8} = 0.125$ day

29. $\frac{1}{14} = 0.0714$ week   **31.** 1, 2, 4, 5, 8

33. $\frac{11}{30}$ **35.** $60(50 - 4) = 3,000 - 240 = 2,760$

Section 9.2   Exercises

1. 8,700 cm   **3.** 16.005 g   **5.** 11,161 L

7. 126 dg   **9.** 5.1 daL   **11.** 0.5 dm

13. 81,060 cg   **15.** 0.03 m   **17.** 4,000 mg

19. 6,000,000 mg   **21.** $\frac{25}{24} = 1\frac{1}{24}$

23. 12,300 (12,344)   **25.** $0.08\overline{3}$ yard

Section 9.3   Exercises

1. 1 foot 4 inches   **3.** 1 hour 25 minutes

5. 2 weeks 3 days   **7.** 15 pounds

9. 6 gallons 2 quarts   **11.** 8 pounds 7 ounces

13. 2 gallons 1 quart

15. 16 liters 300 milliliters (or 1 daL 6 L 3 dL)

17. 15 days 11 hours   **19.** 59 pounds 9 ounces

21. 1 foot 10 inches   **23.** 1 hour 18 minutes

25. 5 days 16 hours 5 minutes

27. 1 ton 1,100 pounds (or 1 T 1,100 lb)

29. 2 weeks 23 hours 29 minutes 53 seconds

31. 1   **33.** $2\frac{14}{275}$ **35.** 126,000 g

Section 9.4   Exercises

1. 21.8 cm   **3.** 38.14 inches   **5.** 0.86 m

7. 87.92 m   **9.** 16.328 cm   **11.** 0.0771 cm

13. 120.78 m   **15.** 21.71 inches   **17.** 43.7 mm

19. 45.68 cm   **21.** 8.5 or $\frac{17}{2}$ or $8\frac{1}{2}$ **23.** 0.875

25. 1 hour 36 minutes 6 seconds

Section 9.5   Exercises

1. 16 sq m   **3.** 1.21 sq mm   **5.** 18 sq in.

7. $(60.5\,\pi + 132)$ sq ft   **9.** 40.8 sq in.

11. 31.0132 sq in.   **13.** 158.2874 sq mm

15. 64.2668 sq in.   **17.** 43.96 sq ft

19. 512 cu cm   **21.** 11.49 cu cm   **23.** $\dfrac{1024}{3}\pi$ cu ft

25. 22.08 cu in.   **27.** 4   **29.** $\dfrac{31}{12}=2\dfrac{7}{12}=2.58$

31. 27.9 m

Chapter 9   Exercise Supplement

1. Measurement is comparison to a standard (unit of measure).

3. 2 pounds   **5.** 6 tons   **7.** 69 feet

9. $\dfrac{2}{3}=0.66\overline{6}$ feet   **11.** 6 pints

13. 80 tablespoons   **15.** 210 seconds

17. $\dfrac{1}{4}=0.25$ L   **19.** 19,610 mg   **21.** 540.06 g

23. 3,500,000 mL   **25.** 6 weeks 1 day

27. 5 gallons 2 quarts 1 pint   **29.** 7 T 1,850 pounds

31. 7 days, 11 hours, 56 minutes, 7 seconds

33. 1 yard 1 foot   **35.** 1 ton 1,100 pounds

37. 1 tablespoon 1 teaspoon   **39.** 3 quarts 1 pint

41. 2.3 meters   **43.** 1.05 km   **45.** 5.652 sq cm

47. 104.28568 cu ft   **49.** 0.18π sq in.

51. 267.94667 cu mm   **53.** 32 cu cm

55. 39.48 sq in.   **57.** 56.52 sq ft

Chapter 9   Proficiency Exam

1. measurement   **2.** 42 feet   **3.** 612 inches

4. 1 foot   **5.** 135 seconds   **6.** 850 cg

7. 0.0058623 kL   **8.** 0.2131062 m

9. 100,001,000,000 mL   **10.** 3 weeks 2 days

11. 29 yards 1 foot   **12.** 2 tons 216 pounds

13. 1 gallon 3 quarts   **14.** 8 weeks 5 days

15. 14 gallons 2 quarts   **16.** 2 yards 5 feet 9 inches

17. 29 minutes 56 seconds   **18.** 3 weeks 5 days

19. 34.44 m   **20.** 36 mm   **21.** 87.92 feet

22. 55.14 miles   **23.** 3.75 sq in.

24. 6.002826 sq cm   **25.** 6.28 sq miles

26. 13 sq in.   **27.** 84.64π sq in.   **28.** 25.12 cu mm

29. 4.608π cu ft   **30.** 340.48 cu mm

CHAPTER 10

Section 10.1   Exercises

1. N, W, Z **3.** W, Z **5.** Z **7.** Z **9.** neither

11.

$$-3 \le x < 4$$

13. yes, 10   **15.** $<$ **17.** $>$

19. $\{-1, 0, 1, 2, 3, 4, 5\}$ **21.** $\{1\}$ **23.** $\{0, 1, 2, 3, 4\}$

25. 4   **27.** 8   **29.** yes   **31.** yes   **33.** yes, -1

35. $\dfrac{9}{5}$ or $1\dfrac{4}{5}$ or 1.8   **37.** 0.3006 m

Section 10.2   Exercises

1. + (or no sign)   **3.** negative seven   **5.** fifteen

7. negative negative one, or opposite negative one

9. five plus three   **11.** fifteen plus negative three

13. negative seven minus negative two   **15.** 2

17. -8 **19.** $7 + 3 = 10$ **21.** $0.\overline{296}$

23. $6,000 + 9,000 = 15,000$ ($5,829 + 8,767 = 14,596$) or $5,800 + 8,800 = 14,600$

25. 0.0025 hm

Section 10.3   Exercises

1. 5   **3.** 6   **5.** 1   **7.** -3 **9.** -14 **11.** 26

13. 4   **15.** 6   **17.** 3   **19.** 6   **21.** 100

23. 92   **25.** -1 **27.** $-\$|-2,400,000|$ **29.** $\dfrac{9}{10}$

31. $3\dfrac{13}{50}$ or $\dfrac{163}{50}$ **33.** 2

Section 10.4   Exercises

1. 16   **3.** -15 **5.** 8   **7.** -25 **9.** -12

11. 24   **13.** -21 **15.** 0   **17.** 0   **19.** 23

21. 328   **23.** 876   **25.** $-1,255$ **27.** -6.084

29. $-\$28.50$ **31.** $\$3.00$ **33.** $\dfrac{17}{36}$ **35.** $\dfrac{62}{100}=\dfrac{31}{50}$

Section 10.5   Exercises

1. 5   **3.** -1 **5.** -14 **7.** -2 **9.** -8

11. 7   **13.** 29   **15.** -324 **17.** -429

19. -71 **21.** 164   **23.** -11 **25.** 16.022

27. 31.25%   **29.** -5

Section 10.6   Exercises

1. 16   **3.** 32   **5.** -36 **7.** -60 **9.** -12

11. 3   **13.** -13 **15.** 9   **17.** -5 **19.** 11

21. 28   **23.** -4 **25.** -12 **27.** -4 **29.** -6

31. 15 **33.** 49 **35.** -140 **37.** -7 **39.** -3

41. -5 **43.** 13 **45.** $\frac{4}{3} = 1\frac{1}{3}$ **47.** $52\frac{1}{2}$

Chapter 10 Exercise Supplement

1. N, W, Z **3.** W, Z **5.** $\{-3, -2, -1, 0, 1, 2\}$

7. $<$ **9.** $<$ **11.** $\{-5, -4, -3, -2, -1\}$

13. $\{1, 2, 3, 4\}$ **15.** none **17.** 4 **19.** 6

21. $+$ (plus) **23.** negative eight

25. negative negative one or opposite negative one

27. one plus negative seven

29. negative one minus four

31. zero minus negative eleven **33.** 15

35. 19 or $1 + 18$ **37.** 9 **39.** 5 **41.** -2

43. 12 **45.** 16 **47.** 16 **49.** 8 **51.** 7

53. -2 **55.** -7 **57.** -14 **59.** 48

61. -42 **63.** 3 **65.** 1 **67.** not defined

69. -4 **71.** -22

Chapter 10 Proficiency Exam

1. $\{-7, -6, -5, -4\}$ **2.** $\{-2, -1, 0, 1\}$ **3.** $=$

4. $<$ **5.** $>$ **6.** $\{-3, -2, -1\}$ **7.** $\{1, 2\}$ **8.** 5

9. 16 **10.** -2 **11.** 16 **12.** 15 **13.** -2

14. -11 **15.** -16 **16.** 42 **17.** 121

18. $\frac{51}{7}$ or $7\frac{2}{7}$ **19.** 6 **20.** 0

CHAPTER 11

Section 11.1 Exercises

1. addition; multiplication **3.** $5x, 18y$

5. $8s, 2r, -7t$ **7.** $7a, -2b, -3c, -4d$ **9.** $-x, -y$

11. m **13.** 5 **15.** -30 **17.** -3 **19.** -14

21. 56 **23.** 64 **25.** 1 **27.** 75 **29.** 24

31. 39 **33.** 5 **35.** -7 **37.** 5

39. $\frac{181}{24} = 7\frac{13}{24}$ **41.** $\frac{20}{27}$ **43.** 0

Section 11.2 Exercises

1. $11a$ **3.** $4h$ **5.** $3m + 3n$ **7.** $17s - r$

9. $5h + 9a - 18k$ **11.** 0 **13.** $16m$

15. $8h + 7k$ **17.** $5\triangle - 3\star$ **19.** $16x - 15y - 9$

21. $2\frac{2}{11}$ **23.** 7 **25.** 6

Section 11.3 Exercises

1. Substitute $x = 4$ into the equation $4x - 11 = 5$.
$16 - 11 = 5$
$5 = 5$
$x = 4$ is a solution.

3. Substitute $m = 1$ into the equation $2m - 1 = 1$.
$2 - 1 \overset{?}{=} 1$
$1 \overset{\checkmark}{=} 1$
$m = 1$ is a solution.

5. Substitute $x = -8$ into the equation $3x + 2 - 7 = -5x - 6$.
$-24 + 2 - 7 \overset{?}{=} 40 - 6$
$34 \overset{\checkmark}{=} 34$
$x = -8$ is a solution.

7. Substitute $x = 0$ into the equation $-8 + x = -8$.
$-8 + 0 \overset{?}{=} -8$
$-8 \overset{\checkmark}{=} -8$
$x = 0$ is a solution.

9. Substitute $8 = \frac{15}{2}$ into the equation $4x - 5 = 6x - 20$.
$30 - 5 \overset{?}{=} 45 - 20$
$25 \overset{\checkmark}{=} 25$
$x = \frac{15}{2}$ is a solution.

11. $y = 11$ **13.** $k = 5$ **15.** $a = -9$ **17.** $x = 11$

19. $z = 0$ **21.** $x = 8$ **23.** $m = 1$ **25.** $k = 19$

27. $m = 10$ **29.** $b = -21$ **31.** $a = 16$

33. $y = 7.224$ **35.** $m = -0.778$ **37.** $k = -9.01$

39. $y = -28.06$ **41.** rate **43.** 3.5

45. $13x + 18y$

Section 11.4 Exercises

1. $x = 6$ **3.** $x = 12$ **5.** $a = -8$ **7.** $y = 14$

9. $m = -31$ **11.** $x = 28$ **13.** $z = 84$

15. $m = -4$ **17.** $x = -1$ **19.** $y = \frac{13}{2}$

21. $k = -2$ **23.** $x = \frac{11}{2}$ or $5\frac{1}{2}$ **25.** $a = -\frac{5}{4}$

27. $m = -1$ **29.** $x = 0$ **31.** $a = 14$ **33.** $x = 6$

35. $a = 4$ **37.** $40(30 - 2) = 1200 - 80 = 1120$

39. 220 sq cm **41.** $x = 4$

Section 11.5 Exercises

1. $x - 12$ **3.** $x - 4$ **5.** $x + 1$ **7.** $-7 + x$

9. $x + (-6)$ **11.** $x - [-(-1)]$ **13.** $3x + 11$

15. $2x - 7 = 4$ **17.** $14 + 2x = 6$

19. $\frac{3}{5}x + 8 = 50$ **21.** $\frac{4}{3}x + 12 = 5$

23. $8x + 5 = x + 26$ **25.** $\dfrac{x}{6} + 9 = 1$

27. $\dfrac{x}{7} + 2 = 17$ **29.** $x - 2 = 10$

31. $x - 3 = 2x - 6$ **33.** $5x - 12 = x - 2$

35. $6 - x = x + 4$ **37.** $9 - x + 1 = 3x + 8$

39. $2x - 7 - 4x = -x - 6$ **41.** $\dfrac{3}{4}$ **43.** $8\dfrac{2}{3}$

45. $x = -8$

Section 11.6 Exercises

  1. 24 **3.** 9 **5.** 10 **7.** 6 **9.** 1

11. length = 12 inches, **13.** 15, 16, 17
    width = 6 inches

15. $-2, -1, 0, 1, 2$ **17.** length is 7, width is 2

19. length is 15, width is 22

21. The perimeter is 20 inches. Other answers are
    possible. For example, perimeters such as 26, 32 are
    possible.

23. 15 meters **25.** $n = 0$ **27.** $n = 1$

29. $-2, -1, 0, 1$ **31.** $-35, -33, -31$

33. . . . because the sum of any even number (in this
    case, 2) of even integers (consecutive or not) is even
    and, therefore, cannot be odd (in this case, 139).

35. 2, 15, 20 **37.** 6 feet tall **39.** 46.3 cg

Chapter 11 Exercise Supplement

  1. $6a, -2b, 5c$ **3.** $7m, -3n$ **5.** $x, 2n, -z$

  7. $-y, -3z$ **9.** -4 **11.** k **13.** 7 **15.** -5

17. -27 **19.** 1 **21.** -5 **23.** 0 **25.** 5

27. 18 **29.** 20 **31.** 11 **33.** -1 **35.** -93

37. 45 **39.** 55 **41.** 10 **43.** -6 **45.** -14

47. $2a + 6$ **49.** $-4n + 4m - 3$ **51.** $-15x + 12y$

53. $3k + 2m - 6s$ **55.** $38h - 7k$ **57.** $x = 4$

59. $x = -2$ **61.** $x = 5$ **63.** $x = 12$

65. $x = -15$ **67.** $x = -3$ **69.** $x = 1$

71. $x = -27$ **73.** $y = -\dfrac{7}{3}$ **75.** $m = -\dfrac{8}{5}$

77. $x = \dfrac{1}{15}$ **79.** $s = 1$ **81.** $x = -3$

83. $x = -4$ **85.** $x = -\dfrac{1}{2}$ **87.** $a = -5$

89. $x = 15$ **91.** $x = 5$ **93.** $a = \dfrac{15}{2}$ **95.** $x = \dfrac{8}{5}$

97. $x = 2$ **99.** $x = 5$ **101.** $y = -12$

103. $w = \dfrac{37}{4}$ **105.** $x = 5$ **107.** $x = -24$

109. $w = 38$ **111.** $x = \dfrac{39}{2}$ **113.** $x = -\dfrac{35}{2}$

115. $y = -3$ **117.** $x = \dfrac{6}{5}$ **119.** $y = \dfrac{2}{3}$

121. $m = -3$ **123.** $a = -24$ **125.** $x = \dfrac{1}{2}$

127. $x = 32$ **129.** $x = -2$ **131.** $x = 4$

133. $x = 6$ **135.** $x = \dfrac{-32}{3}$ **137.** $y = \dfrac{130}{3}$

139. $m = \dfrac{2}{13}$

Chapter 11 Proficiency Exam

  1. $5x, 6y, 3z$ **2.** $8m, -2n, -4$ **3.** -9 **4.** -9

  5. 7 **6.** -15 **7.** 1 **8.** 0 **9.** -5

10. $4y + 6$ **11.** $8a + b$ **12.** $13x - y - 13$

13. $x = 8$ **14.** $y = 8$ **15.** $m = -9$ **16.** $a = 1$

17. $x = 26$ **18.** $y = -4$ **19.** $m = 4$

20. $y = -4$ **21.** $x = -8$ **22.** $y = \dfrac{-9}{5}$

23. $-14, -12, -10$ **24.** $l = 13, w = 6$

25. 6, -9, 0, 1

Index

A

Absolute value
  algebraic definition of, 443, 464
  geometric definition of, 442, 464
Addends, 18, 48
Addition, 48
  additive identity in, 49
  associative property of, 44, 49
  calculators for, 24
  carrying in, 19–20
  commutative property of, 44, 49
  of decimals, 260, 298
    calculator for, 262
  of denominate numbers, 395, 421
  of fractions with like denominators, 202, 238
  of fractions with unlike denominators, 207, 238
  of like signed numbers, 445–446, 464
  of mixed numbers, 215, 238
  on number line, 18
  parentheses in, 44, 49
  process of, 18–19
  properties of, 44–45
  of signed numbers, 445–446, 464
  of unlike signed numbers, 447, 464
  of whole numbers, 18–20
  with zero, 447
Additive identity, 45, 49
Area, 410, 422
  formulas for, 411, 422
Arithmetic, fundamental principle of, 112, 127
Associative property
  of addition, 44, 49
  of multiplication, 86, 89

B

Bar, 101, 127
  fraction, 137, 189
Base, 97, 127, 335, 347
Base ten positional system, 2, 48, 243
Borrowing
  in subtraction, 31
  from zero, 33–35
Braces, 101, 127
Brackets, 101, 127

C

Calculator operations
  for decimal addition, 262
  for decimal division, 278
  for decimal multiplication, 267
  for decimal subtraction, 262
  for roots, 98
  for simplifying computations, 105
Cancelling, of common factors, 156–157

Carrying, in addition, 19–20
Centi-, 389, 421
Circle, area formula for, 411, 422
Circumference, 404, 421
  formula for, 405, 422
Cluster, 363, 374
Clustering, estimation by, 363, 374
Coefficient, 471, 516
Comma, 2, 48
Common factor, 117, 128
Common multiples, 121, 128
Commutative property
  of addition, 44, 49
  of multiplication, 85, 89
Cone, volume formula for, 415, 422
Constant, 434, 464, 471
Conversion
  in metric system, 390, 391, 421
  in United States system, 385
Coordinate, 434, 464
Cross products, 155
Cube, 97, 127
Cube root, 97
Cylinder, volume formula for, 415, 422

D

Deci-, 389, 421
Decimal(s), 249–250, 298
  addition of, 260, 298
    calculator for, 262
  complex, fraction conversion of, 254, 298
  division of, 274–275, 298
    calculators for, 278
    by nonzero decimal, 276–277, 298
    by nonzero whole number, 275
    by powers of 10, 280, 298
  fraction conversion of, 254, 298, 326, 347
  fraction conversion to, 288, 298
  fraction operations and, 294
  multiplication of, 265–266, 298
    calculator for, 267
    by powers of 10, 268, 298
  nonterminating fraction conversion to, 331
  percent conversion of, 326, 347
  reading of, 250
  repeating, 285
  rounding of, 257
  subtraction of, 260, 298
    calculator for, 262
  writing of, 251
Decimal point, 249, 298
Deka-, 389, 421
Denominate numbers, 304, 347
Denominator, 137, 189
Diameter, 404, 421
Dictionary, 496
Difference, 29, 49

Digits, 2, 48
  position of, 2
  value of, 2
Distributive property, 366, 374
  estimation by, 367, 374
Dividend, 66, 89
Division, 65, 89
  by 2, 81
  by 3, 81
  by 4, 81
  by 5, 81
  by 6, 82
  by 8, 82
  by 9, 82
  by 10, 82
  of decimals, 274–275, 298
    calculators for, 278
    by nonzero decimal, 276–277, 298
    by nonzero whole number, 275
    by powers of 10, 280, 298
  of denominate number, 398, 421
  exact, 285
  four steps in, 72
  of fractions, 176, 190
  indeterminant, 68, 89
  with multiple digit divisor, 73
  nonterminating, 284–285, 298
  number comparison by, 304, 347
  with remainder, 75
  of signed numbers, 457, 464
  with single digit divisor, 71–72, 89
  terminating, 285
  of whole numbers, 65–68, 71–76
  by zero, 67, 68, 89
  into zero, 67, 68, 89
Divisor, 66, 89

E

Equal symbol, 18
Equality
  addition/subtraction property of, 482, 516
  multiplication/division property of, 487, 516
  symbol of, 436
Equation(s), 480, 516
  conditional, 480, 516
  equivalent, 482, 516
  solutions of, 480, 516
  solving of, 482
    combining techniques in, 490
Estimation, 356, 374
  by clustering, 363, 374
  by distributive property, 367, 374
  by rounding, 356, 374
  by rounding fractions, 370, 374
Evaluation, numerical, 471, 516
Exact division, 285
Exponent, 96, 97, 127
Exponential notation, 96, 127
  reading of, 97

Expression
  algebraic, 470, 516
    simplifying of, 477
  numerical, 470, 516

F

Factor(s), 110–111, 127, 470, 516
  common, 117, 128
    greatest, 117, 128
      method for, 117, 128
      vs. least common multiple,
        123, 128
  in multiplication, 57, 89
Factorization, prime, 112, 127
Fibonacci, Leonardo, 2
Five-step method
  for equations, 500–501
  for proportions, 316
Formula, 405, 422
  for area, 411, 422
  for circumference, 405, 422
Fourth root, 98
Fraction(s), 136–137, 249–250, 298,
    363, 435
  addition of
    with like denominators, 202, 238
    with unlike denominators, 207,
      238
  bar of, 137, 189
  comparison of, 221, 238
  complex, 226, 238
    simple fraction conversion of, 226
  complex decimal conversion to,
    254, 298
  decimal conversion of, 288, 298,
    326, 347
  decimal conversion to, 254, 298
  decimal operations and, 294
  denominator of, 137, 189
  division of, 176, 190
  equivalent, 154–155, 189
  fractions of, 165
  hyphen for, 139
  improper, 145, 189
    mixed number conversion of,
      148, 189
    mixed number conversion to,
      150, 189, 215, 238
    positive mixed numbers and,
      146–148, 189
  multiplication of, 166, 190
    by dividing out common factors,
      167, 190
  nonterminating, conversion of, 331
  numerator of, 137, 189
  of one percent, 330
    conversion of, 330
  order of operations with, 230–231
  parts of, 137, 189
  percent conversion of, 326, 347
  positive, 137, 189
  powers of, 170
  proper, 145, 189
    mixed numbers and, 147

Fraction(s) *(Continued)*
  raising to higher terms, 158–159,
    190
  reading of, 139, 250
  reducing to lowest terms, 155–157,
    190
    by cancelling, 156–157
    by dividing out common factors,
      157
    by dividing out common primes,
      156–157
  roots of, 170
  rounding of, estimation by, 370,
    374
  simple, 225, 238
  subtraction of
    with like denominators, 203,
      238
    with unlike denominators, 207,
      238
  unit, 385
  writing of, 139, 251
Fraction bar, 137, 189
Fundamental principle of arithmetic,
    112, 127

G

Graph, 4, 434, 464
Graphing, 48
  of whole numbers, 4–5
Greater than, symbol for, 221, 238
Greatest common factor, 117, 128
  method for, 117, 128
  vs. least common multiple, 123,
    128
Grouping symbols, 101, 127
  multiple, 102

H

Hecto-, 389, 421
Higher terms, fraction raising to,
    158–159, 190
Hindu-Arabic numeration system,
    1–2, 48
Hyphen, for word fractions, 139

I

Identity
  additive, 45, 49
  multiplicative, 86, 89
Index, 98, 127
Inequality symbols, 221, 238, 436
Integers, 435, 464

K

Kilo-, 389, 421

L

Least common multiple, 122, 128
  method for, 122–123, 128
  vs. greatest common factor, 123, 128
Length
  measurement unit for
    in metric system, 390
    in United States system, 384
Less than, symbol for, 221, 238
Lowest terms, fraction reduction to,
    156, 190

M

Mass
  measurement unit of in metric
    system, 390
  vs. weight, 391
Measurement, 384, 421
  metric system of, 389–391, 421
    conversion in, 390, 421
  standard unit of, 384, 421
  United States system of, 384–385,
    421
    conversions in, 385
Mental arithmetic, 366, 367, 374
Metric system, 389–391, 421
  prefixes for, 389, 421
Milli-, 389, 421
Minuend, 29, 31, 49
Minus symbol, 29
Missing factor statement, 182–183,
    191
Missing product statement, 181, 191
Multiple(s), 121, 128
  common, 121, 128
    least, 122, 128
      method for, 122–123, 128
      vs. greatest common factor,
        123, 128
Multiplicand, 56, 89
Multiplication
  associative property of, 86, 89
  commutative property of, 85, 89
  of decimals, 265–266, 298
    calculator for, 267
    by powers of 10, 268, 298
  of denominate number, 397, 421
  of fractions, 166, 190
    by dividing out common factors,
      167, 190
  indicators of, 56
  of mixed numbers, 169, 190
  with multiple digit multiplier, 58
  multiplicative identity in, 86, 89
  with numbers ending in zero, 60
  properties of, 85–86
  of signed numbers, 454–455, 464
  with single digit multiplier, 57
  in terms of "of," 269
  of whole numbers, 56–60, 89
Multiplication statements, 181
Multiplicative identity, 86, 89
Multiplier, 56, 89

N

Nonterminating division, 284–285, 298
Notation
  exponential, 96, 127
    reading of, 97
  root, 97–98
Number(s), 1, 48
  composite, 111–112, 127
  denominate, 304, 347, 394, 421
    addition of, 395, 421
    division of, 398, 421
    like, 304, 347
    multiplication of, 397, 421
    simplified, 394, 421
    subtraction of, 395, 421
    unlike, 304, 347
  mixed
    addition of, 215, 238
    improper fraction conversion of, 150, 189, 215, 238
    improper fraction conversion to, 148, 189
    improper fractions and, 146–148, 189
    multiplication of, 169, 190
    positive, 145–146, 189
    subtraction of, 215, 238
  prime, 111–112, 127
  pure, 304, 347
  real, 434, 435, 464
    counting, 435, 464
    double-negative property of, 440, 464
    line graph of, 434, 464
    natural, 435, 464
      prime factorization of, 113, 127
    negative, 435, 439, 464
    opposite, 440, 464
    ordering of, 436
    positive, 435, 439, 464
    rational, 435, 464
    signed, 439
      addition of, 445–446, 447, 464
      division of, 457, 464
      multiplication of, 454–455, 464
      subtraction of, 451, 464
    subsets of, 435, 464
    whole, 4, 48, 435, 464. See also Whole number(s)
Number line, 4, 48, 136, 434, 464
  addition on, 18
Numeral, 1, 48
Numeration system, Hindu-Arabic, 1–2, 48
Numerator, 137, 189

O

Ones position, digits to right of, 243
Order of operations
  with fractions, 230–231
  with whole numbers, 103, 127
Ordered number system, 221, 238
Origin, 4, 434, 435

P

Parallelogram, area formula for, 411, 422
Parentheses, 44, 101, 127
  in addition, 44, 49
Partial product, 58
Percent, 325, 335, 347
  applications of, 334–335, 347
  decimal conversion of, 326, 347
  fraction conversion of, 326, 347
  one, fractions of, 330
Percentage, 334, 347
Perimeter, 402, 421
Period, 2, 48
Pi (π), 404, 421
Plus symbol, 18
Polygon, 402, 421
Power, 97, 127
  of fractions, 170
Prime, relatively, 156, 190
Prime factorization, 112, 127
Prime numbers, 111–112, 127
Product, 56, 89, 110, 470
  partial
    first, 58
    second, 58
    total, 58
Property
  associative
    of addition, 44, 49
    of multiplication, 86, 89
  commutative
    of addition, 44, 49
    of multiplication, 85, 89
  distributive, 366, 374
    estimation with, 367, 374
  double-negative, 440, 464
Proportion, 309, 347
  application of, 316
  five-step method for, 316, 347
  missing factor in, 310–311, 347
  rates with, 312–313, 347

Q

Quotient, 66, 89. See also Division
  nonterminating, denoting of, 285

R

Radical, 98, 127
Radical sign, 97, 98, 127
Radicand, 98, 127
Radius, 404, 421
Rate, 305–306, 309, 347
  proportions with, 312–313, 347
Ratio, 305, 309, 325, 347
Reciprocals, 175
Rectangle
  area formula for, 411, 422
  volume formula for, 414, 422
Reducing a fraction, 155–157, 190
Relatively prime, 156, 190
Remainder, division with, 75

Root, 97, 127
  calculator operations for, 98
  cube, 97
  fourth, 98
  of fractions, 170
  square, 97
Root notation, 97–98
Rounding
  as approximation, 12–13
  of decimals, 257, 298
  estimation by, 356, 374
  of fractions, estimation by, 370, 374
  of whole numbers, 12–14, 48

S

Simplifying, of algebraic expression, 477
Sphere, volume formula for, 414, 422
Square, 97, 127
Square root, 97
Statements
  missing factor, 182–183, 191
  missing product, 181, 191
  multiplication, 181
Subtraction, 48
  borrowing in, 31
    from zero, 33–35
  of decimals, 260, 298
    calculator for, 262
  of denominate numbers, 395, 421
  of fractions with like denominators, 203, 238
  of fractions with unlike denominators, 207, 238
  of mixed numbers, 215, 238
  number comparison by, 304, 347
  process of, 30
  of signed numbers, 451, 464
  of whole numbers, 29–31
Subtrahend, 29, 31, 49
Sum, 18, 48, 470
Symbols
  for greater than, 221, 238
  grouping, 101, 127
    multiple, 102
  inequality, 221, 238
  for less than, 221, 238
  words translated to, 495–496, 517

T

Terminating division, 285
Terms, 18, 48, 470, 516
  higher, fraction raising to, 158–159, 190
  like, combining of, 477, 516
  lowest, fraction reduction to, 156, 190
Time
  measurement unit for
    in metric system, 391
    in United States system, 384
Trapezoid, area formula for, 411, 422
Triangle, area formula for, 411, 422

V

Variable, 434, 464
Volume, 414, 422
  formulas for, 414–415, 422
  liquid
    measurement unit for
      in metric system, 391
      in United States system, 384

W

Weight
  measurement units for, in United
    States system, 384
  vs. mass, 391

Whole number(s), 4, 48, 435, 464
  addition of, 18–20
  cube of, 97, 127
  division of, 65–68, 71–76
  factors of, 111
  fractions of. *See* Fraction(s)
  graphing of, 4
  multiplication of, 56–60, 89
  order of operations for, 103, 127
  reading of, 7, 48
  relatively prime, 156, 190
  rounding of, 12–14
  square of, 97, 127
  subtraction of, 29–31
  writing of, 8, 48

Word problems
  solving of, 500–501
  symbol translations of, 495–496,
    517

Z

Zero
  addition with, 447, 464
  borrowing from, in subtraction,
    33–35
  division by, 67, 68, 89
  division into, 67, 68, 89